CRITICAL TOPON

Re-materialising Cultural Geography

Dr Mark Boyle, Department of Geography, National University of Ireland, and Professor Donald Mitchell, Maxwell School, Syracuse University, USA

Nearly 25 years have elapsed since Peter Jackson's seminal call to integrate cultural geography back into the heart of social geography. During this time, a wealth of research has been published which has improved our understanding of how culture both plays a part in, and in turn, is shaped by social relations based on class, gender, race, ethnicity, nationality, disability, age, sexuality and so on. In spite of the achievements of this mountain of scholarship, the task of grounding culture in its proper social contexts remains in its infancy. This series therefore seeks to promote the continued significance of exploring the dialectical relations which exist between culture, social relations and space and place. Its overall aim is to make a contribution to the consolidation, development and promotion of the ongoing project of re-materialising cultural geography.

Other titles in the series

Cultural Landscapes of Post-Socialist Cities
Representation of Powers and Needs
Mariusz Czepczynski
ISBN: 978 0 7546 7022 3

Fear; Critical Geopolitics and Everyday Life
Edited by Rachel Pain and Susan J. Smith
ISBN: 978 0 7546 4966 3

Towards Safe City Centres?
Remaking the Spaces of an Old-Industrial City
Gesa Helms
ISBN: 978 0 7546 4804 8

In the Shadows of the Tropics
Climate, Race and Biopower in Nineteenth Century Ceylon
James S. Duncan
ISBN: 978 0 7546 7226 5

Critical Toponymies
The Contested Politics of Place Naming

Edited by

LAWRENCE D. BERG
University of British Columbia, Canada

and

JANI VUOLTEENAHO
University of Helsinki, Finland

LONDON AND NEW YORK

First published 2009 by Ashgate Publishing

2 Park Square, Milton Park, Abingdon, Oxon OX14 4RN
711 Third Avenue, New York, NY 10017, USA

Routledge is an imprint of the Taylor & Francis Group, an informa business

First issued in paperback 2016

British Library Cataloguing in Publication Data
Critical toponymies : the contested politics of place
naming. -- (Re-materialising cultural geography)
1. Names, Geographical--Etymology. 2. Names, Geographical--
Political aspects. 3. Toponymy.
I. Series II. Berg, Lawrence D. III. Vuolteenaho, Jani.
910'.014-dc22

Library of Congress Cataloging-in-Publication Data
Berg, Lawrence D.
Critical toponymies: the contested politics of place naming / by Lawrence D. Berg and Jani Vuolteenaho.
p. cm.
Includes index.
ISBN 978-0-7546-7453-5 (hardback) 1. Toponymy--Political aspects. 2
Names, Geographical--Political aspects. 3. Political geography. I. Vuolteenaho, Jani. II. Title.
G100.5.B47 2009
910.01'4--dc22

2009016282

ISBN 978-0-7546-7453-5 (hbk)
ISBN 978-1-138-26775-6 (pbk)

Contents

List of Figures and Tables

Figures

Tables

Notes on Contributors

Terhi Ainiala is a senior researcher and head of the research department at the Research Institute for the Languages of Finland in Helsinki. She is an advanced researcher on proper names and has done her doctoral thesis on changes in place names (1997). In addition, she is adjunct professor of Finnish language at the University of Helsinki. Among her main research interests are socio-onomastics, urban place names and theoretical and methodological questions in onomastics. Between 2007 and 2009 she was in charge of a research project "Transformation of the onomastic landscape in the sociolinguistically diversifying neighbourhoods of Helsinki" (funded by the Academy of Finland). Since 2008, she has been a member of the Board of directors in the International Council of Onomastic Sciences.

Derek H. Alderman is Associate Professor in the Department of Geography at East Carolina University, USA. He studies the intersection of place naming with issues of public commemoration, symbolic resistance, African American identity, and social justice. He also examines the Internet as a place of cultural and political representation. Much of his work examines the politics of naming streets and other places for Martin Luther King, Jr, viewing these toponymic struggles as arenas for larger debates about race and power in America. He is the co-author (with Owen Dwyer) of *Civil Rights Memorials and the Geography of Memory*, published in 2008.

Maoz Azaryahu is Associate Professor of Geography at the University of Haifa in Israel. His research includes the cultural and historical geographies of national myths and public memory in Israel and in Germany, landscapes of popular culture, and the cultural history of places and landscapes.

Lawrence D. Berg is co-director of the Centre for Social, Spatial and Economic Justice at the University of British Columbia, Canada. He teaches in Women's and Gender Studies, Cultural Studies and Human Geography. Lawrence was a founding editor of *ACME: An International E-Journal for Critical Geographies,* and he remains a member of its editorial collective. Lawrence has published widely on the politics of naming places, the cultural politics of knowledge production, masculinities and geographies, and race and space. He is currently undertaking research on the relationship between liberalism, neoliberalisation and white supremacy. His work appears in journals such as *AREA, Antipode, Gender Place and Culture, Geoforum, Qualitative Inquiry, Progress in Human Geography, Social and Cultural Geography* and *Society & Space*.

Douglas Herman is Senior Geographer for the Smithsonian National Museum of the American Indian (NMAI). He holds a doctorate in Geography from the University of Hawai'i, and in addition to his work at NMAI, he is the director of Pacific Worlds, a web-based indigenous-geography education project for Hawai'i and the American Pacific. Since 2001 he has served as Secretary of the Indigenous Peoples Specialty Group of the Association of American Geographers, organising numerous conference sessions on Indigenous Geography. He has published several articles and given numerous scholarly presentations regarding geography and the representation of Indigenous cultures.

Robin A. Kearns is Professor of Geography at The University of Auckland. He has published two books on the links between place and health and numerous articles exploring the links between culture, identity and social wellbeing. Robin is a ministerial appointment on New Zealand's National Health Committee. He is also an editor of two journals: *Health and Place*, and *Health and Social Care in the Community*.

Garth Andrew Myers is Director of the Kansas African Studies Center and Professor of Geography and African/African-American Studies at the University of Kansas. His record includes more than 30 publications, including the books, *Verandahs of Power: Colonialism and Space in Urban Africa* (Syracuse, 2003) and *Disposable Cities: Garbage, Governance and Sustainable Development in Urban Africa* (2005, Ashgate), and articles in journals such as the *Annals of the Association of American Geographers*, *Geographical Review*, *Area*, *Urban Studies*, *Political Geography*, and the *Journal of Modern African Studies*

Catherine Nash is Professor of Human Geography, Queen Mary, University of London. Her research interests are in geographies of identity, belonging and relatedness. Most recently she has explored these themes in relation to ideas of geographical and genealogical roots and relatedness in Ireland, Northern Ireland and in the Irish diaspora. She has written widely on Irish cultural and historical geography and her recent publications include *Of Irish Descent: Origin Stories, Genealogy, and the Politics of Belonging* (Syracuse University Press, 2008). She is currently addressing constructions of human difference and relatedness in human population genetics and geneticised genealogy.

Kaisa Rautio Helander is an onomastic scholar with special interests in indigenous and multilingual place names research. Rautio Helander has worked for a long time with Sámi place name management, especially with questions concerning officialising Sámi toponymy, in the Place Name Advisory Office in Norway. Currently, she is an associate professor in Sámi and Finnish languages at Sámi University College, Guovdageaidnu Norway.

Reuben S. Rose-Redwood is an Assistant Professor of Geography at the University of Victoria, where he teaches courses in urban, social and cultural geography. His current research examines the politics of street naming and the historical geography of street and house numbering as part of a broader spatial history of the "geo-coded world." He has published numerous articles, which have appeared in *Progress in Human Geography*, *Social and Cultural Geography*, *Cartographica*, *Political Geography*, and the *Geographical Review*. His work has also been featured on several occasions in the *New York Times* as well as in documentaries on the Discovery Channel and History Channel. He recently served as a Kluge Fellow at the Library of Congress in Washington, DC.

Brenda S.A. Yeoh is Professor, Department of Geography as well as the Vice-Dean of Research, Faculty of Arts and Social Sciences, National University of Singapore. She is also the Research Leader of the Asian Migration Research Cluster and Principal Investigator of the Asian MetaCentre at the University's Asia Research Institute. Her research interests include the politics of space in colonial and post-colonial cities; gender; migration and transnational communities. She has published widely and her books include *Contesting Space: Power Relations and the Urban Built Environment in Colonial Singapore* (Oxford University Press, 1996; reissued Singapore University Press, 2003); *Toponymics: A Study of Singapore Street Names* (Eastern Universities Press, 2003, with Victor R. Savage), *Theorising the Southeast Asian City as Text* (World Scientific, 2003, with Robbie Goh) and *The Politics of Landscape in Singapore: Constructions of "Nation"* (Syracuse University Press, 2003, with Lily Kong).

Jani Vuolteenaho (Academy of Finland, University of Helsinki) is a cultural and urban geographer who is currently carrying out a research project on the relationships between space, power and language, especially with respect to urban toponymies. The foci of his other publications range from urban renewal projects and everyday spaces of unemployment to disciplinary historical and methodological research. In 2003–2006, Vuolteenaho edited a geographical journal *Terra*, published in Finnish and Swedish.

Acknowledgements

The ideas for this collection of essays first came together at the International Symposium on Place Names and Identities in Multicultural Contexts, held at the Norwegian Sámi Parliament in Kárásjohka, Norway, in the late Nordic summer of 2006. We are grateful to the Norwegian Sámi Parliament for hosting that symposium, and specifically Kaisa Rautio Helander, the key on-the-ground organiser of that symposium. Further discussions of the book project took place at the Placing Names, Naming Places Workshop, held in Greenville, North Carolina in the autumn of 2007. We are thankful to Derek Alderman, Reuben Rose-Redwood and Maoz Azaryahu for organising the workshop and providing us with a space to discuss many of the ideas that are now incorporated in this book.

We thank Pion Limited, London for permission to reprint a revised version of Berg, L.D. and R.A. Kearns (1996) "Naming as Norming: 'Race,' gender and the identity politics of naming places in Aotearoa/New Zealand," *Environment and Planning D: Society and Space*, Vol. 14, pp. 99–122. We thank Wiley-Blackwell Publishing for permission to reprint the following essays: Yeoh, B.S.A. (1996) "Street-naming and nation-building: Toponymic inscriptions of nationhood in Singapore," *Area*, Vol. 28, pp. 298–307; Myers, G. (1996) "Naming and placing the other: Power and the urban landscape in Zanzibar," *Tijdschrift voor Economische en Sociale Geografie*, Vol. 87, pp. 237–46; a revised version of Herman, R.D.K. (1999) "The Aloha State: Place names and the anti-conquest of Hawai'i," *Annals of the Association of American Geographers*, Vol. 89, pp. 76–102; and, a revised version of Nash, C. (1999) "Irish place-names: post-colonial locations," *Transactions, Institute of British Geographers,* Vol. 24, pp. 457–80. We thank Taylor and Francis Publishers (http://www.informaworld.com) for permission to reprint Kearns R.A. and L.D. Berg (2002) "Proclaiming place: towards a geography of place name pronunciation," *Social and Cultural Geography*, Vol. 3, pp. 283–302. We thank Geoscience Publications for permission to reprint Alderman, D.H. (2002) "Street Names as Memorial Arenas: The reputational politics of commemorating Martin Luther King, Jr in a Georgia County," *Historical Geography*, Vol. 30, pp. 99–120. Finally, we thank Elsevier Ltd. For permission to publish a revised version of Rose-Redwood, R. (2008) "Indexing the Great Ledger of the Community: Urban house numbering, city directories and the production of spatial legibility," *Journal of Historical Geography*, Vol. 34(2), pp. 286–310. We are grateful to Guy Puzey for permission to use his photograph of the defaced Gemonio/Gimon road sign for the cover. Finally, we are grateful to Levi Gahman for assistance in formatting the manuscript.

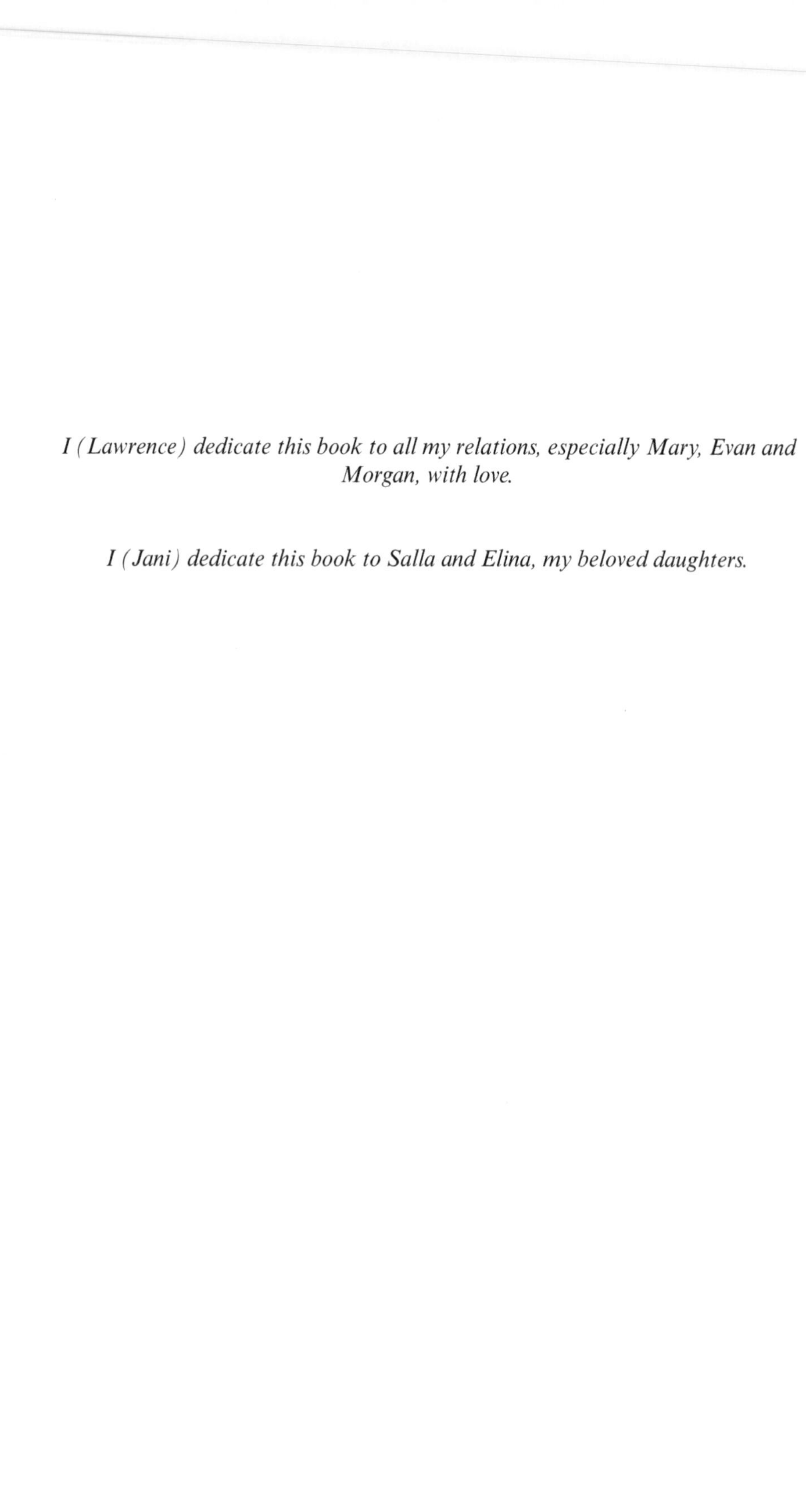

I (Lawrence) dedicate this book to all my relations, especially Mary, Evan and Morgan, with love.

I (Jani) dedicate this book to Salla and Elina, my beloved daughters.

Chapter 1
Towards Critical Toponymies

Jani Vuolteenaho and Lawrence D. Berg

Introduction

Ever since the late 1980s researchers in social sciences and humanities have focussed their efforts on the entanglements of language with power relations. The intellectual and political legitimacy of this "cultural turn" has, however, been called into question for a tendency to focus on theory at the expense of empirical evidence, a reliance on relativism and a "descent into discourse" (Palmer 1990; cited in Jackson 2000) that fails to recognise the importance of material cultural processes. This book attempts to signpost one potential way out of this impasse. Exploring the power of naming in the construction of historical and contemporary landscapes, *Critical Toponymies* aims to show that to take language seriously can bring the material and the discursive together in a productive relationship that avoids binary distinctions between theory and empirics or the material and discursive. In this way, the volume makes a contribution to what at the turn of the century was termed the necessity for "rematerializing cultural geographies" (Philo 2000; Jackson 2000). The chapters in this volume also represent a showcase of recently revitalized "critical" toponymic research, work that extends a field that has traditionally been characterized by political innocence to say the least. This newer critical work on place naming draws on recent social and cultural theories that help to understand the always-already power laden character of naming places (Kearns and Berg, this volume). Whilst there have been many monographs published that discuss toponymy generally, this volume is the first interdisciplinary collection published in English that tackles explicitly place naming as "a political practice par excellence of power over space" (Pinchevski and Torgovnik 2002).

The atheoretical character of traditional approaches to toponymy owes much to the fact that such scholarship has, until recently, chiefly developed through research by map-makers, specialized linguists, and etymologists interested in gaining some understanding about what might be called "bygone landscapes" (see e.g. Eichler et al. 1995; Zelinsky 1997). In contrast, this volume brings together a range of researchers with specific interests in using the insights of social and cultural theories to help us better conceptualize the power relations inherent in geographical naming.

As editors, we have opted to bring together a range of conceptually insightful articles and stimulating case studies that deal with the hegemonic and contested practices of naming. The book's twelve substantive articles focus, inter alia, on the key role of naming in the colonial silencing of indigenous cultures, the canonization of nationalist ideals in the nomenclature of cities and topographic maps, the construction of commodified neoliberal urban landscapes, the contestation over identity and place at the level of street naming, as well as the formation of more or less fluid forms of postcolonial identities. In empirical terms, *Critical Toponymies* guides the reader into struggles over nomenclature through case studies in a multitude of national and local contexts in Africa, Asia, Australasia, Europe, North America, the Pacific, and even the virtual landscapes of the Internet.

Our introduction has three contextualizing goals. In the following section, we discuss certain traditional—and in many respects antithetical—research orientations to naming and toponyms. While excavating philosophical, technocratic and historical-cultural name studies, we also pay attention to the politics of knowledge production behind them, as well as point out some forerunners for the recent rise of multidisciplinary interest in critical analyses of place naming. Next, we focus on the pivotal concepts of place and naming, and elaborate how these concepts are manifest through practices of place naming that are infused with relations of power. Finally, we introduce the articles included in this collection, and try to signpost some of the key theoretical positions drawn upon by our contributors as they attempt to better understand the power relations that inhere in naming places.

Place Name Research Orientations

> The theoretical scene in the study of names leaves much to be desired. Such a situation smacks of the paradoxical, as Klaas Willems has stated, "Proper names are complex signs with specific linguistic, pragmatic, logical, philosophical, semiotic, historical, psychological, social, and juridical properties, and hence represent a vast interdisciplinary field of study" (2000, 86). [...] For whatever reasons (and I have none to present), the number of serious students of names has always been relatively limited. Moreover, onomastics [the study of proper names] has been an academic orphan (Murray 1994). (Zelinsky 2002, 243, 246)

Apparently due to "the vagaries of academic history" (Zelinsky 1997, 465), toponymic studies have been scattered across a spectrum of disciplines. To mention some of these, anthropology, cartography, geography, history, legal studies, literary studies, philosophy, political science, environmental psychology, and especially linguistics are all among the fields in which researchers have endeavoured to make sense of place names and named places around us.

Interestingly, though, as Wilbur Zelinsky (1997, 465) has put it, onomastics has "never managed to gain a firm foothold" in any of these. Indeed, rather old-fashioned philological etymologizing has remained long a foundation for most empirical name studies conducted in the humanities and social sciences. In modern linguistics, for instance, etymological name research has mainly overshadowed positivist (Saussurean or otherwise) approaches otherwise flourishing across its subdisciplines since the late nineteenth century (see Cassirer 1955; Gal and Irvine 1995). Instead of limiting our discussion to the status of name research in specific disciplines, however, we identify generally three conventional (but still influential) research orientations that have preceded the recent rise of critical toponymic approaches. In our somewhat schematic survey of a methodologically compartmentalized intellectual legacy, we will concentrate on typical treatments of (place) names in: i) philosophical, ii) technical-authoritative, and iii) historical-culturalist literature.

To begin with the most sophisticated tradition in conceptual terms, the earliest known roots of philosophically oriented study of names can be traced back to Antiquity. Plato, for example, insisted that the correctness of names for beings is "not the trivial work of chance persons, but rather the task of the community," and the Stoics distinguished between names that "signify a quality proper to an individual" and appellatives that "signify a quality common to a genus" (Summerell 1995, 369–70). Later on, Gottfried Leibniz (1646–1716), along with a number of other early modern thinkers, wrestled with the same proper/common noun distinction, the denotative and connotative qualities of these word classes, and their functions in the "universal grammar" of the innate cognitive apparatus of humans (ibid., 370–71; see also Foucault 1974). Usually characterized by an abstract interest in naming in general (as distinct from place names as such), philosophers' contemplations of this sort have continued in a variety of guises in the course of the last century, too. Thus according to the later Ludwig Wittgenstein's (1968, 13[e]) well-known postulation, all discourses between people are dependant upon the calling into being of real-world or abstract entities: "we name things and then we can talk about them: can refer to them in talk." Notably, the last few decades have also evinced deconstructionist theories that have questioned, in the face of endless semiotic workings of language, the referential capacity of proper names to designate unique presences (e.g. Derrida 1986; 1994, 107–118). Meanwhile, some radical thinkers have stressed the power-related aspects of naming in ways that resonate well with the approaches of this collection. Inter alia, Ernesto Laclau (2005, 93–124) has emphasized the symbolically unifying functions that names enjoy as performative catalysts for popular identities. No less persuasively, Pierre Bourdieu (1991, 47, 239–43) has stressed the symbolic violence involved in the "state monopolies of legitimate naming," producing centralized representations of the social world at the expense of the lower classes and speakers of non-sanctioned languages and dialects.

In contrast to philosophers' broad contemplations upon naming, applied map-makers, linguists and other experts have been interested in gathering and standardizing stocks of toponyms for specific (geopolitical) areas. Indeed, as especially Foucauldian historiographers and geographers have pointed out, the administrative systematization of geographical nomenclature has played its crucial role in the technocratic rationalization of space in the course of modernization (e.g. Harley 2001; Withers 2000; Rose-Redwood 2006). In essence, even if frequently presented in depoliticized guise in hegemonic discourses, the mapping of toponyms has formed an ancillary form of knowledge-production in the service of a wider scientific-geopolitical project of knowing the world as accurately as possible as part of the process of controlling its spaces. In the Renaissance, for instance, the voyages of European explorers yielded "raw material for encyclopaedic works on the world, or parts of it" (Holt-Jensen 1999, 21). In these popular studies, miscellaneous information on both "natural and physical conditions" and "folklore, etymology and history" was organized on the basis of "the jumble of information related to place names." Although encyclopaedic compilations of this kind have since then been relegated "outside the field of science" (ibid.), and place naming itself has been disparaged as an ineffective language of "nominal specification" in positivist inspired circles (Haggett 1972, 83), there is no question that pursuit of the standardization of toponymies has continued to intensify to the present-day. In the nineteenth century, in particular, cartography, statistical record-keeping, and the associated gathering of national toponymies became "the business of the state" (Harley 1992, 244) as various rising nation-states "consolidated their authority and eased their governance through archives and registers of people, places and things" (Nash 1999, 457).

As this technocratic-administrative legacy has continued to expand to the present-day, the current norm is that each country has to have "a national name authority to standardize and disseminate its names," and even "international [toponymic] standardization must of necessity be based on national standardization" (Raper 1995, 263; Kadmon 1997, 169–243). At a multitude of scales from urban subdivisions to global name standardization under the auspices of the United Nations Group of Experts on Geographical Names (UNGEGN), modernization has witnessed countless projects aiming at the systematization of the make-up and orthography of spatial nomenclature. As the gloomy side-effect of topynymic modernization, these projects have frequently resulted in the erasure of the inherited heteroglossia of local names and even definite "toponymic silencing" (Harley 2001, 99) of indigenous cultures (see chapters by Nash and Rautio Helander in this volume). Ironically, a recent use of Internet- and Geographic Information System-based toponymic databases as the tools of cultural empowerment (as with the Inuit or Maori people) also evinces the power that technocratic-authoritative toponymies exert in modern societies.

Occupying a middle position between the above two traditions, the bulk of empirical studies of toponymy in the humanities and social sciences has been traditionally dominated by what might be called historical-culturalist approaches. At least in Europe, philological scholars and clergy engaged in "etymological guesswork" already in the medieval and early modern periods (e.g. Taylor 1896, iv; Stemshaug 1995, 32). In the nineteenth century, however, much more systematic approaches to understanding the origins of place names were adopted. Above all, the foundations for new research were initially laid in German-speaking continental Europe. In this academic and societal context, research became influenced by idealistic philosophy that underscored the organically rooted "spirit" of each national language, as well as the hierarchical *Stammbaum* ("family tree") model of the development of languages (see Cassirer 1955, 117–76; Vološinov 1990, 63–83; Errington 2001, 32–3). In this growing tradition, historical successions of layers of place names used by different native tribes became treated as keys to language and settlement history (see e.g. Grim 1838; Arnold 1875/1983). Instead of focusing on names that had a familiar ring to contemporary speakers, scholars of language in Germany, and soon elsewhere in Europe, were particularly keen on genealogies of past tongues and language contacts (Haubrichs 1995, 62–8). Also, a good number of historians, anthropologists and cultural geographers (such as Otto Schlüter) started to utilize old toponyms as descriptive sources in explicating diffusion routes of settlement types and cultivation habits, traces of cultural and religious values on the landscape, and so on (ibid.; Ratzel 1891, 537–74; Schwarz 1989, 6–9). At least implicitly, the objectives of the then new tradition were tied up with the historical-symbolic legitimatization of modern nation-states, as studies were canonically discoveries of ethno-linguistic pasts in their particular regions. Unsurprisingly, early European toponymists also engaged with research in geopolitically disputed areas such as Alsace-Lorraine/Elsaß-Lothringen, often ending up in scientifically legitimatized yet highly controversial claims over their symbolic ownership (see e.g. Haubrichs 1995, 64–7; 2008).

A central scholar in terms of trans-Atlantic diffusion of the historical-culturalist toponymies was Franz Boas (1858–1942), a German-born anthropologist who is not only regarded as a founder of "Americanist anthropology" (Darnell 2001, 33–67), but also the figure on whose legacy a linguistic scholarship in North America was largely based (Trask 2007, 19). At least partly through his studies on Baffin Land and Hudson Bay Eskimos, and subsequently Kwakiutl Indians,[1] fieldwork based research on Native American place name systems grew into "a viable component of anthropology in the United States" in the first half of the last century (Basso

1 Highlighting the problematic character of Western colonial naming practices and the political importance of naming in anti-colonial and postcolonial contexts, these people refer to themselves as, respectively, *Inuit* and *Kwakwaka'wakw* people.

1996, 43). Besides the influence from Ratzel, Schlüter and other German geographers, Boas and his disciples (see e.g. Kroeber 1916) also very likely oriented many representatives of the culturalist Berkeley School in American geography to consider descriptive place names and other geographical terms—potentially "rich in genetic meaning"—as sources to trace the imprint of past cultures on the landscape (see Sauer 1925/1963, 330; Wagner and Mikesell 1962, 14–15; Zelinsky 1982; Norton 2000, 291; more critically, see Duncan 1980; Jackson 1989, 16–20). Undoubtedly, the internally tensionless "superorganic" view of culture adopted in the so-called Berkeley School of cultural geography has, in many respects, resembled the work of preceding European toponymists. Equally telling in terms of the persistence of historical-culturalist orientation, in the United Kingdom Margaret Gelling's (e.g. 1984) influential interpretations of place names as clues to the bygone "mixture of races" have yielded suspiciously innocent and bloodless accounts of history, as if geographical nomenclatures were akin to the particulate matter within stratigraphic sequences that geomorphologists study (Kearns and Berg, this volume; see also Glennie 2000).

Altogether, it seems to us that traditional name research has not only "been afraid of politics" (Palonen 1993, 103), but it has also failed to honestly reflect on its own complicity in power struggles over toponymies. With notable exceptions such as Bourdieu (1991) or Laclau (2005), the general problem with philosophically oriented reflections on naming has lain in their very abstraction from any place-bound historical context. Conversely, the technocratic-authoritative and historical-cultural strands of toponymic research have typically adopted theoretically (and politically) naïve empiricist foci on the nomenclatures of specific localities, provinces, nation-states or other geopolitical units. Indeed, while historical-cultural toponymists have been occasionally attentive to philosophical considerations on names, in other respects they have often uncritically supplemented bureaucratic institutional standardization programs with (nationally or otherwise) canonized language, political aims, and overtones (on the role of philologists and ethnologists in the anglicization of the toponymy of Ireland, for example, see Nash, this volume; see also Bourdieu 1991, 46–9; Hanks 2005, 75–6). From the viewpoint of marginalized groups, this conjunction of the "practical" with the "intellectual" has had multiple negative implications related to the centripetal development of standard languages and lexicons, suppression of minority tongues, conservation of dialectal nomenclatures as cultural relicts rather than as living entities of linguistic variety, and dissemination of (false) conflict-free narratives about the past linguistic relations and symbolic landscapes of nations.

The contrast is striking, however, if one compares the above three approaches with the recent reorientation of research towards what we call in this volume "critical toponymies." At this point, it is pertinent to refer to a couple of early studies published independently from each other in the 1980s and early 1990s.

Drawing on theorists such as Michel Foucault (e.g. 1980) and Edward Said (1979), J.B. Harley's (for a posthumous collection of articles, see Harley 2001) critical excavations of the history of cartography have been the catalyst for a range of previously understudied questions on "the social processes intrinsic to the authoritative act of naming" (Withers 2000, 533–4). In a similar vein, Paul Carter's book *The Road to Botany Bay* (1988, xxiii–xxiv)—highlighting how naming practices by western explorers and settlers transformed the supposedly "empty lands" (constituted by another name: *terra nullius*) of Australia "into a place, a space with history" in their self-contained cultural imaginations—has provided influential insights for a toponymic scholarship sensitized to power relations in the landscape. In addition to studies in colonial/postcolonial contexts, the interest in toponymic power has also surfaced in the wide-ranging field of urban studies. In this regard, it is important to recognise the significance that Michel de Certeau's (1984) remarks on place names as symbolic armature for both the powerful and the weak, Allan Pred's (1990, 92–142) analysis of blue-collar workers' colloquial, irreverent "words of spatial orientation" in the face of capitalist modernization in late nineteenth century Stockholm, and Maoz Azaryahu's (e.g. 1986; 1996) works on street names as political symbols (see also Azaryahu's article in this volume) have all played in bringing to the fore more critical analyses of naming places. Yet before proceeding to successors of these classic studies, we want to first make some theoretical remarks on the central concepts of this book.

Theorizing Naming, Place-making and Power

> [T]o be a place is to be located (or at least locatable), named, known. (Maskit 1998, 266)

The key concepts of this book involve ideas at the intersection of naming, place-making, and power. As each of these concepts have come to be understood in a variety of ways in different disciplines, our purpose in the following is to clarify the ways that we find these concepts to be most useful for thinking critically about the relationship between social relations infused with power and the naming of places. En route, we will also touch, of necessity, upon the question of why toponyms often appear to people as ideologically innocent rather than power-charged semiotic dynamos for making meaning about places. Indeed, when seen through theories that stress the pervasiveness of power in all socio-spatial relations (e.g. Foucault 1980; de Certeau 1984; Lefebvre 1991; Massey 1991; Simonsen 1996), the mundane taken-for-grantedness of place names (until something "goes wrong" with them), and their politically charged meanings, are elements in the same conceptual tangle. In our view, these seemingly contradictory aspects of place names need to be approached in tandem in order to fully understand their power-FULL function as ubiquitous "cultural stuff" through which geographical knowledges and meanings are constructed, mediated and contested.

Non-linguists have often tended to be rather inexact about what they mean by the concept of naming. In both academic and lay discourses, the term "naming" is often taken to encompass all kinds of linguistic expressions that designate entities or occurrences in the phenomenal world. As opposed to this vague definition, in linguistics and its relatively marginal sub-discipline onomastics (from Greek *onoma*, "name"), names refer to the word class of proper nouns, including not only place names, but also personal names, firm names, names of sport clubs, book titles, Internet domain names, etc. (e.g. Nicolaisen 1976; Kiviniemi 1979; Trask 2007, 197). For linguists, then, place naming is based on specific language conventions and lexical relationships through which the "individuality" of specific areas (British Columbia, Europe), topographic features (Schwarzwald), settlements, passages and other constructions (Alice Springs, L'avenue des Champs-Élysées, Mulholland Drive, Brooklyn Bridge), can be identified apart from other places. As distinct from other spatially designating word classes—such as common names that classify entities to a culturally known category (e.g. continent, badland, alley, boulevard, slum, favela, bidonville), and "elementary spatial terms" like pronouns (I/you, us/them) or spatial demonstratives (here/there, up/down, near/far) (see e.g. Cassirer 1955, 198–215)—"proper" toponyms pinpoint unique places. According to some extremely formal views "the uniqueness and immense practical convenience" of proper (place) names is that they are "meaningless marks" which, as their fundamental *referential* function, "name objects individually without describing them" (Diderichsen 1962; cited in Zilliacus 1997, 14; see also Russell 1912; Searle 1969, 163). Nevertheless, most toponymists in linguistics, and even more so in other disciplines, agree today that place names not only denote single portions of geographical space, but can also act as "pegs on which to hang descriptions," or be full to the brim with connotative meanings (see Azaryahy, this volume; see also Searle 1969, 172).

In our view, there are both theoretical advantages and dangers inherent in this kind of "linguistically aware" approach to place naming. Justifiably, it alerts critical scholars not to ignore the "geographical individuality" (to utilize a concept of Paul Vidal de la Blanche) of places in the face of spatially overarching economic and cultural processes (cf. Cosgrove 1983). From another angle, the onomasticians' insistence upon the referentiality of all proprial expressions also carries the potential for bringing "down to earth" *in toto* conceptualizations of language as *the* model of culture adopted in some textualist versions of constructionism and poststructuralism (see e.g. Lefebvre 1991, 5–7, 16–17; Mitchell 1995). Quite simply, an anti-relativistic view that toponyms literally "make sense" in and of people's lives through pinpointing specific places can be seen as a response to the relativistic argument that "there is no outside-text" (Derrida 1994, 158). Unfortunately, the actual research undertakings by linguistically trained onomasticians have tended to leave such promises unfulfilled because of an unevenly strict foci on isolated *langue* systems (*á la* Saussure)—a methodological trick that brackets everything "extra-linguistic" outside their analyses. Indeed, for a critical researcher of toponymy there always lurks a danger of succumbing too deeply to the minutiae or normalized rules of specific languages, leading to a myopia regarding wider social and cultural processes. Given that naming a place is always a socially embedded act, one that involves power relations, the "pure" linguistic standpoint remains inadequate for the critical study of toponymy. Accordingly, whilst we agree with the need to be specific about what type of "naming" we are talking, we advocate an understanding of place names as "social facts" embedded in intricate cultural interrelations and tension-filled conceptions of space. As such, the practices of place naming are also caught up, in any given society, in the power and possibilities of "making places."

This leads us to the need to define a further concept, that of "place." We suspect that another reason that linguists interested in place naming have avoided its political implications is because they have based their understanding of place chiefly on common-sense understandings rather than any theoretically grounded approaches (Kiviniemi 1990, 44–51). Place, of course, is in itself a contested concept: "sometimes competing, and occasionally confusing, claims ... have been made and continue to be made about" it by geographers, philosophers and others (Entrikin 1991, 14). Interestingly, understandings of place that ignore its embeddedness in uneven power relations have not been uncommon in Geography and other human disciplines. Thus in the 1970s, for instance, philosophically oriented humanistic geographers argued, *contra* the formally geometric conceptions of space by positivist geographers, that a unique essence of each place is rooted in the deeply personal memories and intentions that individuals attach to it (see e.g. Guelke 1974; Relph 1976; Ley and Samuels 1978; see also Light and Smith 1998). A central problem with such theories was, of course, that they failed to take into account the wider societal and cultural processes involved in the production of places. As a result, many humanistic geographers also turned a blind eye to places as particular sites of exclusion and powerlessness for members of marginalized groups (see e.g., the critique in Rose 1993, 41–61).

In contrast to the focus on the *meaning of place* as understood by a trans-historical, universal subject endorsed by humanistic geographers and phenomenological philosophers, more recent interventions in critical geography and related fields of social theory resonate much more strongly with our conceptualizations of place and place naming. In Kirsten Simonsen's (1996, 502–3) conception of social spatiality, for instance, places are "in all scales of social life—from microphenomena, such as daily practices of work or consumption, to macrophenomena, such as the international division of labour"—both caught up in and constitutive of social relations. Similarly,

Doreen Massey (1991, 28) has influentially defined place as a unique conjunction of social relations:

> [W]hat gives a place its specificity is not some long internalised history but the fact that it is constructed out of a particular set of social relations, meeting and weaving together at a particular locus. [...] Each "place" can be seen as a *particular, unique, point of their intersection.* It is, indeed, a meeting place. Instead then, of thinking of places as areas with boundaries around, they can be imagined as *articulated moments in the networks of social relations and understandings* .... (Emphases added)

Nevertheless, neither humanistic geographers nor theorists of socially "produced" space have put much explicit emphasis on naming in their conceptualizations of place. Generally speaking, critical theoretical analyses of the role that naming plays in the construction of places have only recently begun to live up to their promise (for partial earlier exceptions, see Relph 1976, 16–17; Watts 1988; Tuan 1991; Massey 1995; Paasi 2002). Perhaps the key exception in this regard is de Certeau's (1984) book *The Practice of Everyday Life*, which gives insights not only into urban place naming as a normalizing strategy, but also as a contested cultural realm. While sharing a view that "the ideas of 'space' and 'place' require each other for definition" (Tuan 1977, 6), de Certeau otherwise turns the humanistic idealization of place on its head. According to him, place-making strategies by technical experts and other elites essentially entail—besides "imaginary totalizations produced by the eye"—socially sanctioned naming practices. As manifestations of power, official nomenclatures articulate "constellations that hierarchize and semantically order the surface of the city" in line with "proper" rationalities and historical justifications (de Certeau 1984, 93–117). As Carter (1988) has also suggested with regard to colonial naming practices, the hegemonic acts of naming transform polymorphous and uncontrollable "space" into a finite system of neatly isolatable, stabilized and interconnected "places." Interestingly, de Certeau (1984, 104–5) simultaneously maintains that since toponyms are endowed with the "ability to signify," people in marginal societal positions are able to transform the intended meanings of their "first definitions."

Importantly, what de Certeau (1984) generally does with his theoretical analysis of urban place names is to sensitize us to the power and spatiality inherent in Wittgenstein's (1968) notion of naming as an act that "lifts up" something into effective social-discursive existence. In the first place, hegemonic practices of place naming are important aspects of the practical and symbolic governing of social space at various spatial scales. In this role, toponyms pinpoint (individualize) places necessary to the functioning of societies, invest them with variable but politically expedient meanings, and simultaneously totalize space by imposing hierarchical, often monolingual spatial nomenclatures (see e.g., Rose-Redwood, this volume). In this way, for example, toponyms are caught up with spatial processes of governmentality (Larner 2000). Yet, crucially, this should not be taken to imply that naming practices would somehow overshadow other aspects in the production of governable space(s). On the contrary, place naming strategies almost invariably operate inextricably in tandem with other material and discursive processes equally fundamental for the operation of power. In complex ways, *extra-linguistic* (e.g. spatio-temporally routinized practices, ceremonial rituals, boundary-markings and -fencings, cartographic and other representational delineations) and *linguistic* (not only place naming as such, but also coding places through numbering them, calling places by classificatory common nouns, oral and written discourses on places, and so on) *measures of place-making* intertwine and support each other. A crucial viewpoint at this point is that just as the other aspects of the production of space tend to reinforce hegemonic social relations, the official naming of places also frequently yields repressive consequences for groups that do not embrace the "elevated" language(s) and ideological messages of dominant toponymies.

Intriguingly, de Certeau (1984) insists that a lived social space is made practically and symbolically "habitable" through cognitive-orientation (or as he puts it: "constellations of names ... are stars directing itineraries") and the often ideologically evasive affective meanings that place names enjoy in everyday life (see also Roberts 2000, 273). In this regard, his approach paves the way to understanding the socio-cultural "after-life" of the once coined toponyms in ways that are sensitive to power. Once part of the multiplicity of "language-games" among a speech community (Wittgenstein 1968), place names frequently become "shorthands" that articulate and crystallize broader cultural meanings—think here only of Wall Street as a contested symbol of global finance markets and wealthy lifestyles. In this role, place names and their attached meanings operate both as (inter)subjective aids of memory (Rose-Redwood 2008, 433) and semiotic "intertexts" reproduced in everyday conversations and various kinds of media representations (Entrikin 1991, 56). Especially among population segments that support the *status quo* of power relations, as well as in societies that are characterized by a high level of cultural entente, it is well possible that the hegemonic toponymies acquire—as a part and parcel of people's intuitive "common sense"– a largely taken-for-granted status. Conversely, in cases where socio-cultural tensions are paramount, toponymic struggles may surface in a variety of everyday forms: from organized re-naming campaigns to the spontaneous use of alternative names and pronunciations, grouses against the renditions of history in official toponymies, refusals to unlearn marginalized names, and so on. In the following pages of this book, issues revolving around place names as foci of conflicting meanings are explored across a range of historical and geographical contexts.

Approaches in the Volume

Crucially, the chapters of *Critical Toponymies* differ from the mainstream of toponymic research in terms of their explicit interest in the entanglements of place names with power relations and social antagonisms. As we have shown in this introduction, the reluctance to address issues of power related to the capacity of geographical naming to "call something into being, to render invisible visible, to impart a certain character to things" (Tuan 1991, 688) has been an overarching lacuna in the traditional orientations of name research (see also Šrámek 2008, 11). While we believe that none of the authors of *Critical Toponymies* would deny the convenience of toponyms in communicating locational facts, and even less so the affective meanings people attach to them, all of them adopt a wider and most importantly, critical societal perspective on the topic.

In gathering together the book's contributions, we have endeavoured to illuminate toponymic power strategies and contestations especially in different national(ist), colonial/postcolonial and urban contexts. Meanwhile, the organization of the book attempts to give the reader a rough outline of the development of critical toponymic research from the 1990s to the present day. With this goal in mind, we felt it important to include a number of already-published works to illustrate some of the key papers in the relatively recent evolution of this genre of research. In addition, the book elucidates some of the newest work on the politics of naming places, indicating possible future directions for toponymic research that is sensitive to relations of power.

The book's array of case studies begins with an illustration of toponymic contestations in postcolonial contexts, in which Lawrence Berg and Robin Kearns focus on objections to re-instatement of Maori place names in the Otago/Murihiku region of Aotearoa/New Zealand. Their discourse analytic reading unveils an ideology of hegemonic masculinism behind the apparently commonsense reactions against the attempt to rectify iniquities of colonial history.

In a revised version of a now-classic paper on critical urban toponymies, Maoz Azaryahu highlights two fundamental, power-embedded and interrelated toponymic functions: orientation and symbolic commemoration. Based on the analysis of commemorative street names located mainly in European metropolises, he argues that street nomenclature assists people to orient themselves across the urban system, and works powerfully to render natural specific historical narratives that ultimately serve the goals of the ruling socio-political order.

Brenda Yeoh continues the book's analyses of street names by exploring nationalist landscape inscriptions in post-independence Singapore in her chapter. Through combining textual landscape methodology with theories on nationalist and post-colonialist state-formation, she shows how the gradual multi-racialization of the colonial legacy in Singapore's namescape has mirrored substantial shifts in the hegemonic policies of Singaporean nation-building since the mid-1960s.

Garth Myers advocates in his chapter for more research on "the diversity of place names utilized for different purposes by various players" through engaging fieldwork methods. By way of case study, Myers excavates spatial nicknames and associated delineations of socio-spatial otherness by hegemonic and marginal actors in the impoverished Ng'ambo area of Zanzibar City.

In a chapter informed by postcolonial and psychoanalytical theories, Doug Herman emphasizes the ways the Western colonization of Hawai'i was underpinned by the imposition of masculine logos onto the indigenous landscape. Even today, as the safely exotic bilingual name "The Aloha State" suggests, the seeming respect for Hawaiian culture is based on the idealizations of the islands as a "gracious, warm, charming, welcoming" tourism paradise.

In her chapter on the anglicization of the toponymy of Ireland by British Ordnance Survey officers in the nineteenth century, and a more recent campaign against the obliteration of traditional "townland" names in Northern Ireland, Catherine Nash tackles complexities surrounding re-naming projects in postcolonial settings. According to her, a common predilection to idealize the purity of pre-colonial cultures often oversimplifies postcolonial cultural and linguistic relationships.

In their revisit of the cultural politics of recognition in contemporary Aotearoa/New Zealand, Robin Kearns and Lawrence Berg underscore the benefits of taking into account aural aspects of place names. Illuminatingly, their analysis of the sympathetic pronunciations of Maori place names in media, and often hostile reactions towards these by conservative representatives of the *Pakeha* (non-Maori) population, represents an exemplary "move beyond the visual" in the research on the constitution of places.

Derek Alderman elucidates how commemorative street naming becomes frequently embroiled in disputes over what is historically significant and worthy of public recognition. In empirical terms, he focuses on tensions around a failed campaign to name a perimeter highway in Bulloch County, Georgia, after the civil rights leader Martin Luther King Jr.

In an article dealing with the adoption of house (and street) numbering systems in the United States, Reuben Rose-Redwood provides a persuasive example of the importance of the technocratic objective of "spatial legibility" for toponymic modernization. Inter alia, he shows how the numbering of properties was precipitated by nineteenth century engineering journals, celebrating it as a scientific means of "economizing time" with respect to a multitude of urban functions.

Kaisa Rautio Helander's chapter focuses on the contestation between Norwegian nation-building and indigenous Sami people's attempts to gain recognition for their local naming practices in Norway. Rautio Helander draws on linguistic traditions, yet her work illustrates that such approaches might be sensitive to power relations embedded in linguistic changes in the landscape. Her work provides an excellent empirical example of "toponymic silencing" (Harley 2001, 99).

In the penultimate chapter, Jani Vuolteenaho and Terhi Ainiala illuminate the legacy of authority-led name planning in European cities, as well as ideological challenges that the present-day neoliberalization of urban policies have been posing to them. On this basis, they analyze some of the power relations implicated in market-led, image-orientated shifts in the "linguistic landscaping" in a coastal suburb of Helsinki, Finland.

In the final chapter of the book, Derek Alderman opens new directions to critical name research in the light of naming practices in cyberspace. Based on a case study of an internet domain named after Martin Luther King Jr. and used to hoax internet surfers and promote the goals of a white supremacist organization, he convincingly argues against a techno-utopian vision of cyberspace as "the great equalizer" of geographical and racial divisions.

Taken altogether, these contributions provide the reader with a first-of-its-kind introduction to a wide variety of theoretical and methodological approaches to understanding the contested character of naming places. They highlight how toponyms can be theoretically understood as power-embedded, ubiquitous aspects of cultural geography that work in two registers. In one, they betray a complex array of hegemonic power relations that underpin naming and place making. In essence, the hegemonic practices of place naming do some of the heavy work of naturalizing and reinforcing the dominance of existing social orders. In the other register, place names represent a contested cultural realm through which people in marginal societal positions are able to express their own place-bound identities and counter-hegemonic political goals. In bringing this collection of essays together, it is our sincere hope that critical toponymic research gains a wider audience, and critical analyses of the political implications of the naming of places become more the norm than the exception.

References

Allen, J. and Massey, D. (eds) (1995), *Geographical Worlds* (Oxford: Oxford University Press).

Arnold, W. (1875/1983), *Ansiedlung und Wanderungen Deutscher Stämme, Zumeist nach Hessischen Ortsnamen* (Köln: Böhlau Verlag).

Azaryahu, M. (1986), "Street Names and Political Identity: The Case of East Berlin," *Journal of Contemporary History* 21: 4, 581–604.

—— (1996), "The Power of Commemorative Street Names," *Environment and Planning D: Society and Space* 14, 311–30.

Barnes, T. and Duncan, J. (eds) (1992), *Writing Worlds: Discourse, Text and Metaphor in the Representation of Landscape* (London: Routledge).

Basso, K. (1996), *Wisdom Sits in Places* (Albuquerque: University of New Mexico Press).

Blonsky, M. (ed.) (1985), *On Signs* (Baltimore, MD: Johns Hopkins University Press).
Bourdieu, P. (1991), *Language and Symbolic Power* (Cambridge: Polity Press).
Carter, P. (1987), *The Road to Botany Bay* (London: Faber and Faber).
Cassirer, E. (1955), *The Philosophy of Symbolic Forms. Volume 1: Language* (New Haven, CT: Yale University Press).
Cook, I. et al. (eds) (2000), *Cultural Turns/Geographical Turns* (Harlow: Prentice Hall).
Cosgrove, D. (1983), "Toward a Radical Cultural Geography: Theoretical Problems," *Antipode* 15: 1, 1–11.
Darnell, R. (2001), *Invisible Genealogies. A History of Americanist Anthropology* (Lincoln: University of Nebraska Press).
de Certeau, M. (1984), *The Practice of Everyday Life* (Berkeley: University of California Press).
—— (1985), "Practices of Space," in Blonsky, M. (ed.).
Derrida, J. (1994), *Of Grammatology* (Delhi: Motilal Banarsidass Publishers).
—— (1986), *Glas* (Lincoln: University of Nebraska Press).
Diderichsen, P. (1962), *Elementær Dansk Grammatik* (Copenhagen: Gyldendal).
Duncan, J.S. (1980), "The Superorganic in American Cultural Geography," *Annals of the Association of American Geographers* 70: 2, 181–198.
Eichler, E. et al. (eds) (1995), *Namenforschung/Name Studies/Les Noms Propres*, Volumes I–III (Berlin: Walter de Gruyter).
Eller, N. et al. (eds) (2008), Namen und Ihr Konfliktpotential im Europäischen Kontext (Regensburg : Edition Vulpes).
Entrikin, J.N. (1991), *The Betweenness of Place* (Baltimore, MD: Johns Hopkins University Press).
Errington, J. (2001), "Colonial Linguistics," *Annual Review of Anthropology* 30, 19–39.
Foucault, M. (1974), *The Order of Things. An Archeology of the Human Sciences* (London: Routledge).
—— (1980), *Power/Knowledge* (New York: Pantheon).
Gal, S. and Irvine, J.T. (1995), "The Boundaries of Languages and Disciplines: How Ideologies Construct Difference," *Social Research* 62: 4, 967–1004.
Gelling, M. (1984), *Place-names in the Landscape* (London: J.M. Dent and Sons).
Glennie, P. (2000), "Place Names," in Johnston, R.J. et al. (eds).
Golledge, R.G. et al. (eds) (1988), *A Ground for Common Search* (Goleta: Santa Barbara Geographical Press).
Grim, J. (1838), "Über Hessische Ortsnamen," *Zeitschrift des Vereins für Hessische Geschichte und Landeskunde* 2, 132–54.
Guelke, L. (1974), "An Idealist Alternative in Human Geography," *Annals of the Association of American Geographers* 64: 2, 193–202.

Haggett, P. (1972), *Geography: A Modern Synthesis* (New York: Harper and Row).

Hanks, W.F. (2005), "Pierre Bourdieu and the Practices of Language," *Annual Review of Anthropology* 34, 67–83.

Harley, J.B. (1992), "Deconstructing the Map," in Barnes, T. and J. Duncan (eds).

—— (2001), *The New Nature of Maps* (Baltimore, MD: Johns Hopkins University Press).

Haubrichs, W. (1995), "Namenforschung in Deutschland bis 1945," in Eichler et al. (eds).

Haubrichs, W. (2008), "'Germanen oder Kelten?' Die Toponymische Deutungshoheit im Elsass und in Lothringen Zwischen 1870 und 1918 (und danach)," in Eller et al. (eds).

Holt-Jensen, A. (1999), *Geography. History and Concepts* (London: Sage).

Jackson, P. (1989), *Maps of Meaning* (London: Routledge).

—— (2000), "Rematerializing Social and Cultural Geography," *Social and Cultural Geography* 1: 1, 9–14.

Johnston, R.J. et al. (eds) (2000), *The Dictionary of Human Geography*, 4th edn (Oxford: Blackwell).

Kadmon, N. (1997), *Toponymy. The Lore, Laws and Language of Geographical Names* (New York: Vantage Press).

Kearns, R.A. and Berg, L.D. (2002), "Proclaiming Place: Towards a Geography of Place Name Pronunciation," *Social and Cultural Geography* 3, 283–302.

Kiviniemi, E. (1990), *Perustietoa paikannimistä* (Helsinki: SKS).

Kroeber, A.L. (1916), "California Place Names of Indian Origin," *University of California Publications in American Archaeology and Ethnology* 12: 2, 31–69.

Laclau, E. (2005), *On Populist Reason* (London: Verso).

Larner, W. (2000), "Neo-liberalism: Policy, Ideology, Governmentality," *Studies in Political Economy* 63, 5–25.

Lefebvre, H. (1991), *The Production of Space* (Oxford: Blackwell).

Leighley, J. (ed.) (1963), *Land and Life: A Selection from the Writings of Carl Ortwin Sauer* (Berkeley: University of California Press).

Ley, D. and Samuels, M.S. (eds) (1978), *Humanistic Geography: Prospects and Problems* (London: Croom Helm).

Light, A. and Smith, J.M. (eds) (1998), *Philosophies of Place* (Lanham, MD: Rowman and Littlefield).

Maskit, J. (1998), "Something Wild? Deleuze and Guattari and the Impossibility of Wilderness," in Light and Smith (eds).

Massey, D. (1991), "Global Sense of Place," *Marxism Today* (June), 24–9.

—— (1995), "Imagining the World," in Allen, J. and Massey, D. (eds).

Mitchell, D. (1995), "There's No Such Thing as Culture: Towards a Reconceptualisation of the Idea of Culture in Geography," *Transactions of the Institute of British Geographers* 19, 102–16.

Murray, T.E. (1994), "On the Need to Legitimize Onomastics in Academia," *Names* 42, 67–77.

Nash, C. (1999), "Irish Placenames: Post-Colonial Locations," *Transactions of the Institute of British Geographers* 24, 457–80.

Nicolaisen, W.F.H. (1976), "Words as Names," *Onoma* 20, 142–63.

Norton, W. (2000), *Cultural Geography: Themes, Concepts, Analyses* (Oxford: Oxford University Press).

Paasi, A. (2002), "Bounded Spaces in a Mobile World," *Tijdschrift voor Economische en Sociale Geografie* 93: 2, 137–48.

Palmer, B. (1990), *Descent into Discourse* (Philadelphia, PA: Temple University Press).

Palonen, K. (1993), "Reading Street Names Politically," in Palonen, K. and Parvikko, T. (eds).

Palonen, K. and Parvikko, T. (eds) (1993), *Reading the Political: Exploring the Margins of Politics* (Tampere: The Finnish Political Science Association).

Philo, C. (2000), "More Words, More Worlds. Reflections on the "Cultural Turn" and Human Geography," in Cook, I. et al. (eds).

Pinchevski, A. and Torgovnik, E. (2002), "Signifying Passages. The Signs of Change in Israeli Street Names," *Media, Culture and Society* 24: 3, 365–88.

Pitkänen, R.L. and Mallat, K. (eds) (1999), *You Name It. Perspectives on Onomastic Research* (Helsinki: Finnish Literature Society).

Pred, A. (1990), *Lost Words and Lost Worlds* (Cambridge: Cambridge University Press).

Raper, P.E. (1995), "South African Onomastics," in Eichler, E. et al. (eds).

Ratzel, F. (1891), *Anthropogeographie. Zweiter Teil: Die Geographische Verbreitung des Menschen* (Stuttgart: Verlag von J. Engelhorn).

Relph, E. (1976), *Place and Placelessness* (London: Pion).

Roberts, J.M. (2000), "The Enigma of Free Speech: Speakers' Corner, the Geography of Governance and a Crisis of Rationality," *Social and Legal Studies* 9, 271–92.

Rose, G. (1993), *Feminism and Geography* (Minneapolis: University of Minnesota Press).

Rose-Redwood, R.S. (2006), "Governmentality, Geography, and the Geo-Coded World," *Progress in Human Geography* 30: 4, 469–86.

—— (2008), "From Number to Name: Symbolic Capital, Places of Memory, and the Politics of Street Naming in New York City," *Social and Cultural Geography* 9: 4, 431–52.

Russell, B. (1912), *The Problems of Philosophy* (Oxford: Oxford University Press).

Said, E. (1979), *Orientalism* (New York: Vintage Books).

Sauer, C.O. (1925/1963), "The Morphology of Landscape," in Leighley, J. (ed.).

Schwarz, G. (1989), *Allgemeine Siedlungsgeographie*, 4. Auflage (Berlin: Walter de Gruyter).
Searle, J. (1969), *Speech Acts* (Cambridge: Cambridge University Press).
Simonsen, K. (1996), "What Kind of Space in What Kind of Social Theory?," *Progress in Human Geography* 24, 494–512.
Šrámek, R. (2008), "Typologisierendes zu Namenkonflikten," in Eller et al. (eds).
Stemshaug, O. (1995), "Name Studies in Norway," in Eichler et al. (eds).
Summerell, O.F. (1995). "Philosophy of Proper Names," in Eichler et al. (eds).
Taylor, I. (1896), *Names and Their Histories* (New York: Macmillan and Co.).
Trask, R.L. (2007), *Language and Linguistics. The Key Concepts* (New York: Routledge).
Tuan, Y.-F. (1977), *Space and Place. The Perspective of Experience* (London: Edward Arnold).
—— (1991), "Language and the Making of Place: A Narrative-Descriptive Approach," *Annals of the Association of American Geographers* 81, 684–96.
Vološinov, V. (1990), *Kielen Dialogisuus* (Tampere: Vastapaino).
Wagner, P.L. and Mikesell, M.W. (1962), "General Introduction: The Themes of Cultural Ceography," in Wagner, P.L. and Mikesell, M.W. (eds).
—— (eds) (1962), *Readings in Cultural Geography* (Chicago: University of Chicago Press).
Watts, M. (1988), "Struggles over Land, Struggles over Meaning: Some Thoughts on Naming, Peasant Resistance and the Politics of Place," in Golledge et al. (eds).
Willems, K. (2000), "Form, Meaning, and Reference in Natural Language: A Phenomenological Account of Proper Names," *Onoma* 35, 85–119.
Withers, C.W.J. (2000), "Authorizing Landscape: 'Authority', Naming and the Ordnance Survey's Mapping of the Scottish Highlands in the Nineteenth Century," *Journal of Historical Geography* 26, 532–54.
Wittgenstein, L. (1968), *Philosophische Untersuchungen—Philosophical Investigations* (Oxford: Basil Blackwell).
Zelinsky, W. (1982), "By Their Names You Shall Know Them: Toponymic Approach to the American Land and Ethos," *New York Folklore* 8, 85–96.
—— (1997), "Along the Frontiers of Name Geography," *Professional Geographer* 49, 465–6.
—— (2002), "Slouching toward a Theory of Names: A Tentative Taxonomic Fix," *Names* 50, 243–62.
Zilliacus, K. (1997), "On the Function of Proper Names," in Pitkänen, R.L. and Mallat, K. (eds).

Chapter 2

Naming as Norming: "Race," Gender and the Identity Politics of Naming Places in Aotearoa/New Zealand

Lawrence D. Berg and Robin A. Kearns

Introduction

The naming of places is a key component in the relationship between *place* and *the politics of identity* in contemporary societies. In this sense, *naming* is a form of *norming*.[1] Names are part of both a symbolic and material order that provides normality and legitimacy to those who dominate the politics of (place) representation. Saul Cohen and Nurit Kliot (1992), for example, have illustrated the way that the Israeli nation-state selects place names for the administered territories of the Golan, Gaza and West Bank in order to reinforce national Zionist ideologies. Between 1977 and 1992 Biblical and Talmudic place names were introduced by the ruling right-wing Likud bloc to project Israel as the rightful heir to the holy land (Cohen and Kliot 1992, 664–6). Such naming produces a form of "linguistic settlement" that *produces* places through the simple enunciation of intentions to do so. As Paul Carter (1987, 137) observes of the Australian colonial context, "the would-be settler was more than ever obliged to settle the country rhetorically, rather than etymologically: he had, more than ever, to conjure up the object of his desire and, through the act of articulating it, to bring it into being." Wilbur Zelinsky (1983, 1) has directly linked "the intensity and exuberance of nationalism in the young United States' with "the great number and frequency of nationalistic place names to be seen [there]." Zelinsky documented 3,771 counties and minor civil divisions in the United States named after national heroes and notables (1983, 14–15). At the same time, he found only 43 such cases in Canada (Zelinsky 1983, 15). In the South Pacific region, Ron Crocombe (1991) suggests naming places was one form of extending ownership claims. About 600 years ago in Rarotonga, for example, the conqueror Tangiia dominated the indigenous inhabitants by force of arms, but he asserted his hegemony by renaming their lands and places. Crocombe also notes the gendered character of naming places, observing that such activities were "an attribute of masculinity, of power, of control,

1 We are indebted to Anna Yeatman for this phrase.

which were male prerogatives in all cultures at the time the names were given" (Crocombe 1991, 1).[2] The masculinity of naming is certainly reflected in the many New Zealand[3] places that were named for British male military figures such as Hamilton, Edgecumbe, Marlborough, Napier, Nelson, Palmerston, Picton and Wellington (Carter 1987; Pawson 1992).

We argue that naming places reinforces claims of national ownership, state power, and masculine control. Indeed, "affixing names to places is inextricably linked to nation-building and state formation" (Cohen and Kliot 1992, 653). Naming places is thus intricated in the ideological processes involved in the formation of what Benedict Anderson (1983) has called "imagined communities." Such imagined communities arise from the historically-contingent and culturally constructed phenomenon of nationalism. Nationalism results from a complex configuration of emotions, beliefs and attitudes rooted in the perception of singular cultural attributes, myths, and traditions, leading to the belief in a "national community" (Zelinsky 1986, 171–2; also see Blaut 1986). It is a complex combination of nationalist ideologies that lead men (and now women) to go to war to "defend their country."

Nationalist inspired "communities" are *imaginary* for numerous reasons, including the fact that they are represented as universal—that is applying to all members of a socio-geographic polity—when they are actually highly specific to members of the hegemonic class. In the West, this hegemonic class is "White," European, bourgeois and masculine, or what Gillian Rose (1993, 6) terms the unmarked "master subject." In the case of New Zealand, we suggest that nationalist and statist ideologies lead to a representation of New Zealanders in terms of the beliefs, attitudes and cultural practices of a bourgeois Pakeha[4] masculine subject. All those who are marked as Other by this hegemonic "white" masculinity tend to be marginalized.

We make an important distinction here between *men* and a *hegemonic white masculine subject*. Just as not all men will fit the mold of the masculinist ideal, some women (former British Prime Minister Margaret Thatcher, for instance) have strongly masculine (and masculinist) subjectivities (Donaldson 1993). Likewise, not all Pakeha fit the ideal of "whiteness" necessary for acceptance as Pakeha, nor are Maori men (and women) excluded from a Pakeha subject position (New Zealand First Party leader and Member of Parliament,

---

2 See Spender (1985) for a more thoroughgoing account of the gender politics of naming.

3 We shift back and forth between the name "New Zealand" and its Maori equivalent, "Aotearoa," as a way of highlighting the notion of place (names) as sites of contested discourses.

4 We use "Pakeha" here to designate New Zealand born people of "European" descent. Pakeha is a contested term in Aotearoa/New Zealand (Larner and Spoonley 1994; Spoonley 1994) and we acknowledge the tendency inherent in such terminology for obscuring difference.

Winston Peters, for example, seems to cross back and forth between Maori and Pakeha subjectivities, although he appears to do so by articulating a strongly masculinist subject position). As Vron Ware (1992, 237) observes, there is an almost endless number of hierarchically ordered binary distinctions in Western thought and when combined with binary notions of gender, "race" and class, they can articulate in contradictory ways:

> Civilization, for example, is the other side of savagery. In some contexts, white women might indeed be associated with the idea that female nature is inherently uncivilized, primitive when compared to men, and lacking in self-control. In the context of imperialism or modern racism, the dominant ideology would place white women firmly in the civilized camp, in opposition to non-European women whose lack of social and political rights are to be read as a mark of cultural savagery.

With this in mind, then, we argue that men and women, Pakeha and Maori, "are caught in a complex series of (historically and geographically specific) discursive positions, relations and practices" (Rose 1993, 10). The relationship that individual men, women, and racialized persons have to this hegemonic "white" masculinity is thus likely to be ambiguous, contradictory and problematic.

In this chapter we examine some of these complex positions as they are worked out in relation to the politics of identity and place in Aotearoa. We acknowledge that this is a rather too ambitious project, given the nascent body of literature discussing such issues in New Zealand. Nevertheless, we shall attempt, in preliminary fashion at least, to outline some of these discursive politics of/in place. In particular, we consider a specific example of the politics of naming places, and how such politics are intertwined with contemporary discourses of "race" and gender identities. While we accept in principle the notion that there exists an ongoing ideological "struggle" between Maori[5] and Pakeha New Zealanders for ethnic, "community" and "national" identity, we argue that such identity politics tend to be highly complex and nuanced by articulations with notions of gender, class and sexuality. We attempt to illustrate part of this complexity by outlining some the ways that mutually constitutive notions of "race" and gender are implicated in the politics of naming places. Finally, we also recognize our own problematic relationship, and articulation with, these discursive positions. We cannot avoid the power-knowledge of our social position, but we suggest that our partiality gives us a specific perspective from *inside* various discourses of Pakeha racism and

---

5 In *te reo Maori* (the Maori language), plurals are never signified by the addition of an "s" at the end of words, but instead by the context in which they are used. There are some parallels in English usage, for example with words such as moose, elk, or deer.

masculinism. This position is not unproblematically "innocent" and it is not (nor should it be) closed to contestation.

Discourse Analysis

A problem arises when attempting to discuss the "method" of discourse analysis. The notion of "method" is itself the product of a positivist discursive frame which constructs an acceptable set of procedures that lead to the "discovery" of truth. In the type of discourse analysis we utilize, on the other hand, there is no analogous set of codified procedures for getting "results" (Wetherell and Potter 1992). As will become clear, discourse analysis relies upon the skills and perspective of individual analysts, and there is no *singular* way of reading texts nor is there a single truth to be "discovered" by such approaches. Indeed, as Margaret Wetherell and Jonathan Potter (1992: 101) observe, "much of the work of discourse analysis is a craft skill, something like bicycle riding or chicken sexing that is not easy to render or describe in an explicit or codified manner." Furthermore, there can never be a *complete* analysis of texts, for there will always be new interpretations that can be developed. Nevertheless, the craft of discourse analysis does rely on specific understandings of the relationship between language and social practice. With this in mind, the following section discusses our approach to analysing the rhetorics of submissions to the New Zealand Geographic Board (NZGB).

We analysed submissions to the NZGB concerned with an attempt to change three place names in the Otago (*Murihiku*) region of the South Island (*Te Waipounamu*) of New Zealand. We outline the details of the case itself later in the chapter. Our approach was to read submissions with a view to identifying patterns of ideas or constellations of thought relating to notions of "race," "ethnicity," or "culture." From this reading we identified a number of discursive constructs relating to ideas that we group under the descriptive terms "race talk," "culture talk" and "nation talk." As with all representations, our discussion of the rhetorics of submissions to the Geographic Board is necessarily selective and partial. Our analysis of these rhetorics, outlined in a later section of the chapter, centres on attempts to link them to wider discursive patterns circulating—both socially and spatially—in New Zealand. The analysis is thus dependent upon specific understandings of the relationship between language and social practices related to racism, masculinism and nationalism.

We are critical of the role that naming places potentially plays in the legitimation of certain spatialized rhetorics of "race" and gender relations. Our critique is allied partly with the contemporary literature on "race" and the geography of racism. Authors working within this tradition (e.g. Anderson 1991; Berg 1993b; Jackson 1987a, 1992; Jackson and Penrose 1993; Smith 1989; Spoonley 1993) are highly critical of essentialist notions of "race," arguing instead that "race" is a socially constructed historical phenomenon that must itself be explained through historical analysis.[6] Our purpose, then, is not to "uncover" the falsehood of "racial" constructs, but instead "to expose the falseness of our unquestioning acceptance of these constructs from which their legitimacy derives" (Jackson and Penrose 1993, 3). "Race talk" is thus seen as an articulation of locally constructed rhetorics with wider discourses about the immutability of various human "races."

We have already discussed the "imaginary communities" constructed from nationalist imaginings; they also perform important roles in racist rhetorics. The nation-state, as Peter Jackson and Jan Penrose (1993, 9) note,

> is a crucial locus for the articulation of racist ideologies because of the extent to which it embodies the idea of "race" and legitimizes it through the granting or withholding of citizenship, the right to enter and remain within a country, and a host of other entitlements.

"Race" and "nation" gain their power because they are often taken-for-granted as "natural" categories that describe discernible material entities having immutable characteristics. As we have already argued, however, both concepts are imaginaries—socially constructed ideas that arise within historically specific constellations of power-knowledge. As social constructions, both "race" and "nation" can be seen as reifications of social struggles to define group identity. "Nation talk" represents some of the articulations between locally-constructed rhetorics of nationalism and these wider discursive ideas about "The Nation."

We are also interested in the role of "culture" as an integral aspect of the signifying system through which a social order is communicated and (re)produced (Williams 1981, 12–13). In a significant extension of contemporary theorizations of "culture," Don Mitchell (1995, 112) cogently argues that "culture" has no ontological status, and that it should be conceptualized as an *idea* that provides "a means of ordering and defining the world." In this sense, culture can be seen as a reification of social struggles between groups, an idea deployed by powerful social actors. Culture "in itself possesses no explanatory value" (Mitchell 1995, 113). Following Mitchell, then, we argue that—just like "race" and "nation"—"culture" is itself a socially constructed and contested concept that articulates with powerful social actors in society to help in the maintenance of hegemonic social relations. In this way, the "Culture talk" we found in submissions to the Geographic Board can be seen as part of an articulation between local struggles and wider discursive notions of "culture."

---

6 However, our constructivist position is tempered by Diana Fuss's (1989) arguments about the inescapable essentialism of anti-essentialism.

Peter Jackson (1994, 50–51) has recently criticized the tendency in the literature on "race" and racism to treat "race," gender and class as separate analytic variables. Jackson is following the pathbreaking work of feminist historian Vron Ware (1992), who argues that "race" and gender should be seen as *mutually constitutive*. We have already discussed the ways in which "white" women can occupy both sides of binary distinctions when "race" is part of the equation (Ware 1992, 237). Non-white men can also inhabit such contradictory positions. Les Back (1994) argues that working-class white males in South London appropriate black culture because blacks are constructed out of the "fear/desire" couplet. Black masculinities are acceptable to white males because they are hyper-masculine or fearsome in white construction. By contrast, Asian masculinities are unacceptable to white males, because Asian men are constructed as feminine, and they are undervalued. Notions about whiteness, masculinity, and femininity are thus intertwined in highly complex ways that serve to (re)constitute new ideas about both "race" and gender. We concur with these arguments, and we make some tentative steps here toward suggesting a number of ways in which discourses of "race" and gender (and class) might be mutually constitutive.

One key to understanding this complex relationship lies in comprehending the role that discourse plays in the *constitution* of society and space. In this sense, longstanding and widely circulating discourses of "race," "nations" and "culture" articulate with everyday rhetorics of gender in such a way as to produce new ideas about both "race" and gender. As wider discursive notions, such ideas provide taken-for-granted frameworks of *meaning* within which people construct their understanding of everyday relations between different social groups in Aotearoa.

Our analysis is centred on language because of the realization that "language lies at the heart of all knowledge" (Dear 1988, 266). However, the concept of discourse is much broader than language alone. Indeed, discourse might better be characterized as a framework that encompasses "particular combinations of narratives, concepts, ideologies and signifying practices, each relevant to a particular realm of social action" (Barnes and Duncan 1992, 8). It is "a set of capabilities, an ensemble of rules by which readers/listeners and speakers/audiences are able to take what they hear and read and construct it into an organized meaningful whole" (Ó Tuathail and Agnew 1992, 193). Discourse can thus be seen as a *signification system* that governs, controls, and produces knowledge in a culture (Foucault 1984). Finally, discourse is constitutive, it is "implicated in the instantiation and maintenance of social and economic relations" (Wetherell and Potter 1992, 60). Accordingly, discourse should not be seen in any instrumentalist fashion as merely the outcome, or reflection of dominant social relations.

This brings up an important question about the relationship between discourse and the agency (and intentionality) of the subject. We are working within an analytic framework that sees the subject as constituted in and through discourse. Thus, we argue that it is not necessary to discern how far actors are in control of their actions in order to undertake discursive analyses. On the other hand, it is sometimes helpful to treat narrative accounts *as if* they are the result of the active intention of a knowing subject. Accordingly, we adopt an approach to discourse analysis that utilizes "a double movement between styles of reading that emphasize the constitution of subjects and objects, and those that emphasize the ideological work of discourse" (Wetherell and Potter 1992, 93).

Place Names and the Social Construction of Place

As Cohen and Kliot (1992) have observed, place names are integral to the process by which people attach meanings to place. Many geographers have argued cogently that places and landscapes are key aspects of group identity (e.g. Relph 1976; Tuan 1974; Cosgrove 1983). As signifiers of place, place names can evoke powerful emotions within individuals and groups, and they thus "conform to the most classic definitions of symbolism" (Cohen and Kliot 1992, 655). In their ability to transmit symbol, image and meaning, they are an integral component of the "iconography of landscape" (Daniels and Cosgrove 1988). In Auckland, for example, place names like Mission Bay, Parnell and Remuera evoke much different symbolic imagery than do Mangere, Otahuhu or Papatoetoe. The former are associated with elite landscapes of wealthy, "white," Pakeha (heterosexual) families; the latter with landscapes of marginalized poor, working class, Maori and Pacific Islanders (Kearns et al. 1991).

Place imaginings can also be gender coded (Kolodny 1975). Such gender codings are multiple and contradictory in Aotearoa. In rural rhetorics, for example, there is a tendency to code rural landscapes as masculine, and therefore valorize them over the cities which are seen to be effeminate because they are places "of voluptuousness and fashion, of luxury and ease, where men lived a soft life" (Phillips 1980, 228). Cities are also inhabited by (feminized) office workers, not "productive" (masculinized) farmers and labourers. Cities and towns are for women then; farms and bush are the true place of the "kiwi bloke" (Berg 1994). This gendered reading of the landscape has been reinforced in New Zealand through popular novels such as Barry Crump's *A Good Keen Man* (1982) and John Mulgan's *Man Alone* (1949), and it contradicts the types of landscape gender codings found in many other places. Kolodny (1975) notes, for example, how the American rural landscape—associated as it is with nature (not culture)—has usually been coded as feminine. In complementary fashion, the urban landscape has often been coded as masculine. Indeed, as early as the mid-nineteenth century, the dominant discourse on the domestic sphere represented it as a site of feminine virtue, grace and charm in opposition to the masculine public sphere of aggressive competitive business (Poovey

1989; Tosh 1991). More recently the masculine coding of the public sphere has been attached to urban centres of capital accumulation, signified in the urban landscape by the phallic imagery of the skyscraper.

Clearly, these kinds of gendered rhetorics exist in New Zealand cities. Nevertheless, we speculate that alongside these dominant constructions there exists in New Zealand a contradictory set of ideas that code the rural as masculine. These rhetorics exist in the rural areas themselves, which are dominated by small capitalist farming and an attendant frontier culture. Much of this rural frontier culture has developed out of "the concern that as New Zealand became more urbanized, men were no longer exposed to the hard physical struggles of pioneering life and were becoming soft and effeminate" (Phillips 1987, 242). In rural rhetorics of New Zealand, the farm is associated with Man's productive transformation of nature into (horti)culture.

Places and their attendant imaginings are not given, but are instead "made" in the contested process of what James Duncan (1992, 37–9) terms "cultural (re)production." In this regard, places do not exist simply as ontological givens, but instead they are produced through complex constellations of power-knowledge. Space is a social product, but it also "serves as a tool of thought and action ... in addition to being a means of production it is also a means of control, and hence of domination, of power" (LeFebvre 1991, 26). This can quite clearly be discerned in colonial projects of the eighteenth and nineteenth centuries where, as Paul Carter (1987, 41) has argued, "Space itself was a text that had to be written before it could be interpreted." The projects of European colonialism often "wrote space" by naming places and mapping landscapes (Carter 1987).

Topographic maps were—and continue to be—an important form of spatial construction. They gain their rhetorical power from an hegemonic myth of mimesis, where maps are said to unproblematically re-present places in their entirety (Harley 1988a, 1988b, 1992; Wood 1992; Jacobs 1993). As Alison Blunt and Gillian Rose (1994, 8) observe, "Hegemonic claims for the mimetic representation of transparent space underpin both imperial mapping and many subsequent mappings of imperialism." In this way, maps can be seen as the "graphic tools" of colonialism, aiding in the social construction of notions of "empty lands" available for colonization (Blunt and Rose 1994, 9). Place names play a central role in mapping, for they provide the means of filling ostensibly empty space (on maps) with meaningful points of interest. The use (and reproduction) of place names can thus be seen as a means of communicating (often ideological) meaning about place. But these kinds of hegemonic constructions are never uncontested. As Henri LeFebvre (1991, 26) notes, even while produced space provides a means of control and domination, it also "escapes in part from those who would make use of it."

One example of the discursive politics over the social construction of place is provided by Maori (and some sympathetic Pakeha) attempts to re-name and reclaim places. In the sections that follow, we examine some of the local struggles to re-map Aotearoa, to re-appropriate social constructions of place in the name of post-colonial projects. We shall first set the context for our discussion by outlining two brief, and "competing," histories of Maori-Pakeha relations.

Re-presenting People and Place—Two Histories

Subaltern History

Raymond Nairn and Timothy McCreanor (1991, 248–9; also see Nairn and McCreanor 1990) posit the existence of two competing narratives of Maori-Pakeha relations in Aotearoa/New Zealand. First, there exists a revisionist, subaltern history of Aotearoa. Because it resonates with the "truths" of recent revisionist historical research (e.g. Belich 1986; Orange 1987; Walker 1990; Ward 1973), this subaltern history holds more validity for us than the dominant history, which we briefly sketch later in this section. In this subaltern narrative, the current relationship between Maori and Pakeha arises out of Britain's aggressive colonization of New Zealand, and the active opposition to this process by Maori. The British Crown and (some) Maori chiefs signed the Treaty of Waitangi in 1840, guaranteeing *te tino rangatiritanga*, or Maori sovereignty, and *mana whenua*, or control of land (Belich 1986; Orange 1987; Larner and Spoonley 1994; Walker 1990, 1992). It was not long before a collective forgetting of the Treaty began to beset Pakeha, allowing for a British Act of Parliament establishing "responsible" settler government without Maori representation (Nairn and McCreanor 1991). The Pakeha appetite for Maori land drove them to armed conflict in order to establish their dominance (Belich 1986). The settlers were never able to inflict any decisive military defeat, but they managed to grind Maori resources down to the point where they were unable to organize an effective resistance to Pakeha colonization (Nairn and McCreanor 1991). More than 1.6 million acres of tribal lands were confiscated during the land wars, ostensibly as punishment for Maori "rebellion" against the Crown (see e.g. Belich 1986; Orange 1987; Walker 1990; Ward 1974). The Native Land Court system, implemented in the 1860s, was a more successful instrument for divesting Maori of their lands. Between 1840 and 1911, the proportion of New Zealand land controlled by Maori dropped from 100 per cent to just 11 per cent (Pawson 1992, 21; also see Orange 1987; Stone 1973; Walker 1990; Ward 1974). Maori population loss was equally dramatic, dropping from about 80,000 in 1840[7] to just 41,993 individuals in 1891.[8] Conversely, the European settler population increased from a few individuals in 1840 to more than 770,000 in 1901 (Pool 1991, 61).

7 Pool (1991, 76) suggests that this is the "best estimate" of Maori population.

8 Census enumeration (Pool 1991, 76).

Since the early colonial era, Maori have struggled against the colonial state (and recently against international capital) for recognition of their rights.

In more recent times, this struggle has been reflected in a strong resurgence of Maori cultural forms, and increasing calls for self-determination (see, e.g., Awatere 1984; Walker 1990). The response of many Pakeha in general, and the state in particular, has been a number of often minor concessions (Barber 1989), along with a few significant ones such as formation of the Waitangi Tribunal[9] (see Stokes 1992). Maori have won a number of claims put to the Waitangi Tribunal, eventually forcing government recognition of land rights, allocation of greater fishing resources to Maori, and the establishment of numerous statutory bodies to deal with Maori concerns (Larner and Spoonley 1994; Walker 1990, 1992). These victories have, however, been tempered by the New Right monetarist policies implemented from the mid-1980s by both Labour and National Party governments (Britton et al. 1992; Kelsey 1990; Spoonley 1993, 1994). In addition, while some Pakeha actively supported Maori people in their quest for recognition of their rights, there still remains a large number of Pakeha for whom Maori rights are a significant source of anxiety.

Hegemonic History

The above story of Maori-Pakeha relations stands in sharp contrast to the hegemonic discourse circulating in New Zealand. Nairn and McCreanor (1991, 248–9) have best summarized this discourse, so we quote them at length here.

> The standard ideology says that Maori/Pakeha relations in New Zealand are the best in the world, rooted as they are in the honourable adherence to the outcome of a fair fight. Mutual respect for each others' strengths and tolerances of idiosyncrasies has integrated the Maori people into a harmonious, egalitarian relationship with the more recent arrivals; the whole working constructively for the common good.

Maori "failure," in the hegemonic account, is attributable to Maori themselves. Maori are often constructed as incapable of coping with the modern world due to inherent "racial," cultural, or personal flaws (Nairn and McCreanor 1991; Pawson 1992). The outcome of such thinking, is that Maori dissent is cast as resulting from the work of (both Maori and Pakeha)

9 The Waitangi Tribunal is a standing commission of inquiry created in 1975 to examine Maori Treaty grievances. The Tribunal initially was empowered only to examine claims dating from 1975, but the *Treaty of Waitangi Amendment Act, 1985* allows it to adjudicate claims dating from 1840 (see Stokes 1992).

"stirrers" or troublemakers who disturb existing harmonious "race" relations[10] to further their own ends. We shall now turn to a discussion of place naming in New Zealand in order to examine more closely some of these interpretive repertoires concerning Maori people and place names and the ways they might articulate with wider discursive constructs of "race" and gender.

Place Names and the Politics of Identity

The discoverers and original inhabitants of Aotearoa were Polynesian peoples who probably migrated from the region that encompasses the present-day Cook, Society, and Marquesas Islands (Davidson 1992). The date of first settlement is uncertain, with estimates ranging between 800 and 1100 AD (Davidson 1992, 6; cf. Walker 1990, 28). Polynesians had therefore inhabited Aotearoa for at least 500 years before the first Europeans "discovered" the Islands. The *tangata whenua* ("people of the land" or "we who stand here") had developed a sophisticated social structure based upon kinship ties, spirituality and *mana*, or status. *Mana*, based as it was upon lineage systems, involved both men and women equally: "Retaining their own identity and *whakapapa* (genealogy) was of utmost importance and children could identify with the kinship group of either or both parents" (Pere, cited in Larner and Spoonley 1994). Spirituality and *mana* were (and remain today) closely connected to the land.[11] Indeed, the land was more than an economic base: "land, kinship, and individual identity were united for the Maori" (James and Saville-Smith 1989, 17). Moreover, claims to land were not gender-divided; they could be proven through *whakapapa* traced through either the male or female lineage (Kawharu 1977, 294). Maori identity was inextricably linked to land and place and the naming of place played an important role in their way of seeing the world.

The term "New Zealand" was meaningless to its Polynesian inhabitants before about 1840 (Owens 1992). Europeans, on the other hand, had used the name since the seventeenth century, after it was "discovered" by Abel Tasman (see Stokes 1968). The name became more familiar to Europeans after James Cook's best-selling journals were published describing, among

10 See Sinclair (1971) for an example of the "harmonious race relations" myth at work in the academy.

11 This connection to land is exemplified in the Maori term *whenua*, which means both "land" and "placenta." After a child is born, many Maori (and some Pakeha) continue the practice of burying the placenta in their family land, signifying the spiritual connection between *whanau* (extended family) and *whenua*. It is important to remember that this counter-hegemonic narrative on Maori spirituality has the potential (in some circles at least) for becoming hegemonic, and thus producing a new set of exclusions.

other explorations, his circumnavigation of New Zealand between October 1769 and March 1770 (Cook 1955). The inextricable relationship between the symbolic and material orders of society is clearly visible in the politics of Cook's place-naming. In the ten days from the time he made landfall adjacent to the mouth of the Turanganui River on 9 October 1769, Cook and his men were responsible for the death of at least seven natives and the serious injury of five others.[12] As a result of his experiences, Cook named the place *Poverty Bay* because, in his own words, "it afforded us no one thing we wanted …."[13] Cook went on to circumnavigate the islands of New Zealand providing English names for many prominent coastal features, and in so doing, incorporating New Zealand fully "within the rhetorical ambit of a white geo-historical discourse" (Carter 1987, 328). This spatio-temporal discourse was not confined to the land; it extended to the people as well. Cook and his men referred to the *tangata whenua*—who differentiated themselves on the basis of *iwi* (tribe), *hapu* (sub-tribe) and *whanau* (extended family) affiliations—as either "natives" or "New Zealanders." Later, as visits from missionaries and traders became more common in New Zealand the *tangata whenua* came to be called "Maori" (which means "ordinary").[14] This cultural politics of naming people and places was to continue for more than two centuries. In the case of place names, for example, only smaller third order centres in New Zealand retained predominantly Maori names by the early 1980s (Yoon 1986, 101–7).

The process of conveying (primarily) European names to places was part of a larger process of Europeanizing the landscape (Harley 1988b) and seizing control of Maori lands. It was also part of a process of gendering the landscape in Western masculinist terms. In more contemporary times, however—as part of what has been dubbed "the Maori renaissance" which was part of the land rights movement in Aotearoa—Maori have stepped up attempts to rename places, and thereby reclaim (and de-Europeanize) the landscape. Maori attempts to reclaim place names involve them eventually with the New Zealand Geographic Board (NZGB), the statutory body established by Act of Parliament in 1946 with responsibility for naming places in Aotearoa/New Zealand.

---

12 See Cook's journal entries of 9, 10 and 15 October 1769 (Cook 1955, 169, 170 and 178).

13 Journal entry of 11 October 1769 (Cook 1955, 172).

14 Although the term "Maori" is partly the result of a masculinist and racist European desire for categorization, it also has its roots in Maori societies. Indeed, the etiology of the term is a question that has received little attention and would be a worthy study in itself. Furthermore, Maori people have recently reclaimed the label as a signifier of a *political* position from which to oppose racist discourse (for a discussion of how "racial" and "sexual" Others might claim "race" or "queerness" as a speaking position from which to challenge racism and heterosexism, respectively see hooks (1990) and Fuss (1989)).

From her perspective as a member of the NZGB, Evelyn Stokes (1994) provides a comprehensive description of the organization and its activities. The NZGB is comprised of seven persons who, apart from the Surveyor General, are appointed for renewable three-year terms. The NZGB Act (1946) determines that at least two members must be Maori and the present membership includes the prominent Ngai Tahu leader Tipene O'Regan. This level of representation suggests a certain sympathy to te reo Maori. Indeed, the NZGB regularly consults with the Maori Language Commission in recognition of the (at least nominal) status of Maori as an official language of New Zealand. The NZGB has also "... been proactive in its policy on collecting and recording Maori names, and in encouraging a greater interest in them" (Stokes 1994, 9) through the publication and distribution of reports and oral histories. Therefore, while there may be less than one-third Maori membership of the NZGB, its mandate and orientation is cognisant of a Maori perspective, and in Stokes's words "The most important task for the NZGB is to ensure that the rich fabric of Maori names spread over the land, which has meaning and significance to each hapu and iwi, is preserved" (1994, 11).

As Stokes (1994) outlines, the NZGB receives submissions which come either directly from members of the public, or are received more formally from local government departments. Any proposal for new or replacement names is received and considered so long as duplication is avoided and good taste is not breached. Names already in local usage by either Maori or Pakeha are treated more favourably. After receipt, all suggestions are first referred to representatives of the appropriate iwi authority for comment, then subjected to preliminary research by staff of the Department of Survey and Land Information. The NZGB then deliberates and reaches a decision. Should this be favourable, the proposed name change is publicly advertised to allow potential objectors a three-month period in which to register disapproval. If no objections are received the change is simply incorporated into all future maps. However, should significant objection ensue, the Minister of Survey and Land Information becomes involved in accepting or declining the proposal. Members of the public can thus become involved in the process at two stages: as preparers of submissions and, at a later stage, as objectors or supporters of a proposal.[15]

Stokes (1994) summarizes the difficult tasks facing the NZGB:

> It must consider and weigh up the often conflicting issues of popular and customary usage, correct spelling and etymology, orthography, the status of original Maori names against those applied later, and at times cope with

15 See Stokes (1994) for a more complete description of the NZGB and the process involved in naming places in Aotearoa/New Zealand.

> angry and emotive outbursts from a public that does not want anyone to interfere with the place-names that people have grown accustomed to.

Stokes is probably being diplomatic here, as we suggest that there is more to those "angry and emotive outbursts" than the fact that "people have grown accustomed" to certain place names. Indeed, we argue that such responses—which usually arise when Maori attempt to replace European names with Maori ones—are indicative of an explicit cultural politics of identity. This identity politics is inextricably bound up with "race," class and gender politics, and is embodied in exclusionary ideological discourses of community in New Zealand. In this sense, then, place and place names are (material and metaphorical) sites of contested representational discourses (Duncan and Sharp 1993). To illustrate our argument, we shall examine the discursive politics involved in one specific attempt to reclaim Maori place names in the Otago (*Murihiku*) region of the South Island (*Te Waipounamu* or *Te Waka o Aoraki*).

Contesting Place Names in Otago/Murihiku

On 10 November 1989 the NZGB received a submission from John McLachlan proposing the following amendments to three place names on the east coast of the Otago region:

> (a) Wharau Werawera Beach should replace Long Beach …; and
> (b) Whare Akeake Beach should replace Murdering Beach; and
> (c) Purakaunui should replace Purakanui as the more correct form of the name (Lawrence, n.d., 1).

McLachlan invoked the second article of the Treaty of Waitangi in support of his claim, arguing that the proposed names were *taonga* of *te reo Maori* (treasures or very important aspects of the Maori language) (Lawrence n.d., 4). At the time of his submission, Mr. McLachlan was residing in Paraparaumu (located on the North Island near Wellington). However, as part of his submission McLachlan cited his *whakapapa* (genealogy) which linked him to Ngaihuirapa, a *hapu* (sub-tribe) of the Ngai Tahu tribe long-resident in the area where the names were to be changed. He thus invoked a concept of space as "stasis" (Massey 1992) and an essence of place in order to establish his "legitimacy" in the politics of naming these particular places. As we shall see, his strategy did not completely foreclose others from invoking their own place-essences in order to contest the proposed name changes.

Following initial research—reluctantly carried out by the Otago District Office of the Department of Survey and Land Information, whose manager was opposed to the name changes (and subsequently publicly opposed the NZGB's recommendations)—and more in-depth research carried out by

NZGB staff in Wellington, the NZGB decided that there were sufficient reasons to warrant a change of place names. In accordance with statutory requirements, notice of intent to change the place names was outlined in the *New Zealand Gazette* on 1 November 1990, and copies of the notice were placed in the *Otago Daily Times* on 7 and 9 November 1990 (Anonymous n.d.).

Response to the NZGB's notice of intent was swift in coming: on 8 November the *Otago Daily Times* published an editorial critical of the proposed name changes. The first sentence of the editorial is illustrative of some rhetorical tropes that would become common in the representation of this issue by members of the "local community" opposed to the changes:

> The adoption by the New Zealand Geographic Board of *one man's* proposal, and a *stranger* at that, to alter the *well-established local* names of Murdering Beach and Long Beach to *long Maori names* ... has caused a great deal of unfavourable reaction in and around Dunedin. (*Otago Daily Times*, 8 November 1990; emphasis added)

The editorial adopts a viewpoint that putatively represents the view of "the community," but its language gives away the highly specific nature of the Pakeha community it represents.[16] This is an exemplar of one form of *community* discourse, where concrete specificity ("long Maori names") is subsumed by abstract universality ("well-established [European] names"). Traditional concepts of citizenship are tied to the notion of "membership in a community" encompassing various rights and responsibilities (Yuval-Davis 1991, 59). However, in this discourse of citizenship, *community* and *difference* are mutually exclusive terms, because community is constructed as a bounded essence, inhabited by rational, reasonable beings having shared beliefs, a common culture, and unity of purpose (Yeatman 1992; Young 1990a, 1990b). This notion of community has its origins in the Enlightenment, where objective, impartial and putatively universal rationality was elevated above subjective, partial, and particular forms of thought (Seidler 1994). As Iris Young (1990a, 97) has pointed out, the ideal of impartiality produces a whole series of dichotomies: universal/particular, public/private, reason/passion (to name but a few). These dichotomies are aligned with gendered and racialized identities as well. Thus the universal is masculine and "white," while the particular is feminine, "black" ... and so on (see, e.g., Berg 1994; Jay 1981; Massey 1992; Seidler 1994). More important, however, is one of the ideological functions that the ideal of impartiality serves: "it masks the ways in which the particular perspectives of dominant groups claim universality" (Young 1990a, 97).

---

16 Dunedin, where the *Otago Daily Times* is published, has the smallest proportion of person's identifying themselves as Maori of the main urban centres in New Zealand.

Community discourses are far from universal, as one of their primary purposes is to constitute boundaries of inclusion and exclusion. In so doing, they work ideologically to produce an excluded Other. Liberal discourses of community require the individual to bracket out all that is particular, unique or personal in order to enjoy membership in an homogeneous public community. Republican discourses of community rely on a more participatory notion of community, one that is arrived at through rational consensus; however, this rationality is predicated on a notion of universal normative reason that transcends particularist perspectives. Welfare state discourse is predicated on a dualism of independence and dependency, where dependency is the Other against which independence is gauged (Yeatman 1991, 4–6). Western discourses of citizenship and community thus require the production of an Other to bear responsibility for the substantive, private, and particular of social life. In this case of contested place naming, then, this Other—on whom "interdependency conceptualised as dependency" (Yeatman 1991, 6) is projected—comprises those who do not fit within the confining bounds of hegemonic Pakeha masculinity.

The community is also conceived in terms of a bounded essence of abstract space (LeFebvre 1991) which encompasses its members. Legitimacy and illegitimacy are defined by inclusion within (locals), or exclusion from (strangers), this bounded space. Ideas about legitimacy are thus explicitly linked (via the "local names" versus "long Maori names" opposition) to an insider/outsider dichotomy that is, in turn, connected to a racialized masculine/feminine dichotomy. Locals, read as Pakeha insiders, are valorized and masculinized in this discursive construction; strangers, read as Maori outsiders, are devalued and feminized.

We speculate that this gendering process becomes more visible if one understands some of the tensions between North and South Island residents, and the ways that each island has been racialized and gendered by "Mainlanders," as many South Islanders refer to themselves. South Islanders, we assert, have a strong rural identity and frontier mentality—the South Island is less populated, has fewer cities, a harsher climate, and a lower proportion of Maori inhabitants than its northern counterpart. In contrast, the North Island is more populous, more urbanized (and remember, in rural rhetorics the city is an effeminate place), has a milder climate, and has a higher proportion of Maori (and Pacific Island) inhabitants.[17] We suggest that in some South Islander constructions, Pakeha masculinities tend to be rooted in the countryside, associated with farming and taming the frontier—that is, replacing a feminine nature with a masculine (horti)culture, based on food and fibre production. A quintessential part of New Zealand masculine culture is constructed in the physicality of rural activities—and the attendent exposure

17 See Pool (1991) concerning population characteristics of the North and South Islands.

to the elements associated with farming. As one of the 100 New Zealand men interviewed for the book *The Jones Men* put it: "I have a very high regard for New Zealand men. I have a lot of contact with them, in the farming scene. That's where you meet the true blue New Zealander ..." (in Gray 1983, 148). In this way, then, we suggest there is a tendency among Pakeha South Islanders to code the (predominantly rural) South as more masculine and European, while the (more urbanized) North is seen as more feminine and Maori.[18]

This gender coding is likely compounded by the fact that the proponent of the name changes, John McLachlan, was formerly a resident of the Otago area, but waited until he resided in Paraparaumu on the North Island prior to lodging his submission to the Geographic Board. For some in the region, it might have been seen as "manly" of him had he stayed in Otago to "fight" for the name change. But in leaving the area and taking up the battle only once he was at a "safe" distance, we suggest that McLachlan's actions might be perceived by some as having a ring of sneakiness or treachery about them, and they could be closely associated with the feminine.

The politics of naming places in this instance is both a *politics of space* (deciding who names and controls space) and a *spatialized politics* (whereby the spatial defines who has legitimacy to speak) (see Keith and Pile 1993). Moreover, such politics rely on intertwined discursive constructions of "race" and gender. The notion of community as a bounded essence has a long history in Western thought; it can be traced back at least to Aristotle's discourse on politics, where he explicitly linked Reason to those who inhabited the Polis, or city-state (see, e.g., Dossa 1987). Moreover, as we shall discuss later in more detail, this spatialized Reason has long been conceptualized in *masculine* and *European* terms (Seidler 1994).

The *Otago Daily Times* editorial was the precursor to a large number of objections to the NZGB's proposed name changes. By the end of the statutory three-month period for raising objections to its decision the Board had received 95 written objections purporting to represent the views of 1800 people (Lawrence n.d.). Yet in spite of the relatively large number of objections, the objectors themselves relied upon a small number of interpretive repertoires or rhetorical tropes to support their opposition to the changes. We shall now discuss a few of these in more detail. Our purpose is not to be exhaustive (as might be the case with a content analysis), but instead to suggest how some of these interpretations articulate with, and are dependent upon, wider discursive constructs for their rhetorical power.

18 Such gender coding is not uniform, as it tends to be spatially differentiated. Many Aucklanders, for example might valorize urban life and masculinities associated with the "toughness" required of the successful "businessman." For these people, small communities might be devalued as "hick towns" and the masculinities associated with farming similarly criticized as brutish and unintelligent.

A common argument put forth, in one form or another, by most of those who wrote to the NZGB to voice their objection to the proposed name changes centred on the notion that the existing names are well known, historically accurate (in the case of Murdering Beach) and the most appropriate descriptions of the places in question:

> The present names are well established and are meaningful and descriptive. (Submission B C, 14/1/91)[19]

> One cannot change history. Historic events should not be erased from nomenclature. (Submission L P, 23/1/91)

> These beaches in question … in my view are now appropriately named & have been for quite some time, so why introduce the Maori language when the minister of Maori affairs admits difficulty in speaking. (Submission A B, 14 /11/90)

> To change such well known names as Long Beach & Murdering Beach to long, hard to pronounce & hard to spell Maori names is quite unacceptable & unnecessary. (Submission E D, 10/11/90)

> All Maoris speak english and know what Long Beach means but not all New Zealanders speak Maori and consequently the proposed name would mean nothing at all. (Submission D A, 9/1/90 [the year is probably 1991])

Such narratives rely upon a number of often contradictory discursive constructions of "race" and gender for meaning. First, there is the rhetorical devaluation of the Maori language (and therefore culture), which is "hard to pronounce" and "mean[s] nothing at all" (to Pakeha). This is a common maneuver in Pakeha rhetorics, one that posits Maori culture in a subordinate position to "British" culture,[20] such that everything Maori is *always already* "tainted" by inferiority (McCreanor 1993a; Spoonley 1993).

Second, these ideas articulate with wider discourses of masculine Rationality and Reason (Seidler 1994), such that (a valued) English is coded as Rational and Masculine; while (a devalued) Maori is coded Irrational and

---

19 Although the submissions to the NZGB referred to in this chapter are publicly available, we identify them solely by the author's initials and date of submission to protect individual identities. Our purpose is not to impugn those individuals who wrote to the Board to object to Maori names, but instead to illustrate the way they (often unknowingly) draw on a particular set of discursive constructs that work to maintain a hegemonic Pakeha masculinity.

20 Pakeha is a contested localized construction of identity that is nonetheless derivative of British beliefs and practices (see, e.g., King 1985).

Feminine. The links between Pakeha masculinity and Reason are made quite explicit in another objector's admonition that

> New Zealanders must be forward looking not backward looking. This means rather than drag up radical Maori argument about debateable battles a Government body (Geographic Board) must set an example by listening to reason *not emotion*. (Submission L P, 23/1/91; emphasis in original)

The reason/emotion binary has a long history in Western thought (Berg 1994; Jay 1981; Seidler 1994). Indeed, from the very beginnings of Western philosophical thought, masculinity has been associated with Reason; while femininity was associated with everything Reason left behind. This type of binary logic is also linked vertically to other binary concepts, such that racialized Others (Them) are also excluded from Reason (Us). The above comments to the NZGB, then, are reliant upon the wider discursive idea that Reason is masculine and "white," while Emotion is feminine and "black." In this discourse, Maori become the excluded feminine Other, and their claims for justice are *ipso facto*, unreasonable.

This subordination of *taha* Maori (Maori ways) is also reliant upon a Euro-centric conception of what constitutes an appropriate and "historically accurate" description. Indeed, as Michel Foucault has shown, "history" is a slippery and elastic concept. In the case of Murdering Beach, for example, the name is said to have originated from events in 1817, when three Europeans were attacked and killed by local Maori. The Maori had attacked the group only when they recognized one of the Europeans as a person who had stolen a preserved Maori head at Riverton in 1811 (Lawrence n.d., 13). To the Europeans this was murder, but to the Maori it was an appropriate response to the cultural requirement to satisfy the rules of *utu*, or ritual justice.

We would like to suggest, however, that there is something more at work here. This story, and the name "murdering beach" that arises from it, is reliant on a particularly common trope of colonialism. We speak here of the association of violence and savagery with the "native" and peace and civilization with the European—along with their attendant gender-mappings (Ware 1992; Back 1994). The story of murder is thus also the tale of a dangerous and unpredictable (irrational) Maori masculinity, but one that has been subdued (at least rhetorically, via the name) by an appropriately Rational Pakeha masculinity.

Other objectors took their cue from the *Otago Daily Times* editorial, questioning an "outsider's" legitimacy to propose name changes in the first place:

> I do not accept that a John McLachlan living hundreds of miles away thinks … he can demand a name change. (Submission J S, 26/11/90)

> I see no reason why, in response to one person being far away in the North Island, steps should be taken to change the names. (Submission A A, 13/11/90)

> It is impudent of Mr John McLachlan, the mover of these changes, to propose these changes when he is not a current resident of the area or in a position to be aware of the feelings of the visitors to and residents of these beaches. (Submission B W, 13/11/90)

In these narratives, legitimacy is defined in spatial terms, that is, the right to speak is predetermined by one's physical presence in a bounded geographical space—Here (and not There). It seems to us that this notion must be related to that phenomenon Jacques Derrida (1976) criticized as the Western "metaphysics of presence." In this regard, a speaker's *presence* supposedly makes their intentions transparent, thus a key aspect of this metaphysics is the privileging of presence over absence. This, in turn, is related to the tendency of "Westerners" to conflate *seeing* with *knowing* (see, e.g., Berg 1994; Rose 1993; Sayer 1991). John McLachlan's physical presence in the Otago region is required for him to know the Truth; or, to put it another way, those who are Here, know the real Truth. Moreover, in drawing on his *whakapapa* (genealogy) as partial justification for his claims, McLachlan's claim is also temporally more distant than those who oppose him.

As we have already noted, this is not only a politics of space, but also a spatialized politics. Placing boundaries around abstract space is an effective means of policing membership in a community. Elastic notions of membership can then be invoked, depending on whether hegemonic members wish to include or exclude certain bodies or ideas from the space of legitimacy. Thus, despite John McLachlan's strong ties (he grew up in the region, only recently moving away; he is *tangata whenua*; he still owns land and pays property taxes; he continues to visit regularly), he can be excluded because he does not physically inhabit the bounded region defined as the "local community." Some people invoked the fact that they *visited* the Murdering Beach area in order to establish their legitimacy to speak and to oppose the name changes. Interestingly, these people were classed as "locals" by those who wished to construct the notion of a large local community opposed to the name changes. As with the editorial in the *Otago Daily Times*, the local is implicitly gendered and racialized through its articulation with binaries such as Us/Them, Civilized/Savage, and Male/Female. Accordingly the local in this case refers implicitly to a Pakeha masculinity.

Other submissions to the NZGB relied upon much more explicit formulations of "race," "culture" and "nation." These different constructions produce divergent identity politics and result from different rationales, yet they each result in similar outcomes—the maintenance of a hegemonic Pakeha masculinity. We shall examine some of these different forms of identity politics in more detail. We have found it heuristically useful to separate these themes, but it is important to remember they often overlap such that the rhetorics of "race" are subsumed by those of "culture" , and they in turn are also partly "nation" rhetorics.[21]

"Race" Talk

A number of those opposed to place name changes traded explicitly in the rhetorics of "race," relying upon notions of identity rooted in an essential biology:

> Our history of the area is about Long Beach and Murdering Beach not the virtually unpronouncable names suggested by some distant person with a Scottish name who may have a few millilitres of Maori blood. (Submission A K, 7/12/90)

> As for Mr Tipene O'Regan's visit to consult local maoris, this smacks to me of another of his mischievous attempts to disrupt harmony between the two Races and helps to create further ill feeling all in the name of the Treaty of Waitangi. (Submission A L, 14 /12/90)

> Your proposed changes have done nothing for race relations. (Submission G S, 8/1/91)

> As yet we do not have apartheid in New Zealand and it is to be hoped we never will have. Changing the names of these two Otago beaches to their previous Maori counterparts would take us one step closer to the separate development of individual races which is the hallmark of apartheid, so is to be avoided. (Submission A M, 11/11/90)

"Race" talk of this sort relies upon a discursive construction that came to prominence during the late eighteenth and early nineteenth century (see e.g., Baudet 1965; Banton 1977; Bolt 1984). This is the idea that "races" have some biological (thus natural) basis, readily identifiable through physical appearances, and indicative of some underlying essence. "Race" discourse produces biological objects. As Wetherell and Potter (1992, 122) note, "for some, genes are the crucial biological entity; for others the biological division is encapsulated in the shorthand of "blood." The blood possessed by Maoris is said to dictate, in a mysterious way, personality, temperament, ability, outlook on life and social organization." In the case of place name changes, they might somehow "stir the blood" of the two "races," leading ineluctably to ill-feeling and poor "race relations." Perhaps more importantly, despite the rhetoric of

21 We are grateful to an anonymous referee for this point.

some participants in these events, essentialist notions of "race" are clearly tied to complexly intertwined hierarchies of "race" and gender (Jackson 1994; Ware 1992).

There are other interpretive repertoires circulating within the "race" talk of those objecting to the proposed name changes. As we discussed in our survey of Maori-Pakeha relations, many Pakeha believe that current "race" relations are harmonious. In this discursive ordering, then, any attempts by Maori to change their marginal status in New Zealand society are painted as "mischievous attempts to disrupt harmony between the two Races" (Submission A L, 14/12/90). Maori are thus produced as "troublemakers" and "stirrers" whose actions "create ill feeling" (Submission A L, 14/12/90) where none previously existed (McCreanor 1993a, 1993b; Nairn and McCreanor 1991). The actions of such "stirrers" also transgress the bounds of masculine Reason, thereby passing into the feminine territory of Unreason (Seidler 1994).

Similarly, as a result of the modest attempts by the state to redress Maori marginalization (e.g. Davey and Kearns 1994), many Pakeha New Zealanders believe that Maori are the recipients of unfair special privileges (in the same way that unemployed persons are constructed as "dole bludgers"). Turning conceptual somersaults, some Pakeha argue that these "special privileges" are a form of "reverse racism," and many equate them to "separate development of individual races which is the hallmark of apartheid" (Submission A M, 11/11/90). In this rhetorical construction, Maori self-determination—anathema to Pakeha notions of equality—is the inevitable precursor to apartheid (Spoonley 1993, 17).

"Culture" Talk

While "race" talk is still commonly found in the discourse of Pakeha (and Maori) New Zealanders, there has been a subtle shift in emphasis so that now racial discourse might be seen more as "a kind of residual sediment" (Wetherell and Potter 1992, 123). "Race" has been retained as a trope or metaphor of difference, but racial theories no longer provide the sole foundation for taken-for-granted analyses of social relations in New Zealand. Probably as a result of the civil rights movements of the "60s and "70s, as well as Maori pressure for social change, trade in "race" talk has become, at least for some, acutely embarrassing. Racism might thus be seen as a metonymic elaboration in which coded signifiers of race are deployed so as to avoid charges of racism (Solomos and Back 1994; also see Blaut 1992; Spoonley 1994). The most common metonymic substitute for "race" has been "culture":

> The promotion of the proposed names simply because they are maori is insensitive to the popular culture of the Dunedin and Otago area. (Submission B W, 13/11/90)

> If you are going to start changing name how about asking the Maoriori people who were driven off the land by the Maori people they also should be considered as also should the opinion of the Pakeha who helped to educate the people, and who have helped to make N. Zealand what it is ... I think there is too much changing of names to Maori leave the country names as they are, after all why isn't the Pakeha side being considered. (Submission L O, 15/11/90)

> The proposed changes may cause a backlash of ill-feeling between local people. Already graffiti has appeared in a previously quiet and tolerant settlement. (Submission B C, 14/1/91)

> Although we agree that the Maori language should not be lost (and in that we commend to Maoris the example of the New Zealand Chinese), we see these two name changes as a minor contribution towards a cumbersome two language situation which may eventually become exceedingly divisory and dangerous as, for instance, is happening in the Canadian Province of Quebec. (Submission M W, 12/11/90)

In these narratives, "culture" can be implicated in a number of highly complex interpretive repertoires. We can see, for example, reliance upon familiar ideas about the superiority of "the Pakeha who helped to educate the people, and who have helped to make N[ew] Zealand what it is" (Submission L O, 15/11/90). Likewise, notions of Maori privilege are being invoked, "after all, why isn't the Pakeha side being considered?" (Submission L O, 15/11/90). The reference to the "Maoriori" (Submission L O, 15/11/90), signals a shift to another highly questionable, but common argument that Aotearoa was originally settled by the Moriori people, who were defeated in war by the Maori. Pakeha use this "fair fight" (during the Land Wars of the 1800s) argument as a rationalization for their current domination of Maori. This kind of rhetoric has contradictory implications for Maori. On the one hand, it incorporates them within a specific form of Rational Western masculinity, one that is embodied by the military. James Belich's (1986) revisionist history of the Land Wars, which tends to be altogether celebratory of Maori prowess on the battlefield (and thus argues that they were not inferior, and sometimes superior, to the British army), provides an example of this rhetoric. On the other hand, this rhetoric reinforces the binary distinction between a rational, and thus acceptably masculine, form of violence (embodied in the military apparatus of the New Zealand state) and an irrational, unacceptably hyper-masculine violence associated with uncivilized savages.[22] While Maori masculinity is

---

22 Most historians, following the important work of James Belich (1986), now acknowledge that Maori employed rational planning and strategy in defence of their lands, but this idea is not widely accepted in popular (Pakeha) consciousness.

usually incorporated within the rhetorical ambit of an acceptable (Western) masculinity, it is always dangerously close to its "savage" roots.

Maori culture can also be played out within notions of either "heritage" or "therapy" (Wetherell and Potter 1992). Pakeha culture is often invisible, taken-for-granted as the *norm, progress, modernity*, a time and place from which Maori culture is seen to deviate. As "heritage," Maori culture is unchanging tradition, referring back to the past rather than ahead to the future: "New Zealanders must be forward looking not backward looking" (Submission L P, 23/1/91). This is a typical trope of Western modernity, whereby non-Westerners are conceptualized as inhabiting the Then and There as opposed to the Here and Now of the West (Fabian 1983; Geertz 1988). Culture in this sense becomes something you bring out of the closet on weekends for special celebrations (such as Waitangi Day), and for tourists but it is not appropriate for "normal," "modern" life. Maori are thus burdened with the task of saving their culture, while at the same time "using" it sparingly. If Maori do display their culture too much they can be accused of "stirring"—causing "a backlash of ill-feeling … in a previously quiet and tolerant settlement" (Submission B C, 14/1/91). Similarly, too much use of the Maori language "may eventually become exceedingly divisory and dangerous" (Submission M W, 12/11/90).

With "culture as therapy," the emphasis shifts from preservation to identity. Culture becomes a psychological need (Wetherell and Potter 1992). While this might be a less derisive way of conceptualizing culture, it still has its dangers for Maori, who can be discursively produced as non-persons. By "carelessly" losing their culture and identity, Maori become empty containers, awaiting their refill of "culture." Moreover, in addition to being seen as deficient in relation to Pakeha, Maori are also produced as deficient Maori in this discursive frame (Wetherell and Potter 1992, 131). In this sense, loss of identity is seen by Pakeha to be a Maori problem, and it is dissociated from the marginalization of Maori in New Zealand society. This idea was explicitly seized by a number of objectors, who argued that Maori had lost interest in their culture, thus by implication, there was no need to retain Maori place names:

> Of all the permanent residents I can remember only one Maori family—the Mowats whose main mark upon history was the murder committed by one son. In all that time the interest taken by Maoris unheard of … The senior clerk was a Maori by the name of Tamais Paiki which he never used, preferring to be known as Thomas Pike. He was completely disinterested in the land & the history of his race. (Submission R S, 12/11/90)

The reference to the "murder committed by one son" in this passage also relies on the masculinist and racist construction of Maori male subjectivities as dangerous and hyper-masculine (Ware 1992), thus serving to further deny Maori legitimacy. On the other hand, Maori themselves have taken up the idea of culture as therapy in a different way, using it as an explicit oppositional response to Pakeha hegemony. When Nga Tamatoa (an influential urban Maori "political" organization formed in the 1970s) called for the enculturation of young urban Maori in *taha* Maori (Maori ways), for example, this call was tied to claims for social justice and land rights (see Walker 1990, 1992).

"Nation" Talk

There exists yet another powerful discursive ordering of group relations in the politics of place in New Zealand, one that posits group identity in terms of the nation and nationalism:

> I object to the number of Maori place names which are being foisted on us. I do not wish to detract from the Maori culture or destroy their heritage but we are all, white or coloured, New Zealanders and see no reason why either culture should take precedence. (Submission D A, 9/1/90)

> New Zealanders live here not Maoris. (Anonymous submission, 01/91)

> The Board has some pro Maori members. pro to the extent of bias. I have read articles by members. (Submission E O, 11/1/91)

> There should be no change of name to suit a whim of one man there are enough unpronoun[ce]able maori name around now we are all New Zealanders there is too much them & us & it should not be encouraged .... (Submission E P, no date)

"Nation" talk of this sort relies on a dualistic concept of identity: "while arguing at one level that racial considerations ought to be irrelevant, at another level racist distinctions are still made" (Spoonley 1993, 17). The *specific* interests of a hegemonic Pakeha masculinity are asserted as *universal* national interests applicable to all New Zealanders. In this sense, Pakeha masculinity becomes the norm by which all Others are measured—and found wanting. The Other becomes the negatively valued foil by which Pakeha masculinity is positively valued. Thus to be "pro-Pakeha" is to be objective, but if people are "pro-Maori" they are biased.

Nation talk is closely linked to the liberal "community" discourse we spoke of earlier. In this regard, the concrete specificity of ethnic difference is subsumed by the abstract universality of national identity defined in masculine terms. To paraphrase Wetherell and Potter (1992, 141), heterogeneity, revealed as Maori difference and often celebrated at other times [such as Waitangi Day, or during the 1990 Commonwealth Games in Auckland (see Pawson 1992)], becomes restyled as different faces of an underlying homogeneity that is exemplified by Pakeha masculine identity. When people step outside those norms—to protest land rights, to combat patriarchy, or to reclaim place names—they are contradictorily labelled radical, irrational, hysterical (feminine), dangerous (hyper-masculine), divisive, or abnormal. On the other hand, Maori have appropriated "nation" discourses to combat racist and masculinist domination. They argue for a notion of Maori sovereignty and self-determination, invoking the Treaty of Waitangi and its recognition of their "nation" status prior to the formation of New Zealand (see, e.g. Awatere 1984; Walker 1990, 1992).

To complete our narrative, it appears that this particular struggle to reclaim Maori place names was only partially successful. In light of the objections raised, on 26 November 1991 the Minister of Survey and Land Information announced an equivocal decision: the proposed name changes for Long Beach and Murdering Beach were declined, but the changed spelling of Purakanui to Purakaunui was approved (DOSLI 1991). It appears to us that the less controversial proposal (a spelling change) was allowed as something of a palliative to McLachlan and his supporters. It is significant that despite the strong Maori sympathies of the NZGB, the change away from (the offensive) Murdering beach was disallowed by the Minister. In the absence of definitive evidence, we speculate that this outcome relates to the distinctive socio-cultural geography of Aotearoa/New Zealand. As outlined earlier, the proportion of persons identifying themselves as Maori is low in Murihiku/Otago relative to North Island regions. This likely had a bearing upon the naming decision, for within the volatile politics of a small nation, there are limits to how politicians will act in favour of a racialized minority. Furthermore, Otago is one of the longest colonized regions of New Zealand, with Dunedin being known as the "Edinburgh of the south," reflecting the region's Scots influence. As one objector observed in an earlier section, they could only remember "one Maori family" in their part of Otago. Thus we speculate that it might involve too much political risk for a National Party (conservative) Cabinet Minister to approve changing a place name back to Maori within the regional context of a Pakeha majority. We suggest that were the foregoing narrative to have unfolded in a region such as Northland where Maori representation and Pakeha sympathies are stronger, a different outcome and story could be told. In other words, we reiterate the dual importance of the politics of space *and* a spatialized politics at work in the process of naming places in Aotearoa/New Zealand.

Conclusion

Our purpose in discussing submissions to the Geographic Board was not to provide an exhaustive discourse analysis, but instead to suggest some of the ways that the contested politics of naming places in Aotearoa articulate with wider discursive constructs of "race" and gender. We have tried to illustrate some of the ways that place names are important signifiers of meaning, providing symbolic identity to people, place and landscape. Identity, in this sense, is not given. Rather it is "made" in the contested process of cultural (re)production. This complex process is made even more complicated by its articulation with many of the often contradictory but mutually constitutive semiotic constructions of "race" and gender.

Many authors commenting on New Zealand cultural politics posit a binary opposition between a hegemonic Pakeha group and a subordinate, oppositional Maori group (see, e.g. Barber 1989; Belich 1986; Kelsey 1990; McCreanor 1993a, 1993b; Nairn and McCreanor 1990, 1991; Pawson 1992; Ward 1974). These works play an important role in the fight against racism in Aotearoa. However, if one acknowledges that subjectivities are constituted through complexly intertwined identity politics of "race," gender, class, and sexuality, for example, it becomes increasingly more difficult to support this binary construction of cultural politics. Focusing on the mutually constitutive character of "race" and gender, we have argued instead for recognition of a hegemonic Pakeha masculine subject, and numerous subordinate, but contestatory, subject positions. This is different from a simple Maori-Pakeha binary, as the hegemonic subject position can, for example, encompass women and Maori who display an appropriate form of (usually bourgeois) masculinity.[23] In the same way, the numerous subordinate subjectivities can include men and women, working-class and middle class, Maori and Pakeha.

We have used the contested politics of naming places in Otago to illustrate some of our argument. Place names, and the maps used to present them, are the outcome of the appropriation of symbolic production by hegemonic groups, who impose their specific identity norms across all social groups. However, this process concerns more than "merely" symbolic production, because these social symbols are inextricably linked to the production of material conditions. Thus, just as landscapes can be seen as ideological constructs designed to legitimate hegemonic material interests, place names can be seen to legitimate a masculinist colonialism and colonial history in contemporary New Zealand. This legitimation is rationalized by a hegemonic Pakeha masculinity through various, often contradictory, discourses of gender, "race," culture and nation.

References

Anderson, B. (1983), *Imagined Communities* (London: Verso).

Anderson, K. and Gale, F. (eds) (1992), *Inventing Places: Studies in Cultural Geography* (Melbourne: Longman Cheshire).

Anderson, K.J. (1991), *Vancouver's Chinatown: Racial discourse in Canada, 1875–1980* (Montreal: McGill-Queen's University Press).

23 It is important to recognize, however, that those Maori and women incorporated into the hegemonic masculinity will always bear the residual traces of their Otherness as "native" or "woman" (or both).

Anonymous (n.d.), "Report on Assigning and Objections to 3 Names in Otago, Wharauwerawera Beach, Whare Ake Ake Beach, Purakaunui, Purakaunui Bay and Inlet," unpublished report prepared for the Geographic Board, available from the Geographic Board, Wellington.

Awatere, D. (1984), *Maori Sovereignty* (Auckland: Broadsheet Publications).

Back, L. (1994), "The 'White Negro' Revisited: Race and Masculinities in South London," in Cornwall, A. and Lindisfarne, N. (eds), 172–83.

Banton, M. (1977), *The Idea of Race* (London: Tavistock).

Barber, K. (1989), "New Zealand "Race Relations Policy," 1970–1988," *Sites* 18, 5–16.

Barnes, T. and Duncan, J. (eds) (1992), *Writing Worlds: Discourse, Text and Metaphor in the Representation of Landscape* (London: Routledge).

—— (1992), "Introduction: Writing Worlds," in Barnes, T. and Duncan, J. (eds), 1–17.

Baudet, H. (1965), *Paradise on Earth: Some Thoughts on European Images of non-European Man*, trans. E Wentholt (New Haven, CT: Yale University Press).

Belich, J. (1986), *The New Zealand Wars and the Victorian Interpretation of Racial Conflict* (Auckland: Auckland University Press).

Berg, L.D. (1993a), "Racialization in Academic Discourse," *Urban Geography* 14, 194–200.

—— (1993b), "Aboriginal People, Aboriginal Rights and Ideology: The Aboriginal Land Question in British Columbia," in Cant, G., Overton, J. and Pawson, E. (eds), 132–43.

—— (1993c), "Between Modernism and Postmodernism," *Progress in Human Geography* 17, 490–507.

—— (1994), "Masculinity, Place, and a Binary Discourse of 'Theory' and 'Empirical investigation' in the Human Geography of Aotearoa/New Zealand," *Gender, Place and Culture* 1, 245–60.

Blaut, J. (1986), "A Theory of Nationalism," *Antipode* 18, 5–10.

—— (1992), "The Theory of Cultural Racism," *Antipode* 24, 289–99.

Bolt, C. (1984), "Race and the Victorians," in Eldridge, C.C. (ed.), 126–47.

Britton, S.G., Le Heron, R.B. and Pawson, E. (eds) (1992), *Changing Places in New Zealand: A Geography of Restructuring* (Christchurch: New Zealand Geographical Society).

Bunkle, P. and Hughes, B. (eds) (1980), *Women in New Zealand Society* (Auckland: George Allen and Unwin).

Cant, G., Overton, J. and Pawson, E. (eds) (1993), *Indigenous Land Rights in Commonwealth Countries* (Christchurch: University of Canterbury and Ngai Tahu Maori Trust Board).

Carter, P. (1987), *The Road to Botany Bay: An Essay in Spatial History* (London: Faber and Faber).

Cohen, S. and Kliot, N. (1992), "Place-names in Israel's Ideological Struggle over the Administered Territories," *Annals of the Association of American Geographers* 82, 653–80.

Cook, J. (1955), *The Journals of Captain James Cook on his Voyages of Discovery, Volume I: The Voyage of the Endeavour 1768–1771*, ed. J.C. Beaglehole (Cambridge: Cambridge University Press).

Cornwall, A. and Lindisfarne, N. (eds) (1994), *Dislocating Masculinities* (London: Routledge).

Cosgrove, D. (1983), "Towards a Radical Cultural Geography: Problems of Theory" *Antipode* 15, 1–11.

—— and Daniels, S. (eds) (1988), *Iconography of Landscape* (Cambridge: Cambridge University Press).

Crocombe, R. (1991), "Naming and Claiming in the South Pacific," *Journal of the Pacific Society* 14, 1–19.

Crump, B. (1982), *A Good Keen Man* (Auckland: Firstlight Productions).

Daniels, S. and Cosgrove, D. (1988), "Iconography and Landscape," in Cosgrove, D. and Daniels, S. (eds), 1–10.

Davey, J.A. and Kearns, R.A. (1994), "Special Needs Versus 'The Level Playing Field:' Recent Housing Policy for Indigenous People in New Zealand," *Journal of Rural Studies* 10, 73–82.

Davidson, J.M. (1992), "The Polynesian Foundation," in Rice, G.E. (ed.), 3–27.

Dear, M. (1988), "The Postmodern Challenge. Reconstructing Human Geography" *Transactions, Institute of British Geographers* NS13, 262–74.

Department of Survey and Land Information (DOSLI) (1991), "Storey Rules on Maori Place Names," media release, Wellington, 26 November.

Derrida, J. (1976), *Of Grammatology*, trans. G.C. Spivak (Baltimore, MD: Johns Hopkins University Press).

Donaldson, M. (1993), "What is Hegemonic Masculinity?," *Theory and Society* 22, 643–57.

Dossa, S. (1987), "Political Philosophy and Orientalism: The Classical Origins of a Discourse," *Alternatives* 12, 343–57.

Duncan, J.S. (1992), "Elite Landscapes as Cultural (Re)productions: The Case of Shaughnessy Heights," in Anderson, K. and Gale, F. (eds), 37–51.

—— (1993), "Sites of Representation: Place, Time and the Discourse of the Other," in Duncan, J. and Ley, D. (eds), 39–56.

Duncan, J. and Ley, D. (eds) (1993), *Place/Culture/Representation* (London: Routledge).

Duncan, N. and Sharp, J.P. (1993), "Confronting Representation(s)," *Environment and Planning D: Society and Space* 11, 473–86.

Eldridge, C.C. (ed.) (1984), *British Imperialism in the Nineteenth Century* (London: Macmillan).

Fabian, J. (1983), *Time and the Other: How Anthropology Makes its Object.* (New York: Columbia University Press).

Foster, H. (ed.) (1988), *Vision and Visuality* (Seattle: Bay Press).
Foucault, M. (1984), "The Order of Discourse," in Shapiro, M. (ed.), 108–38.
Fuss, D. (1989), *Essentially Speaking: Feminism, Nature and Difference* (London: Routledge).
Geertz, C. (1988), *Works and Lives: the Anthropologist as Author* (Cambridge: Polity).
Harley, J.B. (1988a), "Maps, Knowledge, Power," in Cosgrove, D. and Daniels, S., 277–312.
— (1988b), "Silences and Secrecy: The Hidden Agenda of Cartography in Early Modern Europe," *Imago Mundi* 40, 111–30.
— (1992), "Deconstructing the Map," in Barnes, T. and Duncan, J. (eds), 231–47.
Innes, C.L. (1994), "Virgin Territories and Motherlands: Colonial and Nationalist Representations of Africa and Ireland" *Feminist Review* 47, 1–14.
Jackson, P. (1987a), "The Idea of 'Race' and the Geography of Racism," in Jackson, P. (ed.), 3–21.
— (ed.) (1987b), *Race and Racism* (London: Allen and Unwin).
— (1989), *Maps of Meaning* (London: Unwin Hyman).
— (1992), "The Racialization of Labour in Post-war Bradford," *Journal of Historical Geography* 18, 190–209.
— (1994), "Black Male: Advertising and the Cultural Politics of Masculinity," *Gender, Place and Culture* 1, 49–59.
— and Penrose, J. (eds) (1993), *Constructions of Race, Place and Nation* (London: University College London Press).
Jacobs, J. (1993), ""Shake 'im this Country": The Mapping of the Aboriginal Sacred in Australia—the Case of Coronation Hill" in Jackson, P. and Penrose, J. (eds), 100–18.
James, B. and Saville-Smith, K. (1989), *Gender, Culture and Power: Challenging New Zealand's Gendered Culture* (Auckland: Oxford University Press).
Jay, M. (1988), "Scopic Regimes of Modernity," in Foster, H. (ed.), 3–27.
Jay, N. (1981), "Gender and Dichotomy," *Feminist Studies* 7, 38–56.
Jordan, W.D. (1974), *The White Man's Burden: Historical Origins of Racism in the United States* (New York: Oxford University Press).
Kawharu, I.H. (1977), *Maori Land Tenure. Studies of a Changing Institution* (Oxford: Oxford University Press).
Kearns, R.A., Smith, C.J. and Abbott, M.W. (1991), "Another Day in Paradise? Life on the Margins in Urban New Zealand," *Social Science and Medicine* 33, 369–79.
Keith, M. and Pile, S. (eds) (1993), *Place and the Politics of Identity* (London: Routledge).
Kelsey, J. (1990), *A Question of Honour? Labour and the Treaty, 1984–1989* (Wellington: Allen and Unwin).

King, M. (1985), *Being Pakeha : An Encounter with New Zealand and the Maori Renaissance* (Auckland: Hodder and Stoughton).

Kolodny, A. (1975), *The Lay of the Land* (Chapel Hill, NC: University of North Carolina Press).

Larner, W. and Spoonley, P. (1994), "Postcolonial Politics in Aotearoa/New Zealand," in Stasiulis, D. and Yuval-Davis, N. (eds).

Lawrence, P. (n.d), "Otago Place Names: Report to the New Zealand Geographic Board," unpublished typescript prepared by the Secretary of the New Zealand Geographic Board (available from the New Zealand Geographic Board, Wellington).

LeFebvre, H. (1991), *The Production of Space*, trans. D. Nicholson-Smith (Cambridge, MA: Blackwell).

MacCormack, C. and Strathern, M. (eds) (1980), *Nature, Culture and Gender* (Cambridge: Cambridge University Press).

Massey, D. (1992), "Politics and Space/Time," *New Left Review* 196, 65–84.

McCreanor, T.N. (1993a), "Mimiwhangata: Media Reliance on Pakeha Common Sense in Interpretations of Maori Actions" *Sites* 26, 79–90.

—— (1993b), "Settling Grievances to Deny Sovereignty: Trade Goods for the Year 2000," *Sites* 27, 45–73.

Mitchell, D. (1995), "There's No Such Thing as Culture: Towards a Reconceptualization of the Idea of Culture in Geography," *Transactions, Institute of British Geographers* NS20, 102–16.

Mulgan, J. (1949), *Man Alone* (Hamilton, NZ: Paul's Book Arcade; reprinted Penguin Books, 1990).

Nairn, R.G. and McCreanor, T.N. (1990), "Insensitivity and Hypersensitivity: An Imbalance in Pakeha Accounts of Racial Conflict," *Journal of Language and Social Psychology* 9, 293–308.

—— (1991), "Race Talk and Common Sense: Patterns in Pakeha Discourse on Maori/Pakeha Relations in New Zealand," *Journal of Language and Social Psychology* 10, 245–62.

New Zealand Geographic Board (n.d.), "Report of NZ Geographic Board to the Minister of Lands under section 13 of the New Zealand Geographic Board Act 1946 on objections received to the proposed name change of Mount Egmont," copy available from New Zealand Geographic Board, Wellington.

Nicholson, L.J. (ed.) (1990), *Feminism/Postmodernism* (London: Routledge).

Orange, C. (1987), *The Treaty of Waitangi* (Wellington: Allen and Unwin).

Otago Daily Times (1990), "A Sorry Record," *Otago Daily Times*, 8 November.

Ortner, S.B. (1974), "Is Female to Male as Nature is to Culture?," in Rosaldo, M.Z. and Lamphere, L. (eds), 67–87.

Owens, J.M.R. (1992), "New Zealand before Annexation," in Rice, G.W. (ed.), 28–53.

Padgen, A. (1982), *The Fall of Natural Man: The American Indian and the Origins of Comparative Ethnology* (Cambridge: Cambridge University Press).
Pawson, E. (1992), "Two New Zealands: Maori and European," in Anderson, K.J. and Gale, F. (eds), 15–33.
Phillips, J. (1980), "Mummy's Boys: Pakeha Men and Male Culture in New Zealand," in Bunkle, P. and Hughes, B. (eds), 217–43.
— (1987), *A Man's Country? The Image of the Pakeha Male—A History* (Auckland: Penguin).
Pool, D.I. (1991), *Te Iwi Maori: A New Zealand Population, Past, Present and Projected* (Auckland: Auckland University Press).
Poovey, M. (1989), *Uneven Developments. The ideological Work of Gender in Mid-Victorian England* (Chicago: University of Chicago Press).
Relph, E. (1976), *Place and Placelessness* (London: Pion).
Rice, G.E. (ed.) (1992), *The Oxford History of New Zealand*, 2nd edn (Auckland: Oxford University Press).
Roper, M. and Tosh, J. (eds) (1991), *Manful Assertions: Masculinities in Britain since 1800* (London: Routledge).
Rosaldo, M.Z. and Lamphere, L. (eds) (1974), *Woman, Culture and Society* (Stanford, CA: Stanford University Press).
Rose, G. (1993), *Feminism and Geography* (Minneapolis: University of Minnesota Press).
Said, E. (1978), *Orientalism: Western Conceptions of the Orient* (London: Penguin).
Sayer, A. (1991), "Behind the Locality Debate: Deconstructing Geography's Dualisms," *Environment and Planning A* 23, 283–308.
Seidler, V.J. (1994), *Unreasonable Men: Masculinity and Social Theory* (London: Routledge).
Shapiro, M. (ed.) (1984), *Language and Politics* (Oxford: Blackwell).
Sinclair, K. (1971), "Why are Race Relations in New Zealand Better than in South Africa, South Australia, or South Dakota?," *New Zealand Journal of History* 5.
Smith, S.J. (1989), *The Politics of "Race" and Residence in Britain* (Cambridge: Polity).
Solomos, J. and Back, L. (1994), "Conceptualising Racisms: Social Theory, Politics and Research," *Sociology* 28, 143–61.
Spender, D. (1985), *Man Made Language*, 2nd edn (London: Routledge and Kegan Paul).
Spoonley, P. (1993), *Racism and Ethnicity* (Auckland: Oxford University Press).
— (1994), "Constructing Ourselves: The Post-colonial Politics of Pakeha," in Yeatman, A. and Wilson, M. (eds).
Stasiulis, D. and Yuval-Davis, N. (eds) (1994), *Gender, Race, Ethnicity and Class in Settler Colonies: Against Dichotomies* (London: Sage).

Stokes, E. (1968), "The Naming of New Zealand," *New Zealand Geographer* 24, 201–04.
—— (1992), "The Treaty of Waitangi and the Waitangi Tribunal: Maori Claims in New Zealand," *Applied Geography* 12, 176–91.
—— (1994), "Geographic Naming in New Zealand: The Role of the New Zealand Geographic Board," *New Zealand Geographer* 50: 2, 51–4.
Stone, R.C.J. (1973), *Makers of Fortune: A Colonial Business Community and its Fall* (Auckland: Auckland University Press/Oxford University Press).
Tosh, J. (1991), "Domesticity and Manliness in the Victorian Middle Class: The Family of Edward White Benson," in Roper, M. and Tosh, J. (eds), 44–73.
Tuan, Y.-F. (1974), *Topophilia* (Englewood Cliffs, NJ: Prentice-Hall).
Walker, R. (1990), *Ka Whawhai Tonu Matou: Struggle Without End* (Auckland: Penguin).
—— (1992), "Maori People since 1950," in Rice, R.W. (ed.), 498–519.
Ward A. (1974), *A Show of Justice: Racial Amalgamation in Nineteenth Century New Zealand* (Canberra: Australian National University Press).
Ware, V. (1992), *Beyond the Pale: White Women, Racism and History* (London: Verso).
Wetherell, M. and Potter J. (1992), *Mapping the Language of Racism: Discourse and the Legitimation of Exploitation* (London: Harvester Wheatsheaf).
Williams, R. (1981), *Culture* (London: Fontana).
Wood, D. (1992), *The Power of Maps* (London: Routledge).
Yeatman, A. (1992), "Minorities and the Politics of Difference," *Political Theory Newsletter* 4. 1–10.
—— and Wilson, M. (eds) (1994), *Justice, Biculturalism and Difference* (Wellington: Bridget Williams Books).
Yoon, H.-K. (1986), *Maori Mind, Maori Land* (Berne: Peter Lang).
Young, I.M. (1990a), *Justice and the Politics of Difference* (Princeton, NJ: Princeton University Press).
—— (1990b), "The Ideal of Community and the Politics of Difference," in Nicholson, L.J. (ed.), 300–23.
Yuval-Davis, N. (1991), "The Citizenship Debate: Women, Ethnic Processes and the State," *Feminist Review* 39, 54–69.
Zelinsky, W. (1986), "The Changing Face of Nationalism in the American Landscape," *Canadian Geographer* 30, 171–5.
—— (1983), "Nationalism in the American Place-name Cover," *Names* 31, 1–28.

Chapter 3
Naming the Past: The Significance of Commemorative Street Names

Maoz Azaryahu

Introduction

Drawing on examples from different geographical settings, political contexts and historical periods, this chapter explores the significance of street names as vehicles of commemoration. At a practical level, street names provide the users of the city with spatial orientation. When used for commemorative purposes, street names and the version of history they introduce into the public sphere belong to the semiotic makeup of local and national identity and to the structures of power and authority. In their commemorative capacity, street names communicate official representations of the ruling socio-political order. In particular, they introduce an official version of history into networks of social communication that involve ordinary urban experiences that seem to be separated from the realm of political ideology. The first part of the article expands on the political significance of street names by focusing on commemorative naming and renaming of streets. The second part explores aspects of signification. One is the semantic displacements effected by using street names for commemorative purposes and the shift from history to geography it entails. The other elaborates on reading a spatially configured set of commemorative street names as a city-text of history.

Street names designate locations and pronounce certain thoroughfares as distinct urban units. Street names differentiate: the rule is that not two streets should bear the same name. At a practical level, street names that distinguish between streets provide the users of the city with spatial orientation, and secure administrative control over the city. Notably, street names are not a universal norm. The streets of Tokyo do not have names. Alphanumerical designations, as the case of Manhattan demonstrates, impose a sense of spatial order and orientation in grid-pattern urban layouts with orthogonal intersections and parallel streets. However, since these are not the prevailing urban structure in many cities, street names abound.

Street names conflate urban space and the symbolic realm of cultural signification. Geographical street names (cities, rivers, mountains etc.) bestow prestige or conjure romantic associations. When employed as a vehicle of commemoration, street names intersect between urban space and official ideologies and mediate between political elites and "ordinary" people. They communicate official representations of the ruling socio-political order. In particular, when used for commemorative purposes they inscribe an official version of history onto the cityscape and introduce this version of history into myriad networks of social communication that involve ordinary urban experiences that seem to be separated from the realm of political ideology.

Commemorative street naming demonstrates the effect that administrative-political processes have on semiotic procedures. As the result of commemorative street naming, history becomes a concrete semiotic reality. A primary effect of commemoration is that the commemorated history becomes a culturally shared experience. Though the authorities decide upon what should be commemorated and where, they can hardly regulate the different contexts in which street names are used or evoked. Individual responses are to a large extent governed by preconditioned ideological dispositions and political attitudes, and are also dependent upon the social and psychological contexts in which the evocation of a particular street name takes place.

This chapter examines the significance of commemorative street names in terms of their political significance and their capacity of historical signification. The first part of the article expands on the political significance of street names by focusing on commemorative naming and renaming of streets. The second part explores aspects of signification. One is the semantic displacements effected by utilizing street names for commemorative purposes and the shift from history to geography it entails. The other elaborates on reading a spatially configured set of commemorative street names as a city-text of history.

A General Framework

When used for commemorative purposes, street names and the version of the past they introduce into the public sphere belong to the symbolic makeup of local and national identity and to the structures of power and authority. Edward Shils (1981, 249) observed that:

> The center, when it is effectively expanding, controls the institution of communication: schools, churches and if the society is a literate one, many of the organs for the presentation of representations and interpretations of the world.

From the perspective of those in charge of molding the symbolic infrastructure of society, of primary importance is the integration of representations of the ruling sociopolitical order into networks of social communication. A fundamental distinction is between networks of social communication that operate in a "high symbolic voltage" and such that operate in "low symbolic voltage" in relation to the transmission of symbolic messages and the meanings encoded in them. "High-voltage" refers to a high degree of attentiveness accorded to the transmitted symbolic messages. This is the case with public rituals that are saturated with symbolic messages and are conducted in well-defined ceremonial contexts (Kertzer 1988).

"Low-voltage" refers to networks of social communication that operate with a low degree of attentiveness to the ideological messages they convey. To the latter belong street names, banknotes and postage stamps that, when invested with a commemorative function, introduce political messages into ordinary settings of human experience. Banknotes—"pieces of green paper with pictures of dead presidents on them" as American banknotes were described (Krugman 2007)—and postage stamps are a currency and their ideological message is iconic (Raento et al. 2004; Raento and Brunn 2005); street names provide spatial orientation and are written, read and spoken.

In his treatment of the semiotics of architecture Umberto Eco maintains that architectural objects denote a certain function and connote a certain ideology about this function. He distinguishes between two levels of signification: denotations of utilities and "symbolic" connotations that refer to all other modes of communication associated with an architectural object (Eco 1986). The distinction is between primary functions, such that are denoted (utilitarian functions) and a complex set of secondary functions, which are connoted (symbolic functions). Eco emphasizes that the distinction between primary and secondary functions does not imply hierarchical positioning but rather that the secondary functions are semiotically "supported" by utilitarian functions (Eco 1986, 65).

Aimed at the integration of the past into social communication, the primary function of commemoration is symbolic: to re-present the past. Ostensibly devoid of utilitarian functions, it is the commemorative function of memorials that is publicly celebrated as the purpose and meaning of the monument. Further, the commemorative function is activated as both "high- and low voltage" networks of social communication. Memorials are laden with symbolic meaning and their visual prominence attracts attention. This becomes especially relevant in the course of ceremonial activities that highlight the commemorative function of the memorial. However, memorials also trigger low-voltage transfer of meaning that occurs in the context of the myriad of random encounters between individuals and memorials, especially those woven into the urban fabric.

Unlike memorials that are architectural edifices, street names have a well-defined utilitarian function: to designate thoroughfares and to provide spatial orientation in the city. This is their primary function. Yet they often also have a commemorative function, which is of a symbolic nature and political significance (Azaryahu 1996; Palonen 1993). Importantly, the commemorative function of street names is dependent on the primary function of street names as designations of location. In respect to their commemorative function, beyond the history they denote, street names also connote a certain ideology about what should be commemorated and where.

However, as a conventional element of the urban texture, street names are constantly evoked in contexts that are commonly detached from the sphere of ideology and the realm of politics. In this semiotic capacity they introduce an official version of history into mundane settings of everyday life: it is the ostensible ordinariness of street names that allows them to effectively implicate politics and ideology into the practices of everyday life and ordinary urban experiences.

Political Significance

Naming the Past

Traditionally, street names were vernacular and descriptive: they designated either geographical orientation or referred to a certain peculiarity associated with local topography and history. A case in point is the "street called straight" in first-century CE Damascus reported in the New Testament (Acts 9:11). The modern practice of giving non-local names to streets was introduced in Paris in the eighteenth century. Following the example set by the French Revolution, commemorative street naming has become a common feature of modern political culture.

As a prerogative of elected or nominated authorities and affected by administrative and political agendas competing for control over the public domain, naming streets is an expression of power. The vast majority of commemorative street names perpetuate in the cityscape the memory of historical figures and events found worthy of public honouring by the authorities in charge of street naming. The significance of street names as historical commemorations becomes apparent when the political role of history, as both a version and interpretation of the past, is considered. On the one hand, political regimes and elites utilize history to legitimate their dominance and augment their authority. On the other hand, a sense of a shared past is crucial to the cultural viability and social cohesiveness of ethnic communities and nation-states.

Streets and squares in pre-revolutionary Paris were named after royalty and catholic saints (Ferguson 1988, 390). In the nineteenth century Berlin's street names celebrated the ruling Hohenzollern dynasty and Prussian military glory (Katzur 1968, 5). Since the late nineteenth century commemorating historical heroes and events by street names was increasingly associated with nation-building measures and the efforts of political elites to cast national identity in the mold of a national history. After the unification of Germany in 1871 and the institution of Berlin as a de-facto capital of the German Reich, street names commemorated also German national figures such as medieval emperors and prominent writers and musicians. The street names that appeared in Stockholm in the late nineteenth century commemorated Swedish national history (Pred 1990). The association of commemorative street names with nation-building measures became paramount in the twentieth century in numerous cases of successful ethnic revivals and postcolonial state formations. For instance, street-naming in post-independence Singapore was an aspect of molding a sense of national identity (Yeoh 1992; this volume).

A distinguished case of the relationship between street names and national revival is provided by Tel Aviv. Founded as a Jewish settlement in 1909, Tel Aviv soon emerged as the political and cultural centre of Jewish Palestine. The symbolic linkage between Tel Aviv as the embodiment of a Zionist ideal and the city as urban reality was manifest in its street names. Tel Aviv was the first Jewish settlement to name streets (Bar-Gal 1987). These names represented the Zionist consensus. One visitor to Tel Aviv wrote about the pride he felt while walking in the streets of Tel Aviv, a city where a Jew "[F]inds a little consolation: there is still a corner in the world where the Jew can respect his heroes and bestow their names on what he created!" (Vardi 1928, 21). A new emigrant reported on the pleasure regarding the fact that the streets were "[B]earing the names of people who dedicated the best of their power and their best years to the Zionist endeavor" (*Ha'Aretz* 1935: 6).

The main source of names for the streets of Tel Aviv was "the history of the People of Israel" to the effect that most new street names commemorated Jewish and Zionist history (Ben-Yishai 1952, 37). On 24 March 1954, for instance, Tel Aviv's city council approved almost 300 new street names; according to the official categorization, these names included "names of writers, artists and political activists of the Jewish *Yishuv*," "Jewish personalities from abroad who contributed to the Zionist reconstruction," "names commemorating Jewish communities in the Diaspora (destroyed in the Holocaust)," as well as "names associated with the struggle of the Jewish *Yishuv* for independence and the (Israeli) War of Independence" (Ben-Yishai 1955, 24).

The street names of the Israeli Arab-Muslim city of Umm el Fahm commemorate the glory of early Islam. Umm El Fahm was declared a town in 1985; in 1988 the Islamic Front won the municipal elections. In 1993 the municipality named the streets of Umm el Fahm. The names represented the ideological orientation of the Islamic movement in Israel and its version of Islamic history worthy of commemoration (Azaryahu and Kook-Doron 2002). As a whole, the Umm el Fahm list of commemorations of historical figures includes 65 names. Most are names of historical persons that belong to the early history of Islam. Among these are names of Caliphs, military commanders, religious authorities, poets and scientists. The Islamic aspect is emphasized in a twofold manner. One is through the names of religious authorities, with 21 street names commemorating the names of religious authorities who won fame for their religious learning and dedication. Of special significance are early converts to Islam, with direct association to the Prophet being a distinguished merit.

The theme of early Islamic glory is reinforced by the commemoration of famous military victories of Islam in the period of its emergence and expansion. These names reveal an emphasis upon the early period of Islamic history. Stretching from the sixth to the twelfth century, this was an era of territorial and religious expansion, military conquests (mainly the first two centuries of Islamic history), intellectual and scientific achievements as well as religious accomplishments. Accordingly, these historical commemorations celebrated the Golden Age of the history of Islam.

Revolutionary regimes have utilized commemorative street names to consolidate their power and to legitimate their authority. Street names in Nazi Germany celebrated a National-Socialist version of German history. The extensive use of street names (as well as names of ships, hospitals, schools, and towns) to commemorate Soviet leaders was a prominent feature of Soviet political culture. This practice was introduced after the death of Lenin in 1924, and reached extreme proportions in Stalin's cult of personality. As long as the communist regimes of Eastern Europe were in power, the street names of communist cities represented an ideologically correct version of national history that corresponded to the officially promoted Marxist-Leninist interpretation of history (on East Berlin and Bucharest see Azaryahu 1986 and Light, Nicolae and Suditu 2002, respectively).

In November 1948 the communist party gained administrative and political control of the Soviet sector of Berlin. The institution of a Stalinist regime in East Berlin was also evinced in the commemorative naming of central squares and thoroughfares. Among the Communist heroes commemorated in East Berlin between 1949 and 1951 were Ernst Thälmann, the official hero of the German Communist party; Stalin, after whom a central thoroughfare was renamed in December 1949 Stalinallee on the occasion of his 70th birthday (in 1961, in the course of the de-Stalinization process, the Stalinallee was again renamed); and Lenin, who was commemorated in April 1950, on the occasion of his 80th birthday. The name of Wilhelm Pieck, the first president of the GDR, was bestowed upon a central thoroughfare in January 1951 on the occasion of his 75th birthday. (Azaryahu 1986, 588–9).

Socially constructed and culturally formatted, public memory is continuously re-evaluated as the consequence of changes in cultural codes, social values, ideological premises and political power-relations. In an open society, where public debate is the norm, commemoration of persons and events is potentially controversial. A case in point is the controversies over the commemoration of Martin Luther King Jr in the American South, where issues pertaining to the evaluation of King's historical greatness and the scaling of his memory in terms of selecting an appropriate street to be named after King have figured prominently (Alderman 2000; 2003).

The intentions of national and local elites and the popular support for these elites notwithstanding, challenging the legitimacy of certain street names as well as the historical traditions they represent also belongs to the politics of commemorative street naming. Where political opposition is legitimate, the rejection of commemorative names becomes part of public debate. In the late 1980s a campaign was waged in West Berlin to replace street names that commemorated the colonial past of the German Empire (Luikenga 1988). Though this campaign targeted specific street names, it provided a focal point for public discussion on a fundamental issue: what constitutes German historical heritage and national identity.

The influence of political majorities in the city council is evident in determining commemorative priorities. A pertinent issue is how a change of local administration impacts on the existing register of commemorations. This is especially relevant when the commemorative priorities of a new administration differ from those practiced by its predecessors. When continuity and stability are considered a priority, a possible solution is the commemorative naming of streets in newly built neighbourhoods to compensate for alleged or real past commemorative deficiencies. Another option is to rename existing streets.

Re-Naming the Past

That commemorative street names belong to the symbolic foundations of the ruling sociopolitical order makes them, together with other symbolic expressions of power, vulnerable to changes in the course of a revolutionary phase of national history. A measure of historical revision, renaming streets figures prominently in periods of regime change and revolutionary transformations. On the immediate level, renaming streets is about making street names and the version of history they represent conform to the ideology of the new regime. On another level, it is a distinct "ritual of revolution" (Kertzer 1988, 151–73). Henri Lefebvre (1991, 54) observed that

> A revolution that does not produce a new space has not realized its full potential [...] A social transformation, to be truly revolutionary in character, must manifest a creative capacity in its effects on daily life, on language and on space.

Relatively simple and comparatively cheap, renaming streets and other public spaces in the course of a revolutionary phase of political history has an immediate effect on daily life, on language, and on space. The change of the city's political geography introduces the political change and the ideology of the new regime into mundane spheres of urban experience and even intimate levels of everyday life. When conducted in the context of a regime change, the act of renaming streets, together with the more spectacular pulling-down of monuments, the renaming of streets is an act of political propaganda with immense declarative value and public resonance. Through the renaming of streets, the new regime proclaims the beginning of a new era while demonstrating resoluteness, self-confidence and its power to control the symbolic infrastructure of society. In this semiotic capacity, renaming the past is a message in its own right: an assertion that a radical restructuring of power relations in society and an ideological reorientation have indeed been accomplished, or are underway.

As is often the case in such circumstances, renaming streets is both a celebration of triumph and a mechanism for settling scores with the vanquished regime. As a measure of historical revision, renaming the past in the context of a regime change is a twofold procedure that involves both the de-commemoration of persons and events associated with the old regime and the commemoration of heroes and events that represent the new regime and its version of history. When regime-change is construed in terms of restoration, commemoration may assume the form of re-commemoration, namely, the re-institution of names that had been removed by the former regime. Ostensibly, de-commemoration and commemoration are two sides of the same coin since in principle, renaming the past entails the substitution of one street name by another. However, the dual procedure of de-commemoration and commemoration may involve different ideological emphases and political priorities that reflect specific interests and needs as well as power constellations that influence the actual pattern and direction of the renaming process (Azaryahu 1992).

As mentioned earlier, renaming the past figured prominently in the French Revolution, when royal and religious names were replaced by republican ones. With the collapse of the imperial regime in France in 1815 some 50 streets in Paris were renamed. Most of them re-gained their pre-revolutionary names (Ferguson 1988, 391). Napoleon and his victories were de-commemorated; rue Napoleon was renamed rue de la Paix. The renaming of streets also figured prominently in Paris after the establishment of the Third Republic in 1870. Aware of the propaganda value of street names and determined to erase the traces of the Weimar Republic, the Nazis were swift in their actions to rename all those streets that commemorated the Weimar Republic. National-Socialist political mythology was evident in the commemorations of heroes of the regime, most prominent among them being Adolf Hitler and Horst Wessel, the "Martyr of the Movement" (Giese 1934).

The purge of Nazi street names in German and Austrian cities followed immediately after the surrender of the Third Reich in May 1945. The renaming of the Nazi past was high on the political agenda of the reviving political life of Berlin. The issue was discussed already in the first meeting of the provisional Berlin City Hall that convened on 24 May 1945 (Karwelat 1988; Azaryahu 1990). An important decision was that the districts were responsible for naming streets, yet the right of City Hall to intervene in special cases was formally acknowledged. Even though the need to "democratize" the street signs was a matter of consensus, two different approaches became apparent. The moderate approach advocated by conservative circles was to erase the renamings undertaken by the Nazis and to re-institute the old names. The Communists, on the other end of the political spectrum, supported a radical approach, according to which the democratization of political life should include the renaming of both Nazi and Prussian (namely military and dynastic) traditions in order to accomplish a "true" democratization of public space.

The renaming of the Nazi past was formally concluded in 31 July 1947 Altogether 150 streets were renamed (Verordnungsblatt für Gross-Berlin 1947). Adolf-Hitler-Platz (Charlottenburg) and Hermann-Göring-Platz regained their former names, Reichskanzlerplatz and Ebertstrasse, respectively. The heritage of Prussian glory and German imperial greatness was left intact. The de-commemoration of Nazi leaders, heroes and martyrs evinced the end of the Nazi regime. Notably 34 streets regained their pre-Nazi names. Re-commemorating distinguished leaders of the Weimar Republic such as Friedrich Ebert, Walter Rathenau and Gustav Stresemann was more than a restorative act designed to undo Nazi de-commemorations. It also evinced continuity with the Weimar Republic and the democratic tradition that these names represented. Sign of the times was the commemoration in Berlin of heroes and martyrs of the revolutionary tradition such as Karl Marx, August Bebel and Rosa Luxemburg.

In the Soviet Union de-commemorations indicated major political changes. The end of the Stalinist era in the Soviet Union, as well as in other East European communist states, was sealed in 1961 with the purge of Stalin from public memory, including renaming streets and cities and pulling down monuments. At the height of *perestroika*, the Central Committee of the Communist Party decided to strip Leonid Brezhnev of the public honours bestowed upon him after his death in 1982 and to rename a large square in Leningrad, together with a famed Moscow suburb and a city whose names perpetuated the memory of the former Soviet leader.

Notwithstanding local differences, the cases of post-communist Moscow, Bucharest, Budapest and East Berlin demonstrate that renaming the past also figured prominently in the transition from a communist to a post-communist regime (Azaryahu 1997; Foote et al. 2000; Gill 2005; Light 2004). In these cases, beyond de-commemorating heroes of the communist state the question was also whether and to what extent traditional names erased by the communist regime should be restored. De-commemorating the communist past signified the transition to a post-communist era.

Historical Signification

Semantic Displacements

The primary function of street names is to denote a location. However, as the result of their use in different contexts, street names undergo semantic displacements; as Michel de Certeau (1985, 140) noted, they "... become things apart from the places they were intended to define and turn into imaginary meeting-places in the journeys they map out, having become metaphors, for reasons foreign to their original validity, however known/unknown to the passers by." Anchored as they are in reputations, mythologies, images, and socio-spatial practices, the meanings of a street name are dynamic. Though often determined by personal experiences, they may further be differentiated according to social, ethnic, gender, and generational divisions. Contested as they may be politically, and despite the inherent instability of their meanings in the course of time and for different audiences, the official names provide a fixed point of reference in the geography of the city.

When invested with a commemorative function, the name given to a street also associates the street with the official meaning assigned to the particular historical referent. This meaning figures prominently throughout the naming procedure and dominates the decision making process. According to Michel de Certeau (1985, 140), the naming of streets after "real" historical figures means that the names cease to be "proper." Moreover, a commemorative street naming is bound to trigger a fundamental semantic displacement as a result of the change of referential framework that occurs when a historical reference becomes a spatial designation. Such a semantic displacement disconnects the meaning of the name from the original historical referent without, however, denying the existence of such a referent. The success of commemoration by means of a street name becomes evident in how the name increasingly becomes associated with a location; history becomes geography, and in this geographical capacity it is integrated into various narratives of the city.

The commemorative function is often decisive in naming streets. However, as the result of the semantic displacement that underlies the conversion of history into geography, the geographic denotation takes over while the historical referent becomes increasingly obscure to most users of the city. The "suspension" of history also means that the version of history that street names introduce into social communication is experienced as obvious, part of the "natural order" of things. In this sense, the seemingly weakness of the commemorative denotation actually augments the power of the commemorative street names to render a certain version of history not only familiar, but also self-evident.

Occasionally, however, the meaning of the name as a historical referent is restored ("rediscovered" or "unmasked," depending on the interpretation). This occurs when history is elucidated in an educational context or when certain commemorative names are challenged as ideologically and morally improper. Though exceptional, such occurrences highlight and reaffirm the belonging of the name to the societal discourse of history. Importantly, they also demonstrate that the meaning of a commemorative street name consists of two concurrent layers of denotative signification: the name as both a geographical designation and as a historical reference.

When names are transported from history to geography, they clearly become susceptible to sharing the fortunes of their location. When obscure streets gain in urban significance, their namesake heroes gain in prominence and fame. In the 1980s, when Sheinkin Street in Tel Aviv became a distinguished icon of Israeli popular culture, it also reflected positively on the prestige of a hitherto little known historical figure. The opposite is also true, when famous heroes are associated with the reputation of notorious slums or derelict neighbourhoods.

Semantic displacement also occurs when the history associated with a street name comprises not only the formal historical reference but also the history of the street itself. A case in point is Berlin's Wilhelm Strasse. This street in the centre of Berlin was named after a Prussian monarch. Its international reputation was the result of the location there of the German Foreign Office. The Wilhelm Strasse became a designation for the German Foreign Office and German foreign policy, and hence the imperial aspirations of the German Reich in general. When, in 1964, the East German authorities renamed the street, a traditional symbol of German imperialism was erased from the map of the East German capital. After German reunification in 1990, the "imperialistic" connotation of the old name was a major argument against its restoration. In the fall of 1993, the Senate of Greater Berlin intervened in favour of the restoration of the old name, thereby re-instating not only the traditional name as a historical referent but also its old meanings into the discourse of a German reunification (Dolak 1993).

In a few cases a street name—in its capacity as a geographical designation—also becomes a metaphor for a social milieu, cultural phenomena, modes of behavior and a certain way of life that, at least in their early evolution, were connected with the location. The geographical designation provides the appearance of a specificity and objectivity and solidifies a set of human activities and notions about these activities into a powerful cultural reference that transcends the physical location itself. Built in the 1920s, Tel Aviv's Sheinkin Street was named after a Zionist activist. Celebrated as "the Israeli SoHo or local Village," Sheinkin Street in the 1980s became metaphoric of an "alternative" lifestyle and an enlightened worldview, of an artistic scene and a bohemian milieu (Azaryahu 2006). The process involved two semantic displacements. The first followed the commemoration of Sheinkin by a street name and involved the conversion of the name Sheinkin from history into geography. The second transported the name from geography into the realm of popular culture.

Reading a City-Text

In his discussion of urban sources of signification, Gottdiener observed that "even space itself, such as the city street, is an important object of signification" (1986, 206). The most obvious vehicle of signification are street signs and the ideological messages encoded in street names. The spatial configuration of commemorative street names in a city defines a particular city-text that, as a semiotic feature of the city, presents a particular mapping of urban space and historical time. In connection with street names the term city-text does not mean a reduction of the city into a text (Ferguson 1988), but rather refers to the well-defined ensemble of street names that are integrated into the cityscape and are depicted in cartographic representations of the city.

Constantly written and eventually overwritten, city-text is the sum of additions and erasures; in this sense it is a palimpsest (Ferguson 1988, 392; Crang 1998). Writing a city-text is a prolonged process that conflates urban contingencies, ideological concerns and political interests. As a particular geography of public memory, a city-text represents not only a version of history but also commemorative priorities and hegemonic discourses of former periods. For instance, former communist majorities are still reflected in the street names of Parisian suburbs and Italian towns.

Reading street signs or referring to street names, as is practiced repeatedly in the context of everyday life, mainly involves their practical function as signifiers of location rather than their commemorative function. Moreover, a city-text is not intended to be read as an entirety, and its eventual reading as a text does not involve any obligation to a prescribed order. In principle, a city-text is a spatially configured register of historical figures and events. It is a disorderly representation of history rather than a coherent historical account. However, the semiotic structure of a city-text offers a measure of historical analysis that is pertinent to reading a city-text.

The semiotic structure of a city-text provides a built-in scale which enables the reader to determine the relative status of the commemorated events and heroes. The status hierarchy in an ideal-type text is reflected through the configuration of elements in the urban fabric, for which the underlying rule is that the urban importance of a thoroughfare and the prestige of the associated commemoration are positively correlated and that urban and historical significance conform. Yet deviations from this pattern often occur as the result of the combined effect of urban dynamics and reevaluations of historical significance. One problem is the evaluation of urban prestige. It seems obvious that, from an urban perspective, a central boulevard is more significant than is a small lane on the margin of the city and that the measure of public exposure of the first is much greater than that of the latter. The evaluation of urban prestige, however, also has to take into account that different parameters may have contradictory effects on the production of prestige and status. A newly built highway may be of supreme importance from the point of view of transportation, yet the context in which its name is used is specific and may even reduce its appeal as a vehicle for commemoration in spite of the public exposure it guarantees to the name.

The fact that a city-text is a product of a protracted historical process is a source of further distortions as regards reading a city-text. The history of the text has a twofold effect. One is the possible reevaluation of the significance of historical figures and events, which is an indispensable element of a societal discourse of history. This process is independent of urban developments. The other is the effect of urban dynamics on the prestige of streets and neighbourhoods in general, which is independent of the academic and political discourse of history.

Commemorations reflect prevailing notions of the period concerning the historical significance and greatness of events and persons, respectively. Unless renamings were undertaken, a city-text would conserve notions and attitudes that prevailed in different stages of its emergence. Israeli localities founded in the 1920s and the 1930s commemorated prominent Zionist leaders of the period, yet later reevaluation of Zionist history accorded these figures much less historical significance. The urban expansion of Berlin in the late nineteenth century and the tight control of the Prussian authorities produced a situation in which numerous streets of the former West Berlin further commemorate obscure Prussian generals and officials.

Urban dynamics is a factor to be reckoned with whenever the relative status of commemorations is deduced from the significance of the thoroughfare in the urban fabric. An example is provided by the history of Herzl Street in Tel Aviv, where as the result of the expansion of Tel Aviv to the north, the urban prestige of Herzl Street gradually diminished.

The version of history represented in a city-text is biased in favour of local history; this is evident in the names of local dignitaries, most notably mayors, and local events that are commemorated in the street signs. The recurrence of particular commemorations in different city-texts is a measure of the historical significance assigned to them. The popularity of a commemoration in different city-texts that belong to the same national context is a marker of status and an indicator of reputation. From the practice of street-naming in France it can be deduced that Jeanne d'Arc, Hugo, Pasteur, Jaurès, Clemenceau, and Blum belong to the French national consensus, Napoleon and Richelieu are highly contested, while Danton, Robespierre, and Napoleon III—all salient protagonists of modern French history—have a bad reputation (Milo 1986; Palonen 1993).

In contrast to textbooks, which present both heroes and villains of history and an evaluation of their roles in national history, a city-text mentions heroes and glorious events only. The guiding principle is that villains are not entitled to a place in eternity ostensibly guaranteed by commemorative street names. Hitler and Pétain are absent from contemporary German and French city-texts, respectively, in spite of their paramount significance in twentieth-century German and French history.

The spatial organization of street names does not produce any significant linear order, and the intersection of streets does not necessarily imply any temporal or thematic relationship between their names. Such intersections, however, may suggest imaginary encounters that are sometimes weighed down with irony. For instance, the intersection of the Boulevard Voltaire with rue Saint Sébastien in Paris is ironic in the light of the notorious anticlerical disposition of the eighteenth-century philosopher. In Selma, Alabama, Martin Luther King Street and Jefferson Davies Street intersect, thereby suggesting parity between the martyred hero of the American civil rights movement and the President of the southern Confederacy.

A fundamental property of a city-text is that it lacks a built-in time arrow, and hence chronology. Historical figures and events coexist simultaneously, and one cannot make a distinction between before and after. In his elaboration on the Argentina of the early 1970s, V.S. Naipaul (1981, 140) observed:

> But the country has as yet no idea of itself. Streets and avenues are named after presidents and generals, but there is no art of historical analysis; there is no art of biography. There is legend and antiquarian romance, but no real history. There are only annals, lists of rulers, chronicles of events.

A city-text lacks the narrative structure of the conventional historical account. According to Hayden White, historical narratives—unlike annals and chronicles—weave together events and explanations into a comprehensive plot that conflates temporality and causality (White 1981). As the organizing principle of a narrative account, chronological progression produces a sense of continuity and coherence. Of much significance is "the jointly causal and temporal logic of stories, or narrative's fusion of sequence and consequence" (Herman 2005, 21), namely, that causal explanations correspond to the temporal progression of successive events (Hodge and Kress 1988, 229–30). Importantly, the chronology of a narrative account implies causality: "before" and "after" correspond to "cause" and "effect."

A city-text does not provide its readers with a chronological narrative of history, but rather with an authorized index of a putative narrative, notwithstanding the lack of historical villains. The index of a city map is an exact reproduction of the city-text as it mentions all the names that appear on the map. The alphabetical order of this index implies neither spatial nor temporal order, but is essential for using street names as geographical designations that provide for spatial orientation, which is, after all, considered to be their primary function.

Concluding Remarks

When invested with commemorative function, street names are constitutive elements of particular geographies of public memory and an officially sanctioned vision of history. The encounter with a street name is casual. Unlike commemorative monuments, street signs do not induce pathos and are not charged with the sacred. On the other hand, using street names for commemorative purposes incorporates an official version of history into spheres of social and spatial practices of everyday life that seem to be totally detached from political contexts or communal obligations.

Though spatial orientation and commemoration are two complementary denotative functions of street names, the commemorative function of street names is semiotically supported by the function of street names as markers of location that provide spatial orientation in the city. Normally, the commemorative function does not interfere with spatial orientation, but only associates it with the ideological foundations of the ruling sociopolitical order. However, things are different when the authorities in charge of naming streets consider the commemorative function as the primary one.

This is clearly the case with ideologically motivated renaming of streets. Large-scale renaming procedures evince that for the national and local authorities in charge the commemorative function of street names often overrides their utilitarian function to provide for spatial orientation. In 1882, a British observer accustomed to the stability and continuity characteristic of a map of London, disapproved of the recurrent renaming of streets in Paris (*Dickens's Dictionary of Paris 1882*, 243). The British observer reacted to the aftermath of the establishment of the Third Republic in 1871 and the subsequent change of street names. Less interested in the political context, his judgment was that the authorities in charge showed an "excess of zeal." Intended for a British audience, he noted that "The frequent changing of names of their streets is not to be commended; and under the present Republic the practice seems to have been pushed to an unusual extent" (*Dicken's Dictionary of Paris 1882*, 243).

However, orientation refers to both the geographical and commemorative functions of street names. Commemorative street names denote both local geography and a version of history and in this capacity offer a twofold orientation: both in urban space and in the symbolic space of historical memory. As commemorations, street names transform history into local geography. Michel de Certeau maintains that the naming of streets after "real" historical figures means that the names cease to be "proper" (1985, 140). As street names, Bismarck in Berlin, Martin Luther King in the American south, Herzl in Tel Aviv and Clemenceau in Paris are integrated into ordinary experiences of the city in terms of "where it is" rather than "who is he/she." What distinguishes street names as a vehicle of commemoration is their apparent triviality: everybody uses them but hardly anyone pays attention to their specific historical meaning and to their belonging to the structures of power.

References

Alderman, D.H. (2000), "A Street Fit for a King: Naming Places and Commemoration in the American South," *Professional Geographer* 52:4. 672–84.

—— (2003), "Street Names and the Scaling of Memory: The Politics of Commemorating Martin Luther King, Jr within the African American Community," *Area* 35: 2, 163–73.

Azaryahu, M. (1986), "Street Names and Political Identity: The Case of East Berlin," *Journal of Contemporary History* 21, 581–604.

—— (1990), "Renaming the Past: Changes in 'City Text' in Germany and Austria, 1945–1947," *History and Memory* 22, 32–53.

—— (1992), "The Purge of Bismarck and Saladin: The Renaming of Streets in East Berlin and Haifa. A Comparative Study in Culture-planning," *Poetics Today* 13, 351–67.

—— (1997), "German Reunification and the Politics of Street Names. The Case of East Berlin," *Political Geography* 16: 6, 479–93.

—— (2006), *Tel Aviv: Mythography of a City* (Syracuse: Syracuse University Press).

—— and Kook-Doron, B. (2002), "Mapping the Nation. Street Names and Arab-Palestinian Identity: Three Case Studies," *Nations and Nationalism* 8: 2, 195–213.

Bar-Gal, Y. (1987), "Names for Tel-Aviv's Streets: A Chapter in Urban Cultural History (1909–1933)," *Cathedra* 47, 118–31 (Hebrew).

Ben-Yishai, A.Z. (1952), "Tel Aviv and its Streets," *Yediot Iryat Tel Aviv-Yafo* [*The Municipal Bulletin of Tel Aviv*] 22: 1–3, 37–43.

—— (1955), "New Names for the Streets of Tel Aviv-Yafo," *Yediot Iryat Tel Aviv-Yafo* 23: 1–2, 24–5.

Blonsky, M. (ed.) (1985), *On Signs* (Baltimore, MD: Johns Hopkins University Press).

Crang, M. (1998), *Cultural Geography* (London and New York: Routledge).

De Certeau, M. (1985), "Practices of Space," in Blonsky, M. (ed.), 122–45.

Dickens's Dictionary of Paris 1882 (London: Macmillan).

Dolak, G. (1993), "Ein später Sieg für den Soldatenkönig," *Süddeutsche Zeitung* 16 July.

Eco, U. (1986), "Function and Sign: Semiotics of Architecture," in Gottdiener, M. and Lagopoulos, A.P. (eds), 54–86.

Ferguson, P. (1988), "Reading City Streets," *The French Review* 61, 386–97.

Foote, K., Toth, A. and Arvay, A. (2000), "Hungary after 1989: Inscribing a New Past on Place," *Geographical Review* 90: 3, 301–34.

Giese, A. (1934), "Unsere Helden und ihre Taten in den Strassennamen von Gross-Berlin," *Berlinische Blätter für Geschichte und Heimatkunde* 1: 10–11, 122–3; 1: 12, 134–40.

Gill, G. (2005), "Changing Symbols: The Renovation of Moscow Place Names," *The Russian Review* 64, 480–503.

Gottdiener, M. (1986), "Culture, Ideology and the Sign of the City," in Gottdiener, M. and Lagopoulos, A.P. (eds), 202–19.

—— and Lagopoulos, A.P. (eds) (1986), *The City and the Sign. An Introduction to Urban Semiotics* (New York: Columbia University Press).

Ha'Aretz (1935), 24 October.

Herman, D. (2005), "A Genealogy of Early Development," in Phelan, J. and Rabinowitz, P.J. (eds).

Hodge, R. and Kress, G. (1988), *Social Semiotics (*Cambridge: Polity Press).

Karwelat, J. (ed.) (1988), *Sackgassen. Keine Wendemöglichkeiten für Berliner Strassennamen* (Berlin: Berliner Geschichtswerkstatt, Nishen Verlag).

Katzur, K.(1968), *Berlins Strassennamen (*Berlin: Haude und Spener).

Kertzer, D. (1988), *Ritual, Politics and Power (*New Haven, CT: Yale University Press).

Krugman, P. (2007), "Who Was Milton Friedman?," *New York Review of Books* 54: 2, 15 February.

Lefebvre, H. (1991), *The Production of Space (*Paris: Basil Blackwell).

Light, D. (2004), "Street Names in Bucharest, 1990–1997: Exploring the Modern Historical Geographies of Post-Socialist Change," *Journal of Historical Geography* 30, 154–72.

——, Nicolae, I. and Suditu, B. (2002), "Toponymy and the Communist City: Street Names in Bucharest, 1948–1965," *GeoJournal* 56, 135–44.

Luikenga, M. (1988), "Ein Schuljubiläum und seine Folgen: Streit um die Umbenennung des Dietrich-Schöfer-Weges in Steglitz," in Karwelat. J. (ed.), 46–55.

Milo, D. (1986), "Le nom des rues," in Nora, P. (ed.), 283–315.

Mitchell, W.J.T. (ed.) (1981), *On Narrative* (Chicago: Chicago University Press).

Naipaul, V.J. (1981), *The Return of Eva Peron and Other Essays* (Harmondsworth: Penguin Books).

Nora, P. (ed.) (1986), *Les Lieux de Memoire II: La Nation* (Paris: Gallimard).

Palonen, K. (1993), "Reading Street Names Politically," in Palonen, K. and Parvikko, T. (eds), 103–21.

Palonen, K. and Parvikko, T. (eds) (1993), *Reading the Political: Exploring the Margins of Politics* (Tampere: The Finnish Political Science Association).

Phelan, J. and Rabinowitz, P.J. (eds) (2005), *A Companion to Narrative Theory* (Oxford: Blackwell).

Pred, A. (1990), *Making Histories and Constructing Human Geographies: The Local Transformation of Practice, Power Relations, and Consciousness* (Boulder, CO: Westview Press).

Raento, P., Hamalainen, A. and Ikonen, H. (2004), "Striking Stories: A Political Geography of Euro Coinage," *Political Geography* 23: 8, 929–56.

—— and Brunn, S.D. (2005), "Visualizing Finland: Postage Stamps as Political Messages," *Geografiska Annaler Series B: Human Geography* 87: 2, 145–16.

Shils, E. (1981), *Tradition (*Chicago: Chicago University Press).

Vardi, A. (ed.) (1928), *The City of Wonders: Writers and Statesmen on Tel Aviv—for its Twentieth Anniversar*, Tel Aviv: Lema'an Ha'sefer.

Verordnungsblatt für Gross-Berlin (1947), 3, 21, 1 October.

White, H. (1981), "Value of Narrativity in the Representation of Reality," in Mitchell, W.J.T. (ed.), 1–24.

Yeoh, B.S. (1992), "Street Names in Colonial Singapore," *Geographical Review* 82: 3, 313–22.

—— (1996), "Street Naming and Nation Building: Toponymic Inscriptions of Nationhood in Singapore," *Area* 28: 3, 298–307.

Chapter 4

Street-naming and Nation-building: Toponymic Inscriptions of Nationhood in Singapore

Brenda S.A. Yeoh

Landscape Studies and Toponymic Research

> Illustrating what might be called, after its well-known advocate, the "Humpty Dumpty" position, ... names are rooted neither in reality nor custom, but express instead the power of the namer over the thing named. (Emmerson 1984, 4)[1]

Recent studies of landscape as ensembles of material practices and their symbolic representations have emphasized the power of landscapes to reinforce and "naturalize" prevailing social ideologies (Zukin 1991; Baker 1993). Treated as a "text" laden with a multiplicity of "signs" or a "cultural form" containing a plurality of meanings, the landscape "upon interrogation reveals a human drama of ideas and ideologies, interest groups and power blocs" (Ley 1987, 41; Daniels 1989; Duncan 1990). Notwithstanding cautionary pronouncements on the dangers of privileging the representational realm at the expense of the practical in landscape studies (Jacobs 1993), these insights drawn from textural readings and interpretative analysis have richly informed the craft of cultural and historical geographers working on landscapes ranging from the everyday to the esoteric.

Within this expanding field, and in some ways mirroring the wider trend in landscape studies in moving away from field- or map-based research to discursive analysis (Berleant-Schiller 1991), toponyms have been critically "read" for a variety of purposes, as "signifiers" of wider societal trends. The word toponym is derived from the Greek word *topos* (meaning "place") and *onomia* (meaning "name"), and toponymy refers to the study of place names in a region. Traditionally, etymological inquiry into the naming of places is used to uncover patterns of naming and "these regularities are treated as

1 This observation is based on Alice's (of Wonderland fame) encounter with Humpty Dumpty who believed that words are not rooted in the nature of things but mean just what one chooses them to mean.

beacons for inferring landscape and settlement histories" (Pirie 1984, 51). The analysis of place names, their elaboration and mutation, as instruments for reconstructing the "sequent occupance" of colonizing or immigrant group (Hsieh 1980), tracing the "ecological invasion and succession" of operational territories by urban gangs (Mak 1981), or recovering the localization of trades and tradespeople (Ekwall 1954, 49–57) appears to be a well-honed art if of somewhat antiquarian and esoteric appeal. More recently, rather than treating toponyms as independent clues on environmental change, there is increasing interest in reading them within the wider social context, entrenched within the "social relations of deference and defiance upon which naming is contingent" (Pirie 1984, 51; Yeoh 1992). The toponymic significance of place names honouring public figures has been interpreted as indicators of nationalistic fervour (Zelinsky 1983; Stump 1988) or community identity (Rogers 1995), whilst the renaming of streets in post-colonial societies has been interpreted as an ideological tool to divest the landscape of its colonial associations and achieve political legitimation (Lewandowski 1984). "The power of 'nomination' … is often the first step in taking possession" (Robinson 1989, 160), and street names are "among the first to undergo a refurbishing to commemorate new regimes" (Richman 1983, 16). In a different vein but again reflecting the power of dominant others to shape landscape meanings, Katz and Kirby (1991, 266) argue that place names often provide "particularly sardonic example[s]" of "salvage toponymy" wherein "historic" place names are retained and recycled as signifiers of heritage, obscuring the loss of cultures and practices in which the "text" was originally grounded.

The naming of place, whether as a conscious, deliberate event or a more informal process of evolution, is in varying degrees a socialized activity. As Pirie (1984, 43) has argued, this is so "either by virtue of it involving joint decision-making and/or in respect of it occurring within a given social milieu in which there are formal or informal conventions of name selection, assignment and adoption." The naming process is hence not only of toponymic significance but also embodies some of the social struggle for control over the means of symbolic production within the built environment.

Aims

In this chapter, I am concerned with the way in which toponymic inscription in the landscape reflects and is shaped by broader socio-political conditions in post-independence Singapore. In particular, I examine the changing bases underlining the naming of streets as part of the newly created nation-state's bid to foster a sense of "nation" and "national identity" in Singapore.

As Jackson and Penrose (1993a, 8) have argued, the term "nation" is a social construct, an "imagined community" (Anderson 1983) which draws on notions of common interest between heterogeneous groups. By grounding it in physical space and delimiting a recognisable territory, the belief in the existence, legitimacy and inviolability of the entity is reinforced. Central to the construction of a "nation" is the articulation of a "mystical bond between people and place," an immutable relationship between citizens and their country (Penrose 1993, 29). Newly emergent from under the wing of colonial tutelage in the 1960s, Singapore, with a diverse population drawn mainly from immigrant groups and sojourners, lacked established ideological boundaries and common cultural ballast to bind newly enfranchised citizens. The state thus had to engage in a constant struggle for political dominance, state power and ideological hegemony through a range of tactics (Chua 1995, 43; Hill and Lian 1995) including, crucially, reshaping the built environment in the attempt to forge a radical discontinuity with the colonial past, a "nation" of people. Rewriting the landscape involved a multitude of strategies combining ideological and material effects on a broad range of spaces (Kong 1993; Yeoh and Kong 1994; Yeoh and Tan 1995). This chapter examines one of these strategies, attempts at toponymic inscription, in the social production of a "nation." In as much as there is often slippage between the ideological intents of the state and the effects on the people, this chapter is also concerned with the people's reactions to the naming of the city's streets.

Context

One hundred and fifty years of colonial rule in Singapore conferred on its landscape an official network of street and place names which reflected the mental images and ideological purposes of the dominant culture (Yeoh 1992). As part of a "realized signifying system" (Agnew et al. 1984) embedded in everyday practice, street-names provided a tangible (though not necessarily unproblematic) record of colonial imaginations. The predominant pattern of street-names, for example, commemorated mainly European city fathers, public servants and "deserving" citizens (for example, Farrer Road after a municipal president, Onraet Road after an Inspector-General of Police and Owen Road after the secretary of the Cricket Club) who were considered to have contributed significantly to public work and urban development as well as Governors (for example, Anderson Road and Shenton Way) and British royalty (for example, Victoria Street and Connaught Drive); recalled linkages with other places within the British Empire including Britain itself (for example, Bristol Road and Sussex Garden); and endowed places with racial connotations, reflecting the colonial tendency to order society by separating the colonized into distinct, recognizable containers (for example, China Street in Chinatown; Baghdad Street in the Muslim Quarter and Hindoo Road in Little India). Most street-names thus honoured the perceptions and priorities of powerful European namers rather than those of the people living in the places so named.

For Singapore, the age of independence only dawned at the end of a slow and uneven road following the Japanese Occupation (1942–1945) which shattered the myth of white superiority and British invincibility; internal self-government and the installation of the People's Action Party government in 1959; and after a failed attempt at merger with Malaysia (1961–1965), the rude thrust into independence on 9 August 1965. Independence as a consequence of expulsion from Malaya was an unexpected one and the parameters which engendered the mood of uncertainty at the time were stark; the island had no natural resource base to depend on, external relations with immediate neighbours Malaysia and Indonesia were strained, the communist challenge was "crippled but not crushed" (Chew 1991, 363), and Singapore's vulnerability was further underscored by the withdrawal of British military forces from the island from 1967. Indeed, in the words of Goh Keng Swee, then Minister of the Interior and Defence, Singapore was "a complex, multiracial community with little sense of common history, with a group purpose which is yet to be properly articulated, in the process of rapid transition towards a destiny which we do not know yet" (quoted in Chew 1991, 363). National survival became the dominant preoccupation of Singapore's political leaders, who lost no time in urging the local population to "buckle down to the job of 'nation-building'" (Benjamin 1988, 20). The state's vision was to integrate the "nation" on the basis of principles to create a "multiracial, non-communist, non-aligned, and democratic socialist state" (Chan 1991, 158). In particular, multiracialism (along with the other "3Ms:" multilingualism, multireligiosity and multiculturalism), the ideology that each of the major races—Chinese, Malays, Indians and "Others"—is separate but equal have since been promulgamated as a social formula to forge a single identity out of the heterogeneous population riven by racial, religious, language and cultural lines (Benjamin 1976; Siddique 1989, 365).

Rewriting the Toponymic Text: Inscriptions of Nationhood

Malayanizing the Colonial Landscape

In mounting a programme of nation-building, one of the projects was a concerted effort to sever colonial apron strings and foster and assert a sense of *local* identity, a sense of place, through a rewriting of the everyday landscape. Changes in street-naming policies clearly reflected these imperatives. In March 1967, ground rules laid down to guide the decisions of the Street-Naming Advisory Committee[2] instructed the committee to steer clear of "old colonial

2 Appointed by the Minister of Finance in February 1967, the Committee comprised the Director of Inland Revenue (chair); the Postmaster-General; the Head of Urban Renewal, Housing and Development Board; a Research Officer from the

nuances, British snob names, towns, and royalty," and give priority to local names (MMACNRS 14 March 1967). Preference was to be given to a fuller utilization of Malay names to signal Singapore's allegiance to the Malay as opposed to colonial world; and where English names were to be used, names should reflect "the historical background of the area," "current affairs of public interest in Singapore," local flora and fauna, and the physical nature of the area (MMACNRS 14 March 1967). In addition, as a departure from the liberal toponymic "canonization" of prominent public figures which had preoccupied street-naming decisions in the prewar colonial days, "no person should have a street named after him [sic] during his [sic] life-time, except where the individual ha[d] performed some outstanding public service" (MMACNRS 14 March 1967).

In the wake of these decisions, while the majority of street-names of colonial origins remained inscribed in the landscape, the post-independence era spawned a whole generation of indigenous toponyms which not only drew inspiration from local material but also substituted the Malay equivalent *jalan* or *lorong* for the word "road" or "street" (for example, Jalan Layang-Layang after the Malay name for the swallow and Jalan Silat Gayong after a Malay martial art form). One of the earliest proposals was to change "colonial" names (such as Queensway, Margaret Drive and Commonwealth Avenue) in Queenstown (named after the British monarch), the first satellite town built in the 1960s by the newly created Housing and Development Board, to Malay ones (MMACNRS 26 June 1967). While this was "strongly resisted" by the Board so as to avoided "confusion and inconvenience" and also because "Queenstown [was] well-known throughout the world and should be preserved," the policy to name all streets in new housing estates in the National Language (Bahasa Melayu) took effect in various smaller estates such as the Kallang Airport estate where streets were named using Malay numerals (for example, Jalan Satu, Jalan Dua) (MMACNRS 26 June 1967; 6 July 1967). Conversely, street-names which perpetuated colonial imagery were avoided at all cost.[3]

The experiment to rework the toponymic text to assert local identity and represent the Malayan vision of independence soon encountered constraints. The available Malay vocabulary relating to Malayan fauna, flora and material culture was soon exhausted and painstaking scraping of the bottom of the

Institute of National Language and Culture; and a representative from the History Department of the University of Singapore (MMACNRS 2 February 1967).

3 Twenty years down the road, the debate over the place of "colonial" names in the landscape was resurrected, this time over the naming of the newly built Mass Rapid Transit (MRT) stations. Two stations, originally christened Victoria (after the British monarch) and Maxwell (after a prominent colonial family), were renamed Bugis and Tanjong Pagar (local names), because, to quote the chairman of the MRT Corporation, while "we have no hang-ups about our colonial past, ... we thought we'd have a few more local names" (*The Straits Times* 1986).

lexicographical barrel produced names which the Deputy Prime Minister described as "comic and unintelligible" (MMACNRS 26 June 1967). The ideological significance of introducing intricate Malay toponyms was also unappreciated by the predominantly Chinese, dialect-speaking inhabitants of the housing estates who found street-names such Pesiaran Keliling (meaning "Circular Drive") "tongue twister[s] of absolute horror" (MMACNRS 26 June 1967). Developers of private housing estates and residents protested against the proliferation of Malay street names (for example, names after birds such as Jalan Chiak Padi, Jalan Chiak Raya etc. in Clementi Park estate and names after colours such as Jalan Merah Saga, Jalan Hitam Manis, etc. in Chipbee Gardens), describing these concoctions as "confusing," "ridiculous" and "prejudicial to the dignity [of their estates]" and, in turn, appealing for the reinstatement of English names (MMACNRS 22 August 1967).[4] Even ostensibly "meaningful" names encapsulating local pride such as Jalan Khairuddin (after a well-known local composer of Bangsawan plays) in Opera Estate (where street-names are connected to the operatic world) met with the disapproval of residents who petitioned for the street to be renamed Fidelio Close (after a character in western opera) (MMACNRS 10 October 1967). Apparently, the exercise to use Malay street-names to signify independence and provide a common toponymic language to draw together the plurality of local worlds did not succeed; ironically, people preferred road signage and residential addresses in English, the language of the colonial masters, which they perceived as neutral if not superior.

"Multiracializing" the Landscape

While the failure for street-names in one language to be properly understood and pronounced by fellow citizens of different ethnicities was nothing new (indeed, this was the social reality for a long time under the colonial regime when English was the official toponymic language), it became unacceptable in the newly forged context of multiracialism where the state must not be seen to privilege any one "ethnic" language but to strike a balance between the interests of

4 Not all petitions met with success but the case of Mount Sinai, a private housing estate where street-names had been named in Malay after different types of cloths (for example, Jalan Kain Limau, Jalan Kain Telepok and Jalan Kain Matsuli), is illustrative of how residents pressed for the renaming of the streets (as Mount Sinai Drive, Crescent, View, etc.) First, they pointed out that "not everyone [could] pronounce [Malay] names with the correct accent;" few knew the meaning of these names and thus could not remember them; and that adjacent estates had "nice-sounding" names such as Moonbeam Walk, Moonbeam Terrace, Holland Grove Drive and Holland Grove Terrace. Next, they solicited the help of a powerful individual, in this case the Education Minister who lived within the estate, to support their case and bring it to the attention of the authorities (MMACNRS 12 July 1968).

"separate but equal" "racial" groups. By early 1968, newly proposed guidelines for street-naming specified that street-names should "reflect the multi-lingual, multi-racial and multi-cultural context of the society," and "be easily translated or pronounced in the other official languages [English, Mandarin, Malay and Tamil]" (MMACNRS 17 January 1968). A few months later, the about-turn became even more explicit: the Street-Naming Advisory Committee was to "stop naming streets after flowers, fruits, animals etc. in Malay," avoid the use of "Jalan" for "Road" and "carefully review any objection to the existing street names in Malay" (MMACNRS 29 May 1968). In short, the attempt to create unity on the basis of a National Language signifying identity with the mainland (Peninsular Malaysia) soon collided with another nation-building framework, the political ideal of multiracialism, a contradiction intensified by the painful experience of separation from Malaysia in 1965.

The policy directive to use the National Language for street-naming was truncated and in turn, a new proposal to use "mathematical naming," a strategy with much less racial overtones, put on trial (MMACNRS 4 July 1967). As Singapore roads were not laid out in a grid or "cobweb" (radial) fashion, a fully-fledged numerical naming system was not feasible. Instead, an adapted version whereby streets in newly built private and public housing estates were sequentially named Lorong 1, 2, 3 etc. was proposed (MMACNRS October 1967). This scheme, however, was censured by the Finance Minister (under whose purview the Street-Naming Committee came) for "sterility and lack of imagination" (MMACNRS 29 December 1967). Under the Minister's direction, a number of housing estates which had already acquired numerically-inspired street-names were brought back to the drawing boards: Lorongs in Goldhill Gardens, a middle-class private estate, for example, were renamed Goldhill Avenue, Rise, View and Drive (MMACNRS 2 May 1968). Numerical nomenclature, however, continued to be used for streets and roads in several generations of public housing estates from Toa Payoh, a New Town of the 1970s designed using the ring road system, to the newest estate today, Pasir Ris.

It was in the industrial rather than residential landscape that the logic of multiracialism and the integral part it played in nation-building were most conspicuously inscribed. In naming some 30 new roads in Jurong Industrial Estate (Figure 4.1), Singapore's pioneer estate developed by the Jurong Town Corporation (JTC) and the lynchpin of the nation-state's fledgling industrialization programme, the Street-Naming Committee was tasked to place "special emphasis on names in four official languages … to reflect the multi-lingual, multi-racial and multi-cultural content of the society in Singapore" (MMACNRS 2 May 1968). In addition, chosen names should suggest "industry" and "progress" to reflect constant striving after economic success, an ingredient critical to the survival of the "nation" (MMACNRS 2 May 1968). These two strands of state ideology, multiracialism and an industrial work ethic, were communicated through the toponymic text in a systematic and none-too-subtle manner: permutations of words which suggested "industry" were chosen in turn from the four official lexicographic repositories, resulting in names such as Enterprise Road; Quality Road; Tractor Road; Fan Yoong Road (Mandarin, meaning "prosperity"); Soon Lee Road (Mandarin, meaning "progressing smoothly"); Jalan Tukang (Malay, meaning "skilled craftsman"); Jalan Pesawat (Malay, meaning "machinery"); Neythal Road (Tamil, meaning "to weave," a reflection of the textile factories located along this road) (MMACNRS 29 May 1968; 12 July 1968; 7 August 1968; 6 September 1968).

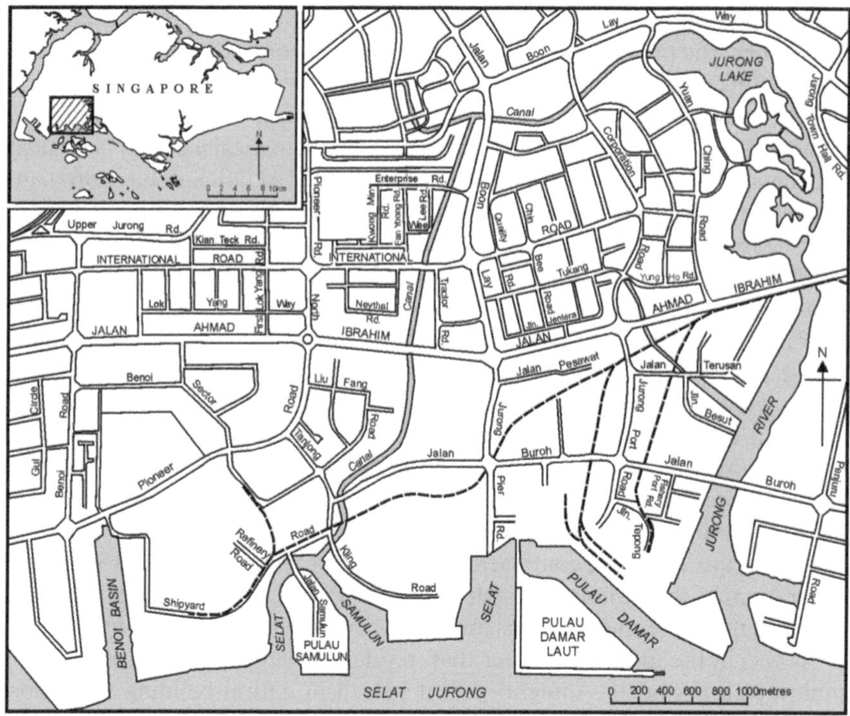

Figure 4.1 Jurong Industrial Estate: combining multiracialism and the industrial work ethic in the naming of streets

In inscribing the language of multiracialism in the landscape, the process was not a straightforward and uncontested one. For example, there were differences of opinion as to how multiracialism was best represented toponymically. JTC favoured creating "districts" of Chinese, Malay and Indian names as "a means of easy identification" (MMCNRS 6 September 1968). In contrast, the Street-Naming Committee felt that such a form of "segregation" was not "in line with the Republic's multi-racial character" and that names in the four official languages should be interspersed throughout the estate (MMACNRS

6 September 1968). JTC also preferred a preponderance of Chinese names as "most of the investors [were] Chinese" (MMACNRS 12 July 1968; 6 August 1968). The Corporation provided an alternative list of mainly Chinese street-names and while the Street-Naming Committee rejected the principles on which this was based as they were not congruent with the logic of multiracialism, it tacitly acceded to JTC's request by assigning a larger proportion of Chinese street-names while still retaining a "multi-racial" mix (MMACNRS 6 August 1968). The inscription of hegemonic meanings in landscape text is hence more akin to an uneven, negotiated process of constant mediations rather than a static consensual once-and-for-all translation of a monolithic ideology into material form.

Standardization, Pinyinization and Bilingualism

The power of landscape text to render ideology more concrete and therefore more real and unquestioned will only have maximum effect if people actively encountered and drew upon the text in daily practice. Up to the late 1960s, while "multiracializing" the landscape required assigning street-names using the four official languages, only Romanized script was used in road signs and street directories. For the majority of the population unschooled in English, Romanized street-names were not comprehended and in fact often transliterated beyond recognition.[5] In order for the multiracial toponymic text to move from an ideological statement of intent to an internalized social reality in the lives of the citizens of the "nation," the people must be able to understand and make use of the available repertory of street names. It was towards this end that in 1967, the state appointed a committee to provide and disseminate all authorized Chinese translation of street-names (RCSSNC 1970). The Committee on the Standardization of Street Names in Chinese was charged with the responsibility of systematizing and simplifying "existing renderings of street names in Chinese so as to avoid confusions and unhappy transliterations" and ensuring that translations were not only "faithful rendering[s] by sound but should also be elegant and meaningful" (*The Straits Times* 1970). The ardous task of translating over 500 street-names, particularly given the various practical and linguistic difficulties such as transliterating multisyllabic Malay street-names into monosyllabic Chinese characters, took a full three years to complete (RCSSNC 1970, 5). While the Committee achieved a measure of standardization in Chinese street-names, this did not close the debate as to how street-names could be made more meaningful within a multilingual society. In the 1980s, the state's attempt to apply the *hanyu pinyin* system (a universally accepted Mandarin system of romanizing

5 One Chinese transliteration of Raffles Place meant "Raffles was impolite" while Woodbridge Hospital, a mental hospital, was transmogrified into "the House of Devils" (*The Straits Times* 1970).

Chinese characters) to street and place names triggered off heated discussions in Parliament and on the streets. Those who supported the "*pinyin* revolution" pointed out that apart from the advantages of standardization, *pinyinized* names were more acceptable than the old romanized dialect names as the latter were no more than haphazard translations bequeathed by the British colonial administration (*The Straits Times* 1991). Those who appealed against the move argued that the old, familiar dialect place names were "part of our heritage," "of strong historical meaning for Singapore" and that the new *pinyin* code of pronunciation would "baffle" older generation Singaporeans and even "come across as insensitive to minority race[s]" (*The Straits Times* 1987a; 1987b;1988; 1991). Tek Kah market, a Hokkien name which meant "market under the shade of the bamboo trees" when *pinyinized* to become Zhujiao market sounded like "pigs' legs market" while the Nee Soon Road area, named after Lim Nee Soon, a well-known Chinese rubber and pineapple magnate, transformed into the dialect version, Yishun, bore no obvious etymological relation to the illustrious pioneer (*The Straits Times* 1987a; 1988). The debate only subsided after March 1987 when the National Development Minister announced that old place names including Chinese dialect ones would be kept although the changes (including Zhujiao and Yishun) which had already been cast in stone would not be rectified (*The Straits Times* 1987a).

In the 1990s, the debate over street-names was resurrected once more, this time in the wake of the Prime Minister's suggestion that road names and public signs in "historic districts" such as Chinatown, Little India and Geylang Serai (Malay heartland) should carry two languages (*The Straits Times* 1992a). Over 60 bilingual signs embellished with "appropriate" ethnic motifs (the lotus for signs in Little India, the pagoda in Chinatown and the Minangkabau house in Geylang Serai) were put up in "a pilot scheme to gauge public response (*The Straits Times* 1992b). The aim of the state was to "remind Singaporeans of the historical significance of these areas," to underscore the multiracial character of the city and to give it "an Asian feel" (*The Straits Times* 1992b). Akin to an exercise in "salvage toponym," these road signifiers were intended as reminders of lost heritage in a modern city where much of the original cultural practices had already been eroded (Yeoh and Kong 1994; Yeoh and Lau 1995). In effect, however, these street-names attracted little public response and the one comment expressed in the main paper's forum page indicated disappointment that Romanized and not Jawi script was not used alongside English names in the signs erected in Geylang Serai (in contrast, Chinese and Tamil characters were used in Chinatown and Little India respectively) (*The Straits Times* 1992c).

Evidently, while the state invested much ideological weight in Singapore's toponymies and constantly tried to configure and reshape landscape text to both systematize and popularize its usage, this process is not without negotiated encounters as people questioned, challenged or came up with alternative readings of both the forms and meanings of street-names. The "web of signification" (Ley and Olds 1988, 195) spun by the state is not totally hegemonic or compelling as landscape contains a "duplicity" (Daniels 1989, 206), being illusory and at the same time impervious.

Conclusion

In as much as the landscape under colonial rule was racialized (Yeoh 1992), the landscape of nationhood also bore the imprint of equally salient ideologies such as multiracialism since all systems of authority draw on some form of landscape text to legitimize their rule (Duncan 1985). The multiracial landscape of independence was as much a social construct as the racialized landscape of colonial rule. At the same time, in as much as "landscape serves as a vast repository out of which symbols of ... ideology can be fashioned" (Duncan 1985, 182), this process is one in which hegemonic meanings are inscribed in the manner of a "moving equilibrium" which is "always contested, always changing" (Warren 1993, 183). The mapping of nationalist ideologies onto Singapore's street-names was an uneven process, reflecting the contradictions and swings in the policies of nation-building and at the same time incorporating to some extent the reactions and resistances of its citizens. The result is a "littered" landscape, one reflecting the co-existence of different systems of signification (colonial names; Malayanized names; numerical names; names in different languages; dialect names; *pinyinized* names; bilingual names), each with its own pattern of names, serving to underline the view that the "tapestry" of a landscape is constituted by an "unglamorous dishevelled tangle of threads" (Eagleton, quoted in Baker 1993, 9).

References

Agnew, J.A., Mercer, J. and Sopher, D. (eds) (1984), *The City in Cultural Context* (Boston: Allen and Unwin).

Agnew, J.A. and Duncan, J.S. (eds) (1989), *The Power of Place: Bringing together Geographical and Sociologic Imaginations* (Boston: Unwin Hyman).

Anderson, B. (1983), *Imagined Communities: Reflections on the Origin and Spread of Nationalism* (London: Verso).

Baker, A.R.H. (1993), "Introduction: on Ideology and Landscape," in Baker, A.R.H. and Biger, G. (eds), *Ideology and Landscape in Historical Perspective* (Cambridge: Cambridge University Press), 1–14 .

Berleant-Schiller, R. (1991), "Hidden places and Creole Forms: Naming the Bardudan Landscape," *Professional Geographer* 43, 92–101.

Benjamin, G. (1976), "The Cultural Logic of Singapore's 'Multiracialism'," in Hassan, R. (ed.), *Singapore: Society in Transition* (Kuala Lumpur: Oxford University Press), 115–33.

—— (1988), "The Unseen Presence: A Theory of the Nation-state and its Mystifications," National University of Singapore, Department of Sociology, Working Paper No. 91.

Chan, H.C. (1991), "Political Developments, 1965–1979," in Chew, E.C.T. and Lee, E. (eds), 157–81.

Chew, E.C.T. (1991), "The Singapore National Identity: Its Historical Evolution and Emergence," in Chew, E.C.T. and Lee, E. (eds), *A History of Singapore* (Singapore: Oxford University Press), 357–68.

Chua, B.H. (1995), *Communitarian Ideology and Democracy in Singapore* (London: Routledge).

Daniels, S. (1989), "Marxism, Culture and the Duplicity of Landscape," in Peet, R. and Thrift, N. (eds), *New Models in Geography: The Political Economy Perspective*, vol. 2 (London: Unwin Hyman), 196–220.

Duncan, J.S. (1985), "Individual Action and Political Power: A Structuration Perspective," in Johnston, R.J. (ed.), *The Future of Geography* (London: Methuen), 174–89.

—— (1990), *The City as Text: the Politics of Landscape in the Kandyan Kingdom* (Cambridge: Cambridge University Press).

Duncan, J. and Ley, D. (eds) (1993), *Place/Culture/Representation* (London: Routledge).

Ekwall, E. (1954), *Street-names of the City of London* (Oxford: Clarendon Press).

Emmerson, D.K. (1984), "'Southeast Asia:' What's in a Name?," *Journal of Southeast Asian Studies* 15, 1–21.

Hassan, R. (ed.) (1976), *Singapore: Society in Transition* (Kuala Lumpur: Oxford University Press).

Hill, M. and Lian, K.F. (1995), *The Politics of Nation Building and Citizenship in Singapore* (London: Routledge).

Hsieh, C.M. (1980), "Sequent Occupance and Place Names," in Knapp, R.G. (ed.), *China's Island Frontier: Studies in the Historical Geography of Taiwan* (Honolulu: University Press of Hawaii), 107–14.

Jackson, P. and Penrose, J. (1993) "Introduction: Placing 'Race' and Nation," in Jackson, P. and Penrose, J. (eds), *Constructions of Race, Place and Nation* (London: UCL Press), 1–23.

Jacobs, J. (1993), "The City Unbound: Qualitative Approaches to the City," *Urban Studies* 30, 827–48.

Johnston, R.J. (ed.) (1985), *The Future of Geography* (London, Methuen).

Katz, C. and Kirby, A. (1991), "In the Nature of Things: the Environment and Everyday Life," *Transactions of the Institute of British Geographers, N.S.* 16, 259–71.

Knapp, R.G. (ed.) (1980), *China's Island Frontier: Studies in the Historical Geography of Taiwan* (University Press of Hawaii, Honolulu).

Kong, L. (1993), "Ideological Hegemony and the Political Symbolism of Religious Buildings in Singapore," *Environment and Planning D: Society and Space* 11, 23–45.

Lewandowski, S.J. (1984), "The Built Environment and Cultural Symbolism in Post-colonial Madras," in Agnew, J.A., Mercer, J. and Sopher, D.E. (eds), *The City in Cultural Context* (Boston: Allen and Unwin), 237–54.

Ley, D. (1987), "Styles of the Times: Liberal and Neo-conservative Landscapes in Inner Vancouver, 1968–1986," *Journal of Historical Geography* 13, 40–56.

—— and Olds, K. (1988), "Landscape as Spectacle: World's Fairs and the Culture of Heroic Consumption," *Environment and Planning D: Society and Space* 6, 191–212.

Mak, L.F. (1981), *The Sociology of Secret Societies: A Study of Chinese Secret Societies in Singapore and Peninsula Malaysia* (Kuala Lumpur: Oxford University Press).

MMACNRS (Minutes *of Meeting of the Advisory Committee on the Naming of Roads and Streets*) Files 2/67, 70/59 and 515/59, Dewan Bahasa Files, Ministry of Culture (Singapore: National Archives).

Peet, R. and Thrift, N. (eds) (1989), *New Models in Geography: The Political Economy Perspective* (London, Unwin Hyman).

Penrose, J. (1993), "Reification in the Name of Change: The Impact of Nationalism on Social Constructions of Nation, People and Place in Scotland and the United Kingdom," in Jackson, P. and Penrose, J. (eds), *Constructions of Race, Place and Nation* (London: UCL Press), 27–49.

Pirie, G.H. (1984), "Letters, Words and Worlds: The Naming of Soweto," *African Studies* 43, 43–5.

RCSSNC (*Report of the Committee on the Standardisation of Street Names in Chinese*) (1970), Singapore.

Richman, J. (1983), *Traffic Wardens: An Ethnography of Street Administration* (Manchester: Manchester University Press).

Robinson, D. (1989), "The Language and Significance of Place in Latin America," in Agnew, J.A. and Duncan, J.S. (eds), *The Power of Place: Bringing together Geographical and Sociological Imaginations* (Boston, Unwin Hyman), 157–84.

Rogers, A. (1995), "Cinco de Mayo and the 15th January: Contrasting Situations in a Mixed Ethnic Neighborhoods," in Rogers, A. and Vertovec, S. (eds), *The Urban Context: Ethnicity, Social Networks and Situational Analysis* (London: Berg), 117–40.

Sandhu, K.S. and Wheatley, P. (eds) (1989), *Management of Success: The Moulding of Modern Singapore* (Singapore: Institute of Southeast Asian Studies).

Siddique, S. (1989), "Singapore Identity," in Sandhu, K.S. and Wheatley, P. (eds), *Management of Success: The Moulding of Modern Singapore* (Singapore: Institute of Southeast Asian Studies), 563–77.

The Straits Times (1970), "When Mental Hospital was 'House of Devils'," 13 June 1970, Singapore.

—— (1986), "3 MRT Stations to be Renamed," 23 November 1986, Singapore.

—— (1987a), "HDB to Keep Old Names: Dhana," 19 March 1987, Singapore.

—— (1987b), "Keep the Old Names," 20 March 1987, Singapore.

—— (1988), "Preserve Names of Places that are Part of Our Heritage," 25 June 1988, Singapore.

—— (1991), "Govt Should Keep Promoting Use of Pinyin Names," 27 December 1991, Singapore.

—— (1992a), "Bilingual Road Signs Go Up," 3 April 1992, Singapore.

—— (1992b), "PWDs 'Sound' Solution to Tamil Street Sign Poser," 11 April, Singapore.

—— (1992c), "Why Not Jawi Signs for Geyland Serai," 21 April 1992, Singapore.

Stump, R.W. (1988), "Toponymic Commemoration of National Figures: The Case of Kennedy and King," *Names* 36, 203–16.

Warren, S. (1993), "'This Heaven Gives Me Migraines:' The Problems and Promises of Landscapes of Leisure," in Duncan, J. and Ley, D. (eds), 313–22.

Yeoh, B.S.A. (1992), "Street Names in Colonial Singapore," *The Geographical Review* 82, 313–22.

—— and Kong, L. (1994), "Reading Landscape Meanings: State Constructions and Lived Experiences in Singapore's Chinatown," *Habitat International* 18, 17–35.

Yeoh, B.S.A. and Lau, W.P. (1995), "Historic District, Contemporary Meanings: Urban Conservation and the Creation and Consumption of Landscape Spectacle in Tanjong Pagar," in Yeoh, B.S.A. and Kong, L. (eds), *Portraits of Places: History, Community and Identity in Singapore* (Singapore: Times Editions), 46–67.

—— and Tan, B.H. (1995), "The Politics of Space: Changing Discourses on Chinese Burial Grounds in Postwar Singapore," *Journal of Historical Geography* 21, 184–201.

Zelinsky, W. (1983), "Nationalism in the American Place-name Cover," *Names* 31, 1–28.

Zukin, S. (1991), *Landscapes of Power: From Detroit to Disney World* (Berkeley: University of California Press).

Chapter 5
Naming and Placing the Other: Power and the Urban Landscape in Zanzibar

Garth Andrew Myers

Introduction

Names for places, and boundaries around them, are markers of the spatiality of power relationships embedded in the landscape. In this chapter, I suggest that the social construction of space and place consists of a multifarious and ever-changing process of positioning oneself and others in the matrix of power relations, and not a spatial fixing of identities based solely on domination. I utilize examples from the Ng'ambo section of Zanzibar.

Ng'ambo literally translates as "the Other Side," and throughout its history it has been seen as a place where "the Others" live—slaves, dockworkers, peasants, and marginalized peoples within Zanzibar's political economy. Yet naming Ng'ambo's landscape and marking boundaries within it are integral facets of identity politics in a manner more suggestive of the complex spatial discourse on power than of any fixed categorizations.

The remainder of the chapter is divided into five sections. The first argues for reconnecting place name analysis to the study of power, drawing on recent geographical studies. The second introduces Zanzibar city before examining naming and bounding actions by the colonial and post-colonial state in Ng'ambo. Residents' responses to these actions are also considered. In the fourth and fifth sections, I delve beneath the surface of official naming and bounding to demonstrate the core of my argument: that spatial discourses of "Othering" are highly nuanced, contextualized and dynamic. The conclusion links this contention to broader debates in geography.

Place Names and Power

Intersections between geography and power have been on the minds of geographers lately, as have questions of geographic language (Agnew and Duncan 1989; Barnes and Duncan 1992; Cosgrove 1993; Duncan 1990). Debates on place, space, power and language have occurred on a fairly abstract plane of philosophical inquiry, punctuated by forays into empirical analysis. For instance, only modest connections have been made between theories on power or language and empirical studies of place names (but see Carter 1988).

This limited analysis of naming/power is unfortunate, because place names provide a rich source of discussion on space and power through varied strategies for contestation embedded in their use. In part, this is due to what Entrikin terms the "performance aspects" of place names in discourses: "in both everyday conversation and in literary narratives, place names have a semantic depth that extends beyond the concern with simple reference to location or to a single image" (Entrikin 1991, 55–6). Both the "symbolic role of place names as expressions of ideological values" (Cohen and Kliot 1992, 653) and the capacity for "understanding indigenous systems of knowledge and ways of ordering and classifying the world" (Berleant-Schiller 1991, 93) through place names are tied into this semantic depth.

Place name study has largely languished in atheoretical caverns of geographical inquiry. Geographers have categorized "vernacular" names to describe and provide boundaries for regional cultures (Zelinsky 1955; Jackson 1970). This interesting but uncritical procedure fostered the widespread belief that place name analysis has "more antiquarian appeal than geographic relevance" (Berleant-Schiller 1991, 92). A critical appreciation of power and ideology is often far from the centre of concern in toponymy studies (Sopher 1978; Lewis 1982; Sales 1991).

One reason why place name study has been limited in recent discussions of social theory may be the decline in geographic fieldwork (Nast 1994; Katz 1994). There is simply no way to appreciate the performance aspects of names without hearing them called, played with as tactics of empowerment, or used as vehicles of derision. The most creative analyses of place names in recent years have all resulted from a deep engagement with the places named (Soja 1989, 224–47; Cruikshank 1990).

A few widely scattered recent pieces have aimed to rejuvenate place name study and connect it with concerns for power and language. Cohen and Kliot's work on Israel is perhaps the most explicit of such study. The authors note that "affixing names to places is inextricably linked with nation building and state formation," and we can trace "ideological upheavals" and the coming or going of "revolutionary values" in toponymy (Cohen and Kliot 1992, 653). But their analysis is an account of how names have been used to reinforce Zionism; little is made of Palestinian resistance. It is crucial to look to all sides in order to truly see the "fields of power" at work in the production and reproduction of space (Katz 1994, 68).

In my study, a sample is shown of the diversity of place names utilized for different purposes by various players, and the divergent ways of placing boundaries on "otherness" in the Zanzibar cityscape. The focus is on Ng'ambo because it has been the key site for constructions of otherness in Zanzibari society and for reconstructions of Zanzibari identity. Before moving to the

Other Side, however, Ng'ambo's development is placed in the context of the growth of the city as a whole.

Othering the Other Side

Zanzibar is an urban area of nearly 200,000 people on the island of Unguja (called Zanzibar island by Westerners) off the East African coast. It is the United Republic of Tanzania's second city politically, and the country's third largest urban district (Tanzania 1991). It rose from a small fishing community in the 1700s to the capital of an extensive commercial empire by the 1840s. Clove, coconut and spice production, based on Unguja and neighbouring Pemba island, supplemented the large inflow of cash to the city from duty collection control over ports from Mogadishu to Mozambique (Sheriff 1987).

The Zanzibari Sultanate's coastal empire gradually slipped into European hands, and by 1890 the Sultan was forced to hand over the islands to a British Protectorate. With Britain's mandate over Tanganyika after World War One and the growth of Dar es Salaam, a period of relative stagnation ensued on Unguja, economically and demographically (Sheriff and Ferguson 1991). Since independence and union with Tanganyika in 1964, Zanzibar city has grown dramatically in size; recently its moribund economy has been on the upswing (Sheriff 1995).

The Kiswahili word *ng'ambo* may be defined as both "an embankment or side of a river or creek" and as "the second or other side of a place" (Johnson 1991, 334). Historically, "Ng'ambo" in Zanzibar referred to lowlands on the eastern side of what was locally termed the Pwani Ndogo (little shore, tideland), named Funguni Creek in colonial documents. Ng'ambo was also the second side of Zanzibar, since it began in the 1850s, long after settlement had taken place across the creek in the "Stone Town" (Clark and Horton 1985).

From its inception, Ng'ambo has been a community of fishing families, domestics, day-laborers, home-makers, porters, artisans and displaced peasants. Ng'ambo did not exist prior to British imperialism; thus, unlike many "other sides" on the Swahili coast, it is as much a colonial creation of otherness as a Swahili or Omani one (Fair 1994; Menon 1978; Sheriff 1987; Myers 1994a, 1994b and 1995). Hence it is crucial to analyze colonial naming and placing in some detail. Nineteenth-century European travel writings set Ng'ambo apart from the city itself in their descriptions (Burton 1872; Christie 1876; Baumann 1897). Colonial building and planning laws. though, quite literally set the division of the *city in stone*.

The 1929 Building Rules established a legal separation between Stone Town, which was called "Town," and Ng'ambo, which was considered a "native location" (see Robinson 1990, for comparison to South Africa). The distinction was predicated on a perception that there were simply two kinds of dwellings, a "stone house" and a "native hut." The former belonged in Stone Town, the latter in Ng'ambo.

Building rules, in Zanzibar as in Mombasa (Cooper 1987), Cairo (Mitchell 1988), or Cape Town (Western 1981), were about defining who could be where. The laws made it difficult to change the character of opposing sides by building a house that fit the definition of a "native hut" in Stone Town or a stone house in Ng'ambo. Rent laws passed in the same period solidified this difficulty by imposing stiff rent and rating differences between stone houses and native huts (Zanzibar National Archives, hereafter ZNA, files BA 13/3 and AB 36/23). The differential was so severe that improving walls to include more stone came to be seen as an illegal act, when it technically was not.

The application of building regulation through the decades has imparted to many contemporary Zanzibari Africans this colonial bifurcation of the city. Enforcement of a legal division with clear racial underpinnings helped establish an enduring differentiation between Stone Town and Ng'ambo in many residents' spatial and racial conceptions of the city. The Stone Town/Ng'ambo divide had a physical dimension prior to the British arrival. As that physical divide was steadily eroded (the British filled in Pwani Ndogo with garbage), the colonial state strengthened the legal and conceptual break between the sides as a means of impressing the spatial order of empire onto the cityscape. For many older residents especially, this boundary is alive in the city today even without its legal foundations.

It often seemed that the real concern of the British in Ng'ambo was simply having things under control. To use Mitchell's (1988, 33, 44) words, having the place "contained" meant translating it into a spatial language they could "read like a book." Their most important translation of Ng'ambo's geography into planning language is quite literally a book: a set of 20 maps covering the entire area for use in building control, planning, rent collection and sanitation. One significant problem the British had in attempting to establish their sense of geographical order in Ng'ambo was that they got lost in its alleys. This irritated the police, who claimed they could not tell one "Juma Ali" from another without numbered hut addresses (ZNA file AB 39/36). The colonial power needed a system of neighbourhoods, streets and numbered houses to "containerize" the Other Side for social control. This map-book, the Ng'ambo Area Folder, was produced between 1934 and 1943 and guided a formal hut-numbering system for the colonial police and health office in 37 separate divisions of the Other Side (ZNA file AW 2/100).

Few records exist of how boundaries between divisions were drawn. In codifying local neighbourhood units as "administrative areas" many commonly used area names in Ng'ambo were bypassed and merged into one another (Hussein 1991). Curiously, two areas on the Stone Town side of the Creek (Funguni and northern Malindi) were mapped and included with the Ng'ambo Folder. Ng'ambo was the "Other" part of town; even if they were not on the "Other" side of the Creek, Funguni and northern Malindi were areas settled by African fisherfolk and poor outsiders. The Empire's geographic "structure of attitude and reference" (Said 1993, xxiii), telling its subjects where they belonged, was codified by placing Funguni and Malindi with the Other Side (Myers 1994b).

Naming the Ng'ambo landscape played a key role in British efforts to secure legitimacy. The British named the protectorate's sports stadium for Sultan Khalifa and propped up the Sultan's retinue elsewhere through toponymy. Since many Ng'ambo residents descended from slaves, from the British point of view legitimacy in Ng'ambo rested on the legacy of having ended slavery. Hence the model village in their reconstruction effort in Ng'ambo was named Holmwood, to honor Frederick Holmwood, a nineteenth-century anti-slavery crusader. Another model neighbourhood was named Mji Mpya (New City), its orderly lanes articulating the linkage of spatial and social order in the post-war Empire. The civic centre established with the colonial reconstruction was named at its ribbon-cutting by the regime's spatial tactician, Eric Dutton: "This building—in fact the whole of this little scheme here [the reconstruction] – is an earnest of what we are trying to do in Ng'ambo … I profoundly hope that this place will not become known by some such sloppy and pretentious designation as the Civic Center … but by some lively Zanzibar name, like 'RahaLeo' [HappinessToday]" (ZNA file AB 9/9). Unfortunately for Dutton, underneath the surface of official names, nicknames emerge which often convey markedly different place meanings from official ones. "Holmwood" slipped out of colonial hands and became "Hamoud" in local parlance, the name which survives in post-colonial revolutionary party files (ZNA file AK 26/18). The neighbourhood that Dutton had demolished, from which Hamoud's first residents came, was called "Poland" by a populace being shown footage of war devastation and drawing parallels with their surroundings. At the time that homes were being cleared to create "Raha Leo" (Happiness Today) illegal squatters on the outskirts of town formed a neighbourhood they called "Raha Kesho"—Happiness Tomorrow (ZNA file AE 9/9).

Postcolonial projections of power onto the landscape fared no better below the surface. Khalifa Stadium became Mao Tse Tung Stadium officially, and was supplanted by the new Amani (Peace) Stadium. The postcolonial redevelopment of Ng'ambo went by the name Zanzibar Mpya (New Zanzibar), the physical manifestation of the new socialist order (Myers 1994c). In names, Ng'ambo residents quickly subverted the new order. The cheap seat side of Amani stadium became "Urusi" (Russia), and was identified with poverty and oppression. "Playing at Mao" connotes playing in the minor leagues. New Zanzibar gathered a litany of new names, among them: New Berlin Street, The Trains, and Kikwajuni GDR. The name attached to the finished blocks, Michenzani, became the "Watergate" of Zanzibar. Like the Washington office complex, Michenzani is a codeword for scandal (see Figure 5.1).

Yet if we stop analysis with the colonial construction of Ng'ambo as Other, the post-colonial state's reconstruction of it as New Zanzibar, or Ng'ambo

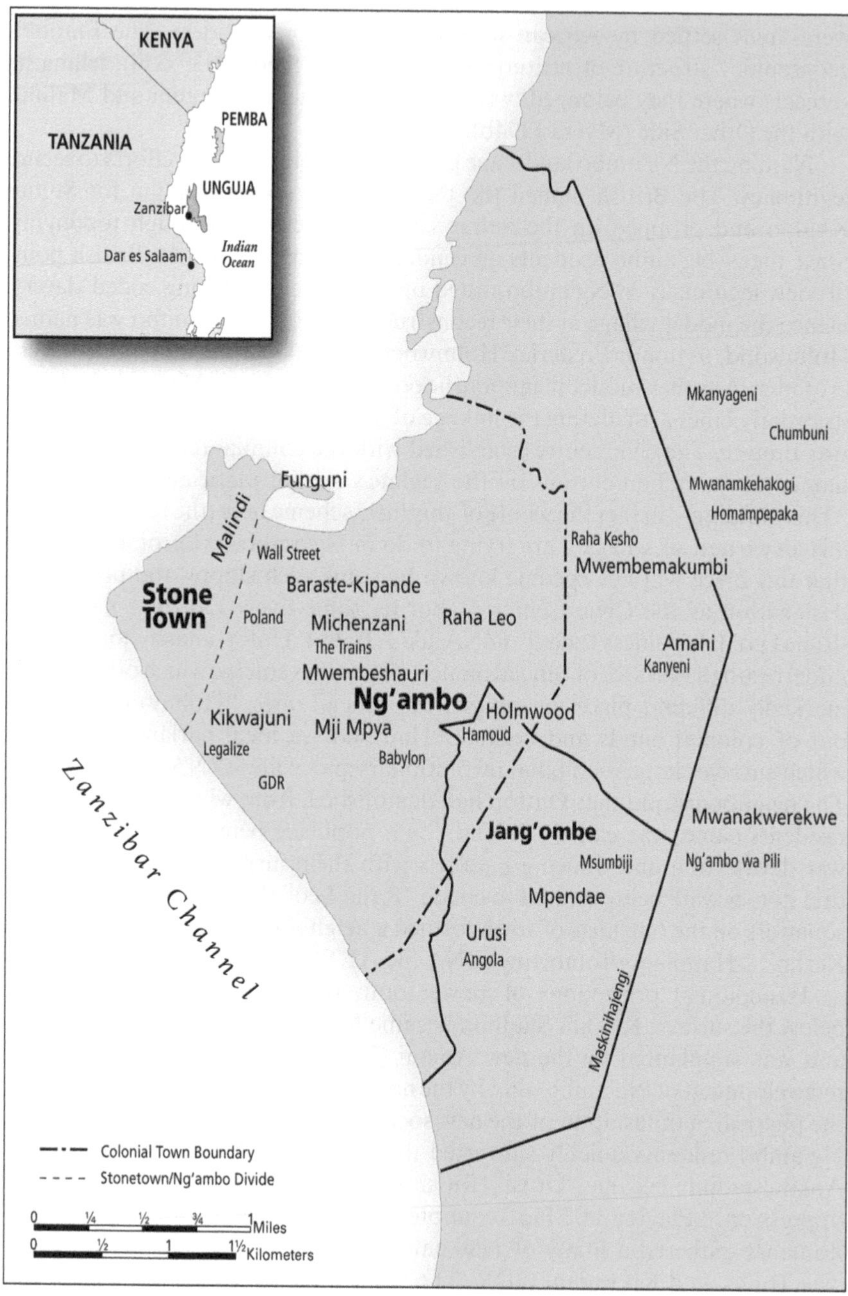

Figure 5.1 Place names on the other side of Zanzibar

residents' clever subversions of these efforts in nicknames, we still miss a rich tapestry of naming and placing done by Ng'ambo's people themselves. Their processes of spatial production under elite noses created "a study in African town architecture unequalled by anything to be seen on the coast" (Zanzibar Protectorate 1912, 18). In the following section, the evolution of this unequalled community and the names of its constituent parts are explored.

Placing and Naming

> They passed through Ng'ambo where the buildings were laid out chaotically, in bits and pieces. They passed alongside one neighborhood after another, every few feet or so. The city is small, but its names are as many, and as different, as the roofs of its houses. (Mohamed 1978, 105, my translation from the Kiswahili)

There are many levels of socio-spatial organization in Swahili cities below the two commonly identifiable "sides." The most vital of these is the *mtaa*, usually translated as "ward" or "quarter" (pl. *mitaa*). The Tanzanian Swahili dictionary defines *mtaa* as "a living area or ward within a city, a section that divides itself from other sections in the city owing to its origins and those of the city itself" (Tanzania Swahili Institute 1989, 135). In my opinion, the most appropriate English word is neighbourhood, because the custom of neighbourliness (ujirani) is so intrinsic to any *mtaa* (Allen 1993; Middleton 1992; Hino 1971).

Most *mitaa* have around 1,000 people. *Mtaa* boundaries, like those of neighbourhoods worldwide, can be flexible; other authors have used the vagueness of distinctions to dismiss the significance of *mitaa* for analysis of spatial organization (Horton 1984, p. 160; Garlake 1966, p. 87). Yet there are many illuminating social codes embedded in the spatiality of *mitaa*, and the most crucial of these is the name.

Other studies (Menon 1978; Allen 1974) have noted that *mtaa* names provide clues for who owned or controlled the land on which it was built (such as Kwa-Bi-Jokha, Land of Bi Jokha), or for settlers' places of origin (Kwa-Buki, Place of the Malagasy). Actually, many more *mtaa* names have no relation to these factors at all. Most Ng'ambo *mtaa* names are given to the place by residents based on the everyday, customary world around them: names of trees; a specific tree paired with an activity underneath it; local land uses; natural features; or well-known local personalities. Of 77 *mtaa* names used on maps of Ng'ambo since 1870, only 10 have any relation to land ownership, and many of these are actually names of Wakf (Islamic inheritance) trustees, not "owners." Only two names refer to places of origin. Sixteen *mtaa* names are based on trees, another 16 on local activities or uses, eight more on a tree plus a use, and 10 on natural features (Williams 1949; ZNA file AW 2/ 100).

Mtaa names must be considered clues to the lives, thoughts and customs of Ng'ambo residents. Ng'ambo was heavily influenced by Zanzibari culture, where land control and family wealth is measured in trees, where jini (spirits) and shetani (devil-spirits) live in trees, and where medicines to treat cases of possession are all derived from specific trees (Larsen 1992). Some named trees are economically valuable, but others are historically prominent. This prominence can be based on size, as is the case with the *mtaa* Mzambarauni (By the Damson Tree). Shade trees, especially mangos, provide extensive clearings amidst Ng'ambo's densely-packed homes; hence these become prominent centres of community activity. Here, the *mtaa* bears the name of both the tree and the activity: Mwembetanga, Mwembeshauri and Mwembemakumbi are *mitaa* named for mango trees underneath which sails are mended, advice is given and coir rope made, respectively (see Figure 5.1).

Many prominent tree-based *mtaa* names have spiritual roots. Trees valued for healing properties are commemorated, as in Mwarubaini (By the Tree of Forty Medicines, the Neem Tree). Other times, the tree is the *mtaa*'s spiritual core. For example, the large *mtaa* Kikwajuni (By the Small Tamarind) is actually subdivided by older residents into more than 10 *mitaa*. What binds these *mitaa* into one is an area known as Mzimuni (By the Spirit). Mzimuni is where that small tamarind tree, inhabited by a genuinely feared spirit, is situated. Similarly, many *mitaa* are named for mango trees not for activity-based reasons, but because mangos are favoured residences of spirits. That over 30% of all *mtaa* names have a tree in them shows that these aspects of custom take precedence over landlords' names.

And what activities receive the attention of Ng'ambo's namers? Sail-mending, cattle-keeping, groundnut-selling, rope-making, advice-getting, and dancing: the everyday activities of Ng'ambo's people. In the toponymic feast of Ng'ambo *mitaa*, there is a Well of Millipedes, a Conceited Tap (it won't give water), a Pregnant Mango (by the maternity hospital), a place By the Shopkeeper Who Burps. There are places you will have to move to, places the Will of God will move you to, places to practice the trumpet for the army, places to wash the clothes, places to spread them out to dry, places for "happiness tomorrow." Ng'ambo's *mtaa* names are the everyday politics and customs of its people written into the landscape, because that has been one of the few spheres where they have had the power to write their stories.

A wry sense of humor pervades the naming of places in Ng'ambo by residents. Under colonialism, many sardonic nicknames became official. Because few colonial officials spoke Swahili, the absence of broad official knowledge of place name meanings created wonderful opportunities for Ng'ambo's disenfranchised poor to make political statements, then listen to their sentiments repeated unwittingly by colonizers. A case in point is Baraste-Kipande. When the British Consul cut a ribbon on a Ng'ambo alley in 1913 and proclaimed it the starting place of a new road, and the road was never built, the poor to whom the land had been dedicated (in Wakf) by its pre-colonial owner made certain the story of colonial inaction would endure. They called the *mtaa* Baraste-Kipande to mean "A Piece of a Road;" older residents will still point out the exact place where markers were laid.

In contemporary times, names and nicknames utilized in such a manner often derive from English. Sometimes these involve an acknowledgement of illegal activities associated with that place. The main illegal money-changing houses, prior to their 1992 de facto authorization, lay along Ng'ambo's business thoroughfare, marked in graffiti as "Wall Street." The practice field of a Ng'ambo sports club across from Zanzibar's House of Representatives is widely known as "Legalize:" the club's fans smoke marijuana in plain view of legislators and a thousand passersby every afternoon. The youth of Ng'ambo who call the sprawling police barracks in the centre of Ng'ambo "Babylon" had no trouble naming the new police lorry: "Babylon By Bus."

The Other Side's Other Sides

This "performance aspect" of place names or nicknames in Ng'ambo often involves oppositional subtexts, like the Babylon or Legalize examples. Just as often, though, the performance of naming and bounding acts by Ng'ambo residents shows a more complex dynamic at work, not some easily identifiable, "us vs. them" consciousness. This complexity is best exhibited by shifting lines figuratively drawn by Zanzibaris to separate an eastward moving "Ng'ambo" from older western parts of the historic "Other Side." These shifts can be examined by considering residents' analyses of Zanzibar's spatiality in three areas—Kikwajuni on Ng'ambo's western edge, Jang'ombe straddling the colonial town boundary, and Chumbuni over the new town line.

Kikwajuni residents make great distinctions between neighbourhoods near Stone Town and those at the town's outer edge. To most western Ng'ambo residents, the oldest part of the Other Side is not the Other Side any more: it is *mjini*, the downtown. Outer districts, especially Jang'ombe, are said by Kikwajunians to be in Ng'ambo; in fact to many, Jang'ombe literally is Ng'ambo. Kikwajuni residents seldom remember the names of Jang'ombe's six separate wards, but nearly everyone knows where it is: somewhere "past the Post Office; once you pass the post office, you have left civilization" (Mohammedridha 1991). Calling someone "cassava from jang'ombe" is a common insult in "downtown" Ng'ambo. Holmwood and nearly all *mitaa* just east of it are perceived as the city's most dangerous. Jang'ombe is seen by those closer to Stone Town as far away: in Kikwajuni, one uses Jang'ombe as a point of reference somewhere near the other side of earth.

Residents of older *mitaa* in western Ng'ambo are not alone in distancing themselves from Jang'ombe. Its residents are among the first to point up its problems. The Other-worldly perceptions of Jang'ombe by many Zanzibaris are linked partly to its history, its boundaries, and its living conditions.

During the 1920s and 1930s, this was the most popular illegal drinking site for Ng'ambo residents, as much of it then lay outside of town police boundaries (Fair 1993). It is, perhaps as a consequence of this history of subversion, bounded on the west by the central prison and the manorial police "Babylon." The mental hospital occupies its northern boundary. To the south, its extent is curtailed by a swamp and the Zanzibar Headquarters of Tanzania's army. The town cemetery at Mwanakwerekwe lies due east.

Jang'ombe constituency now comprises a large percentage of the city's population. As Table 5.1 shows, Jang'ombe has grown in less than a half-century to a district twice the size of Stone Town. The stunning urbanization of the outer edge has as its companion the difficult living conditions experienced there. Jang'ombe is extremely poor, has some of the most severe flooding and erosion in Zanzibar, a chronic clean-water shortage, and almost no electricity, sewerage or solid-waste disposal.

Table 5.1 Demographic shift in Zanzibar City, 1948–1988

Zone of the city	1948	1958	1978	1988
Stone Town	16,698	18,179	15,493	15,854
Historical Ng'ambo	28,578	31,423	59,378	85,913
Total population outside colonial-era line	2,008	8,321	35, 798	65,551
Jang'ombe area*	<100	1,803	14,902	30,986
Total urban population	47,284	57,923	110,669	167,318
Stone Town as % city population	35	31	14	9
Ng'ambo as % city population	60	54	54	51
Outer areas	4	15	31	39
Jang'ombe as % city population	–	3	13	19

* Jang'ombe figures are included in the outer area figures.

The toponymic responses of residents to the experience of living in Jang'ombe demonstrate their acute recognition of these conditions. Two unofficial *mtaa* names given official status in 1985 are Urusi and Mpendae. Urusi derives from the distance, both perceived and real, between this *mtaa* and downtown: it is like living in Russia, people say. Residents explain the name, Mpendae—You Have to Love It Here—by completing the implicit thought of the imperative: because you have no choice. Names which remain unofficial for Jang'ombe *mitaa*—like Maskinihajengi (A Poor Person Can't Build: a government development area where many destitute land recipients were pushed aside by local officials)—convey similar collective responses to impoverishment.

In spite of their forthright means of coping with life there, many Jang'ombe residents do not accept their district as "the Other Side" without making reference first to their eastern neighbours. Beyond Jang'ombe, one enters what Jang'ombe residents call "Ng'ambo wa Pili," the Second Ng'ambo, the Other Side's Other Side. This is the illegal squatter housing that invaded the urban fringe in the 1990s. The *mitaa* of Ng'ambo wa Pili, such as Chumbuni, are rapidly becoming the next century's version of Jang'ombe, and Jang'ombe-dwellers are quick to distance themselves from the newcomers to these settlements.

The government formed a commission to give *mitaa* in areas like Chumbuni official names because they worried about the ones people used. But once again the place names on the official maps do not adequately give the flavor of these places in the way names people really give them do. In and around Chumbuni are the *mitaa* Mkanyageni (Place for Trampling), Homampepaka (Give a Cat the Fever), Mwanamkehakogi (The Woman Doesn't Bathe), and Kanyeni (Go Take a Shit) (Makame 1992; Mohammed 1992). The squatter camps of the Other Side's Other Side are places where the trampled live, those who beg for cats to get malaria rather than their children, who have no water to wash, who live where people farther uphill used to defecate.

Yet, as the enduring custom of imaginative place naming suggests, the newest of Ng'ambo's poorest people are remarkably resilient, creating an Other Side that re-interprets the order of state and elite, remaining what Cosgrove (1989) calls an "alternative landscape." In effect, these alternative geographies are engagements in indirect negotiations with officialdom over the character of poor places. Names and neighbourhood boundaries are both symbolic and material points of contention in contesting the "right to the city" (Friedmann 1992).

Conclusion

"For power to 'speak' socially," writes Constantin (1987, 219), "power must 'speak' with space." I have endeavoured to show here how, for state power, "speaking with space" often begins with naming places and putting boundaries around them. The extent to which the oppressed are "speaking with space" is often left unsaid in geographic analyses. Naming and boundary-making can often be part of what Scott (1990) calls the "arts of resistance," particularly through performance and manipulation. Moreover, they embody the degree to which many marginalized residents hold the "power of definition" (Western 1981, 8) in their hands and exercise their defining rights in diverse ways. In Zanzibar city, there has been an active, conscious contestation of the naming process and the establishment of boundaries. Speaking with space is not simply a matter of the powerful othering the world's Other Sides. "Others" have tactics and strategies for defining the spaces and places of their lives, too.

In recent years, theorists in geography have debated the nature of intersections between place, space, power, language and identity. Foucault's thought is quite influential in this literature. As Harris (1991, 674) suggests, Foucault, with help from other social theorists, has "redrawn the map of power." In post-structuralist thought, "power is less a property than a strategy and is widely distributed" (Harris 1991, 671). Accordingly, "social power is no longer conceived apart from its geographical context. Such power requires space, its exercise shapes space, and space shapes social power" (Harris 1991, 678).

Many mainstream geographers view this political turn in cultural studies with suspicion (Tuan 1991; Price and Lewis 1993). Tuan (1991, 685) contends that "the literature on power" in geography "deals more with techniques of control, social hierarchy and inequality, domination and its contestation, than with the creative acts of making and building." Both Foucault devotees and Tuanian cultural geographers should accept, though, that balancing humanistic and materialistic concerns when studying place naming and boundary-making must begin with the acts of naming and placing. These acts are the most basic of Foucault's (1980, 77) "tactics and strategies" for "control of territories and organization of domains." Yet naming and bounding strategies are exercised both by those having a great deal of social power and by those comparatively lacking in it. As Soja (1989, 134) argues, "emplacement is a passionate process that links subject and object." There is an "active consciousness and assertiveness" of social forces with varying capacities for exercising power through the contestation over definitions for "territorial and social enclosures" (Soja 1989, 134). As I have attempted to show with reference to "the Other" in Zanzibar, the socially constructed boundaries around—and names for—"Other" places are far from fixed. A fluid negotiation of capillary *positions* characterizes toponymic discourse more than a dominative artery of *imposition*.

Acknowledgements

This chapter is the product of research made possible by grants from the Social Science Research Council, the Fulbright-Hays Doctoral Research Program, the UCLA ISOP Center, and the UCLA Geography Department. The research was carried out under the supervision of the Tanzania Commission for Science and Technology and the Zanzibar Ministry of Information. I gratefully acknowledge the resources and guidance that each of these institutions provided. Dr Leo van Grunsven, Edward Soja and several anonymous reviewers made helpful comments on earlier drafts. Marvin J.

Barton was responsible for the original cartography. Darin Grauberger did the cartography for this version. I would also like to dedicate this piece to the memory of the late Juma Maalim Kombo, Chickie-Boy, my neighbour, interviewee and friend.

References

Agnew, J. and Duncan, J. (1989), *The Power of Place: Integrating Sociological and Geographical Imaginations* (Boston: Unwin Hyman).

Allen, J. de V. (1974), "Town and Country in Swahili Culture," in Allen, J. de V. (ed.), *Perspectives des Etudes Africaines Contemporaines* (Cologne: Deutsche UNESCO-Kommission), 298–316.

—— (1993), *Swahili Origins* (Athens, OH: Ohio University Press).

Barnes, T. and Duncan, J. (1992), *Writing Worlds* (New York: Routledge).

Baumann, O. (1897), *Die Insel Zanzibar-Archipel* (Leipsig: Verlag von Duncker und Humblot).

Berleant-Shiller, R. (1991), "Hidden Places and Creole Forms: Naming the Barbudan Landscape," *Professional Geographer* 43, 92–101.

Burton, R. (1872), *Zanzibar: City, Island and Coast* (London: Tinsley Brothers).

Carter, P. (1988), *The Road to Botany Bay* (New York: Knopf).

Christie, J. (1876), *Cholera Epidemics in Eastern Africa* (London: MacMillan).

Clark, C. and Horton, M. (1985), *The Zanzibar Archeology Survey* (Zanzibar: Ministry of Information, Culture and Sport).

Cohen, S. and Kliot, N. (1992), "Place-Names in Israel's Ideological Struggle over the Administered Territories," *Annals of the Association of American Geographers* 82, 653–80.

Constatin, F. (1987), "Condition Swahili et Identite Politique," *Africa* 57, 219–33.

Cooper, F. (1987), *On the African Waterfront: Urban Disorder and the Transformation of Work in Colonial Mombasa* (New Haven, CT: Yale University Press).

Cosgrove, D. (1989), Geography Is Everywhere: Culture and Symbolism in Human Landscapes," in Gregory, D. and Walford, R. (eds), 118–35.

—— (1993), *The Palladian Landscape* (University Park, PA: Pennsylvania State University Press).

Cruikshank, J. (1990), "Getting the Words Right: Perspectives on Naming and Places in Athipaskan Oral History," *Arctic Anthropology* 27, 52–65.

Duncan, J. (1990), *The City as Text: The Politics of Landscape Interpretation in the Kandyan Kingdom* (Cambridge: Cambridge University Press).

Entrikin, J. (1991), *The Betweenness of Place* (Baltimore, MD: Johns Hopkins Press).

Fair, L. (1993), "A House of Mud, Stories of Stone: Tarab as an Expression of Gender, Class and Sexuality in Zanzibar, 1922–1972," paper presented to the African Studies Association, 4 December, Boston, MA.

—— (1994), "Pastimes and Politics: A Social History of Zanzibar's Ng'ambo Community," PhD thesis, University of Minnesota.

Foucault, M. (1980), *Power/Knowledge* (Brighton: Harvester).

Friedmann, J. (1992), "The Right to the City," in Mofse, R. and Hardoy, J. (eds), 98–109.

Garlake, P. (1966), *The Early Islamic Architecture of the East African Coast* (London: Oxford University Press).

Godlewska, A. and Smith, N. (eds) (1994), *Geography and Empire* (Oxford: Blackwell).

Gregory, D. and Walford, R. (eds) (1989), *Horizons in Human Geography* (Basingstoke: Macmillan Education).

Harris, C. (1991), "Power, Modernity, and Historical Geography," *Annals of the Association of American Geographers* 81, 671–83.

Hino, S. (1971), "Neighborhood Groups in African Urban Society: Social Relations and Consciousness of Swahili People of Ujiji, a Small Town of Tanzania, East Africa," *Kyoto University African Studies* 6, 9–23.

Horton, M. (1984), "The Early Settlement of the Northern Swahili Coast," PhD thesis, University of Cambridge.

Hussein, A. (1991), personal interview, 11 December.

Jackson, J. (1970), *Landscapes: Selected Writings of J.B. Jackson* (London: Allen and Unwin).

Johnson, F. (1991), *A Standard Swahili-English Dictionary* (Oxford: Oxford University Press).

Katz, C. (1994), "Playing the Field: Questions of Fieldwork in Geography," *Professional Geographer* 46, 67–72.

Larsen, K. (1992), "People and Masheitani in Zanzibar Town: On the Perception of Cultural Difference," unpublished manuscript, Department of Social Anthropology, University of Oslo.

Lewis, P. (1982), "The Politics of Iranian Place Names," *Geographical Review* 52, 99–102.

Ley, D. and Samuels, M. (eds) (1978), *Humanistic Geography: Prospects and Problems* (Chicago: Maaroufa Press).

Makame, A.H. (1992), personal interview, 10 February.

Menon, R. (1978), "Zanzibar in the Nineteenth Century: Aspects of Urban Development in an East African Coastal Town," MA thesis, University of California, Los Angeles.

Middleton, J. (1992), *The World of the Swahili: An African Mercantile Civilization* (New Haven: Yale University Press).

Mitchell, T. (1988), *Colonizing Egypt* (Cambridge: Cambridge University Press).

Mofse, R. and Hardoy, J. (eds) (1992), *Rethinking the Latin American City* (Washington: Woodrow Wilson Center Press).

Mohamed, S. (1978), *Asali Chungu* [Bitter Honey] (Nairobi: Shungwaya).

Mohammed, S. (1992), personal interview, 2 April.

Mohammedridha, A. (1991), personal interview, 7 December.

Myers, G. (1994a), "Eurocentrism and African Urbanization: The Case of Zanzibar's Other Side," *Antipode* 26, 195–215.

—— (1994b), "From 'Stinkibar' to 'The Island Metropolis': The Geography of British Hegemony in Zanzibar," in Godlewska, A. and Smith, N. (eds), 212–27.

—— (1994c), "Making the Socialist City of Zanzibar," *Geographical Review* 84, 451–64.

—— (1995), "The Early History of the Other Side," in Sheriff, A. (ed.), 21–34

Nast, H. (1994), "Women in the Field: Critical Feminist Methodologies and Theoretical Perspectives," *Professional Geographer* 46, 54–66.

Price, M. and Lewis, P. (1993), "The Reinvention of Cultural Geography," *Annals of the Association of American Geographers* 83, 1–17.

Robinson, J. (1990), "'A Perfect System of Control?' State Power and 'Native Locations' in South Africa," *Society and Space* 8, 135–62.

Said, E. (1993), *Culture and Imperialism* (New York: Knopf).

Sales, P. (1991), "The Origin of Country Names in Africa," *Nomina Africana* 5: 2, 10–31.

Scott, J. (1990), *Domination and the Arts of Resistance: Hidden Transcripts* (New Haven, CT: Yale University Press).

Sheriff, A. (1987), *Slaves Spices and Ivory in Zanzibar: Integration of an East African Commercial Empire into the World Economy, 1770–1873* (London: Heinemann).

—— (ed.) (1995), *The History and Preservation of Zanzibar Stone Town* (Athens, OH: Ohio University Press).

—— and Ferguson, E. (1991), *Zanzibar Under Colonial Rule* (Athens, OH: Ohio University Press).

Soja, E. (1989), *Postmodern Geographies: The Reassertion of Space in Critical Social Theory* (London: Verso).

Sopher, D. (1978), The Structuring of Space in Place Names and Words for Place," in Ley, D. and Samuels, M. (eds), 251–68.

Tanzania Swahili Institute (1989), *Kamusi* (Dar es Salaam: Government Printers).

Tanzania, United Republic of (1991), *The 1988 Census: Preliminary Report* (Dar es Salaam: Tanzania Government).

Tuan, Y.-F. (1991), "Language and the Making of Place: A Narrative-Descriptive Approach," *Annals of the Association of American Geographers* 81, 684–96.

Western, J. (1981), *Outcast Cape Town* (Minneapolis: University of Minnesota Press).

Williams, R.O. (1949), *The Useful and Ornamental Plants of Zanzibar and Pemba* (Zanzibar: Government Printers).

Zanzibar Protectorate (1912), *A Handbook of Zanzibar* (Zanzibar: Government Printers).

Zelinsky, W. (1955), "Some Problems in the Distribution of Generic Terms in the Place Names of the Northeastern United States," *Annals of the Association of American Geographers* 45, 319–49.

Zanzibar National Archives Files:

AB 9/9: Civic Centre at Ng'ambo, 1943–1954.

AB 36/23: Ground Rents in Ng'ambo, 1940–1963.

AB 39/36: Numbering of Huts in Native Locations, 1940.

AK26/18: Hamoud (sic) ASP Branch, 1966–1969.

AW 2/100: The Ngambo Area Folder, 1938.

BA 13/3: The Laws of Zanzibar Volume 111, Cap 79 the Towns Decree and Cap 85 The Town and Country Planning Decree.

Chapter 6
The Aloha State: Place Names and the Anti-conquest of Hawai'i

Douglas Herman

In 1778, when James Cook in the ship *Resolution* rested offshore the islands of Kaua'i and Ni'ihau, he named this archipelago the "Sandwich Islands" in honour of the prominent sponsor of his third voyage, the Earl of Sandwich. By this appellation—as recorded in Cook's journal and subsequently engraved onto maps, charts, globes, and geography texts of the time—did this group of islands become known to the Westernized world. An identity imposed from without, this designation sprang from the process of inscription and classification that was producing the modern scientific and geographical order. Though over subsequent decades this name would yield to the epithet "Hawaii", this too is a colonial by-product; there was no overarching name for the entire archipelago prior to Western contact. Occupied by up to 1 million Polynesians who had been in residence for at least 1000 years, it was not, in effect, one place, but several places. It is supposed that because most early contact took place on the island of Hawai'i, this name became synonymous in Western discourse with the entire group (Blaisdell 1989).[1]

In 1993, an oppositional *resolution* sailed by the Islands—one introduced to the Honolulu City council by Native Hawaiian leaders to change the name of Thurston Avenue. Lorrin A. Thurston, after whom the street was named, was denounced in the resolution as a "radical insurgent who was the early leading force behind the 1893 overthrow of Queen Lili'uokalani." Thurston and a Committee of Thirteen *haole*[2] (white) businessmen, backed by troops from the USS Boston, forced the abdication of the Queen and declared themselves

1 For this reason, throughout this text I tend to refer to the "Islands," rather than to "Hawai'i." Some scholars (e.g. Sahlins and Valeri) use "Hawaii" for the group and "Hawai'i" for the island, reinforcing that "Hawaii" is an imposed, "English" name (hence the spelling without glottal stop). Readers should note that "Hawaiian" is an English word, so does not take the glottal stop either.

2 *Haole*, believed originally to have meant "foreigner," has historically come to designate persons of European ancestry. The connotations of this term was augmented by the historical class difference between "white" persons as owners of plantations and businesses, and other persons (Hawaiians, Asian and Portuguese immigrants) as labourers.

the new government of the island nation. This government, recognized as illegitimate by then-President Cleveland, five years later negotiated a treaty of annexation to the United States with President McKinley. The 1993 resolution asked that the city's Department of Land Utilization rename the street "Kamakaeha Avenue" based on Queen Lili'uokalani's birth name, "Lili'u Kamakaeha." Continuing to have a street named in Thurston's honour was denounced as "especially anachronistic" in light of the January 1993 centennial observance of the overthrow (Waite 1993).

The range of opinions on the resolution varied. It was immediately assaulted by some as revisionism and "political correctness." Critics pointed out that nine of the thirteen men who engineered the overthrow had streets named after them, and that the resolution's logic, if applied across the board, would lead to many more street name changes. Even McKinley High School would have to be renamed. Other views were reported in the Honolulu *Star-Bulletin* (Vickers 1993). One resident liked the idea of changing the name, "just because it seems more Hawaiian." A resident of Vietnamese ancestry felt that the name change was "disrespectful for the residents here now" noting also that she prefers "English street names. At least you can pronounce them." A third stated that "Hawaiians need to realize there's more than one ethnic group involved now. It's part of history too, good or bad. I think we need to spend more time trying to blend together and become one nation. Not just in Hawaii, but in the whole country." Lilikalā Kame'eleihiwa, assistant professor of Hawaiian studies at the University of Hawai'i, countered with "They're lucky Hawaiians don't rise up and burn the street sign."

Inasmuch as the changing of place names, especially country names, has been an ongoing part of decolonisation throughout the Pacific (Crocombe 1990) and forms part of the greater legacy of colonization world-wide, the debate over a street name seems rather trivial. Indeed, it was treated as such by Honolulu newspapers. But the comments by residents above indicate the ongoing struggle over identity that is at the heart of the postcolonial condition. Combined, these comments suggest that things Hawaiian have some value or cultural capital, making for a distinct sense of place; while at the same time, pointing out that this archipelago is not the Hawaiians' place anymore, but rather belongs to us all. It is part of America. Speak English, please.

This study examines the relationship between place names and colonialism in the Hawaiian Islands. While this relationship is a complex and often subtle one—its true magnitude easily overlooked—this study excavates ways in which American hegemony is embedded in the place name code. This hegemony consists primarily in the shift of human-environment discourses and their political-economic contexts towards a capitalist understanding of space that served Western [*haole*] interests. Close examination reveals that place names explicate the relationship between conquest and what Pratt (1992) has called "anti-conquest," while psychoanalysis allows this relationship to be more clearly understood.

Anti-conquest involves glorifying the Other at the same time that the Other is denied real power. Anti-conquest constitutes a seemingly contradictory practice interlaced into the broad field of colonial discourse, posing itself disingenuously as antithetical to overt colonisation. Behdad (1994) and Jacobs (1994) expand on this idea as being the colonizer's attempt to recapture the disappearing Other out of a nostalgia for the lost exotic. It is the positive representation of Others that serves to displace blunt modes of racial differentiation and to rearticulate them in ways that make Otherness seem natural rather than constructed and imposed. Inasmuch as glorification of "Noble Savages" has not precluded their being colonised, anti-conquest poses itself as a benign paternalism that puts the Other on a pedestal—a gesture of respect that is also an exclusion, an isolation, and a fixing of the Other into a historical space separate from the modern. In this case, anti-conquest manifests as the promotion of things Hawaiian at the same time that Hawaiians themselves are excluded from power. Anti-conquest is, therefore, a part of conquest in no way antithetical, but only masquerading as different by operating backwards.

In the matter of Hawaiian place names, "conquest" is most easily seen in the inscription of Western family names onto Honolulu's streets, occurring primarily in the decades before or soon after annexation by the United States. "Conquest" also extends to those practices which deny authority to Hawaiians. It includes the imposition and fixity of a Western order by Western authorities—the cultural colonisation in which Hawaiians themselves as authorities, and Hawaiian worldview as dominant, are overinscribed by their Western counterparts. The participation of the US government and military in this activity heightens the imperial connection to this process. Pacific Explorations from Cook onwards had a distinct, combined military-and-scientific character: well-armed naval vessels, loaded with scientists, on missions to both gather data and, where possible, to acquire territory. Both processes involved acquisition, and together they helped form what became a dominant worldview and world order, both geographically and politically.

Anti-conquest, however, manifests in the veneration and manipulation of the Hawaiian names within the process whereby they are catalogued, promoted by law, and ordered into American systems of geographical knowledge and land control. This takes place after US annexation, when Hawaiians no longer pose a major political obstacle to American hegemony. Within the resultant texts, there is always an implied respect for the integrity of the Hawaiian names. Yet a re-ordering of place is revealed, not just in how these texts are compiled and produced, but in who did so and for what purpose.

Before embarking on an analysis of Hawaiian place names and colonialism, it is worthwhile to consider *why* anti-conquest occurs, which will shed greater light on *how* it works, and what broader contexts and interactions are involved. Unlike most forms of colonialism, anti-conquest is never a conscious process. Colonizers usually see it as paying genuine respect to the local culture, and would take offence if one were to confront them by suggesting their "gracious acts" were in fact modes of power. Anti-conquest is, rather, a subconscious act, best understood with the aid of tools from psychoanalysis. The key point to be made here is the role of language in establishing meaning, and how the assertion of meaning—hence order—is a tool of domination.

The (re)naming of places is what Lacan (1977) and Grosz (1990) would call a *phallic* practice whereby *logos* is imposed. Indigenous place names are ignored and place names in the language of the colonizer are instated, suggesting that the indigenous systems do not constitute order, any more than indigenous ways of knowing were said to constitute "science" or "rationality." Cixous remarks that the feminized Others "always inhabit the place of silence, or at most make it echo with their singing" (1981, 49). Hawaiian place names, and Hawaiian language, have come to play this role and no more. Any effort to bring Hawaiian language and place names into the realm of logos—to allow them to have meaning—meets with fear and opposition by the colonizer. Thus anti-conquest silences the Other by allowing her to "talk" without allowing her to "speak", to control meaning: "she" does not speak, she only chatters, or sings. Her words form a pleasant background noise. Hawaiian place names thus form what Shapiro (1988, 92) calls a code of those who lost the struggle to maintain their practices and have them be the ones that are intelligible—the "Others" whose practices no longer control prevailing understandings on their old turf, but whose discourses remain in a nonpolitical code.

A New World Order

In the case of land, attaching a family name to one's abode or estate—however egotistical or patriarchal it may seem—still contains a certain geographic sense. It reflects a genuine relationship with the land. By comparison, the honorific attachment of names to land otherwise unconnected with the person involved is "commemorative" (Stewart 1970, xxx) as it establishes a monument, maintains a reminder, usually of someone designated as a "Great Man."[3] Inasmuch as language is used to command and control, to render—in this case—Western military-cum-science as the subject of culture and "natives" (lands and peoples) as voiceless objects, so much more does imposing names of Great Men on the land assert the *name of the father* by which this order is assured.

Waterman's (1922) study of indigenous naming practices in Puget Sound points to differences in worldview between Western and Indigenous systems. Waterman notes that "it may be stated as a rule that there is a large series of names for small places, with astonishingly few names for the large features of

3 Crocombe (1990, 2) remarks that he is unaware of a single instance of place naming by a woman—Polynesian, *haole*, or otherwise—anywhere in the Pacific.

the region" (Waterman 1922, 178). There are, for example, names for places on the mountain, but not for the mountain itself, much less the mountain range as a whole. Likewise there are names for specific beaches and canoe landings, but not for "bays" or "islands" as entities. In addition, Waterman goes on to say, "there is always the difficulty of telling absolutely and finally which out of a number of explanations is correct … there is an element of uncertainty which no scientifically minded person would deny" (Waterman 1922, 177).

The differences noted by Waterman reflects a larger aspect of the cultural contrast between the European explorers (and later colonizers) and indigenous islanders. Pre-capitalist island societies were characterized by what Cosgrove (1984) calls an "analogic" context—having a fluidity between culture and environment and an absence of land "ownership" per se. In this context, individuals understand their position of the world in terms of relationships to larger processes. Starting with the British, however, Europeans were experiencing the shift to private property and a resultant "atomistic" context in which individuals understand themselves as isolated entities. Land is commodified, the environment secularized and atomized into "resources," rather than being understood as related aspects of a holistic system. The difference between these two worldviews is profound, and continues in varying degrees to characterize the struggle between indigenous (primal) peoples and their colonizers, European or otherwise.

Atomization is essential for capitalism: private property and the individual entrepreneur (and consumer) are, within this system, understood as "rational." In the Hawaiian Islands, the colonial forces (missionaries, merchants, and military) sought to intentionally institutionalize this shift towards capitalism. One major result was the 1848 *Mahele* [partition]: the privatization of land brought on by Western geopolitical pressure and in the service of Western commercial interests. The subsequent rise to power of Western-owned plantations constitutes the de facto colonisation of the Islands, first by wresting control from the indigenous monarchy, and later, after a short period as an independence, achieving annexation to the US. Place names have come to show, among other things, the transformation of worldview that accompanied this shift.

A Hawaiian Worldview

Despite a social class structure rigidly defined by birth and genealogy into chiefs [*ali'i*] and commoners [*maka'āinana*], Hawaiian geography is characterised by fluidities. These include geographic mobility (not being legally bound to the land); a fluidity of access to resources and resource areas rather than being limited to discreet plots only; a fluidity between the human, divine, and natural realms; and a fluidity of sound, meaning and a metaphor regarding place names. Chiefliness is rooted in sacredness or divinity [*mana*], verified through genealogical connections to ancestral deities and origins of the world itself. The highest ranking *ali'i*—with the most divinity—are linked most closely with Hawaiian deities. These deities, simultaneously, are intimately linked with every aspect of the environment, and back to humanity in a circular linkage between gods, nature and human society (Dudley 1990; Johnson 1983; Herman 1988).

The administrative hierarchy (Figure 6.1) was mirrored in spatial order. Hawaiian land divisions ran from the islands' mountainous centres to the sea like slices of a pie. Large sections (*moku*) were divided into finer slices called *ahupua'a* (Figure 6.2). Transecting the land's altitudinal zonation as well as reaching out into the sea, the *ahupua'a* encompassed the full range of natural resources necessary for subsistence. Though all land was "held" by

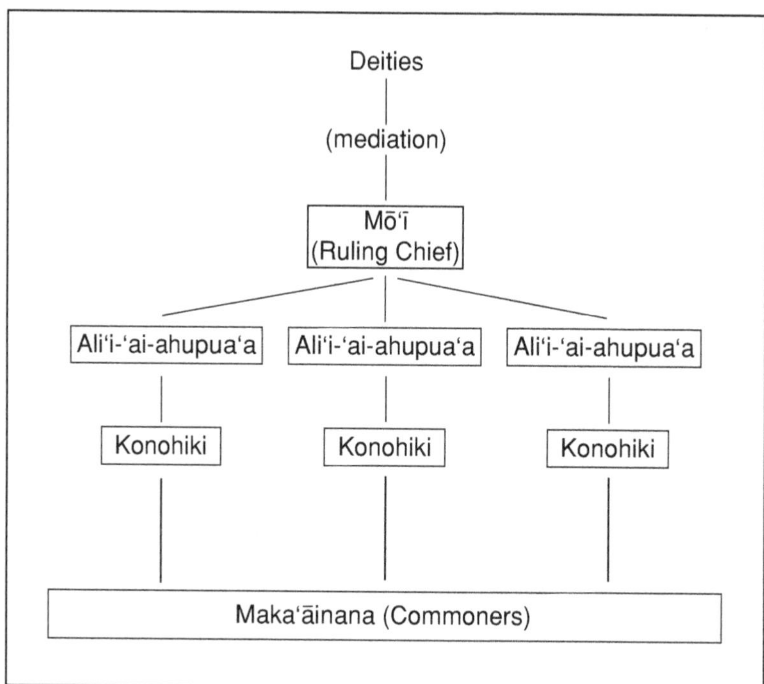

Figure 6.1 Social and political structure of traditional Hawaiian society

Source: after Kirsch (1985).

Note: The ruling chief mediated with the gods on behalf of the people. Lesser chiefs [ali'i] were put in charge of individual *ahupua'a*. They in turn could appoint *konohiki* [administrators] to manage the land and resources. Commoners [*maka'āinana*] mostly fished or worked the land. The goal of maximum fruitfulness produced checks and balances to discourage both poor farmers/fishermen and poor administrators.

the supreme chief, he did not own it, but served as a "trustee" under the gods Kane and Lono. He placed *moku* under the supervision of other high chiefs, who further subdivided administrative responsibility to lesser, warrior chiefs or supervisors [*konohiki*]. The final allotment was made by these retainers to the common people (Handy and Handy 1972, 41).

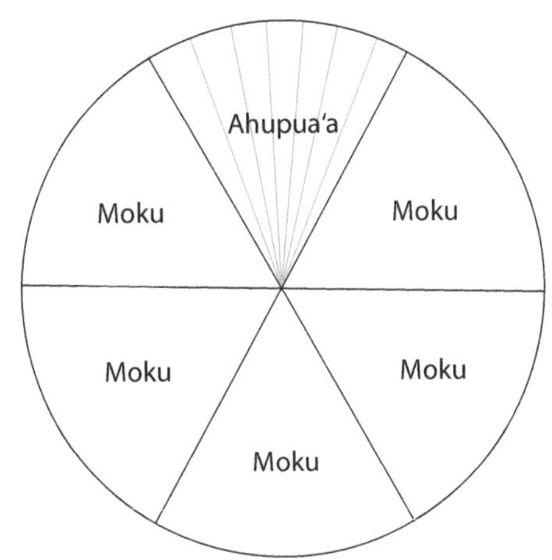

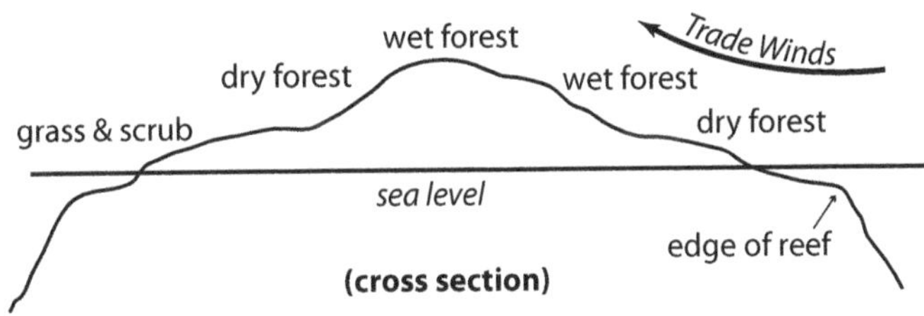

Figure 6.2 Hypothetical island, divided into large districts [*moku*] and administrative units [*ahupua'a*]

Source: after Wise (1965)

Note: With mountains roughly in the island centre (ranging as high as 13,000+ feet on the island of Hawai'i), this method of dividing land assured the full range of environmental resources was available in each *ahupua'a*. The Hawaiian word for "politics" is *kālai'āina*, or "land carving."

"Trusteeship" over the land ranged upwards from the cultivators themselves, to their local chief or supervisor, to yet higher chiefs and finally to the gods themselves. The products of the land likewise filtered upwards through the social-administrative hierarchy through taxation. Yet cutting across this vertical organization of power was a horizontal system of common access to resources. In part this formed a second economy wherein produce was exchanged within the extended family residing in an *ahupua'a*. The fruits of fishermen were exchanged with those of the planters. But most importantly, people within an *ahupua'a* had gathering rights to all the necessary resources within that land division, from the waterfront into the mountains. Thus for any family, land use was spread out to include not just individual holdings, which themselves might be spatially disconnected, but grass areas for thatch, forest areas for timber and medicinal herbs, beach access, and other resource areas.

Within the complex, diverse, and variable materials from which Hawaiian world-views may be drawn (Dudley 1986, 71), it is possible to identify an overall analogic character that differs critically from a generalized modern worldview. Three different aspects of this analogic relationship include Hawaiian "natural science" as elaborated in the *Kumulipo*, a Hawaiian creation chant; the kinship relationship between the land and people, outlined in a subportion of the *Kumulipo*; and a "spiritual ecology" wherein energies flow across the boundary between the manifest and unmanifest worlds.

The *Kumulipo*, a chant of 2,077 lines believed to have been composed around 1700 CE (Beckwith 1970, 310–11) presents the evolution of life in a natural sequence not unlike Darwinian evolution and modern natural science. Species emerge in pairs from the primordial ooze, increasing in complexity from the smallest visible organism, the coral polyp. Johnson (1981, i) states that "when life appears in the Kumulipo, it is the product of active, *natural* forces" though "Supernatural forces are not excluded from that process." With each new set of lifeforms, "*akua* enters in." The term *akua*, used elsewhere as the general term for "gods," has been interpreted here as meaning "active consciousness" (Dudley 1990; Herman 1988). One might say that increasingly complex species are increasingly intelligent and intuitive, hence increasingly godlike. In this interpretation, beings differ quantitatively, but not qualitatively, in their degree of *akua*. At the same time, the universe is pervaded by *mana*, a living force that infuses all things and that, like its inanimate counterpart electricity, manifests more strongly in persons, places, and things that are more "godlike" (better conductors, in the electrical metaphor). And like electricity, this spiritual energy can be stored, and channelled or transferred. The Hawaiian word for "prayer", *ho'omanamana*, literally means "to cause [to move] *mana*." So this energy pervades—hence unites—all things, and circulates among them.

The story of Wakea and Papa, the Sky and the Level Earth, is a sub-plot within the *Kumulipo*. In this oft-repeated story, these two deities united to produce some of the Hawaiian Islands, as well as a beautiful daughter. Within the story, Wakea has an affair with the daughter, producing first a stillborn foetus that becomes the first taro plant (the staple food of Hawaiians), then producing a child that is the first human being. Later, Wakea and Papa come together again and produce the remaining islands. This story demonstrates the *kinship relationship* between the gods, the islands, the taro plant, and human beings. All are related—the land, the people, the gods, and the taro—in a kinship bond that is primary to Hawaiian environmental understanding and comparable to other worldviews held by primal peoples elsewhere.

Finally, this worldview is integrated through the concept of *kinolau* ("many bodies"). This is the manifestation of deities as natural phenomena—plants, weather, animals. These manifestations, like the bread and wine of Holy Communion in some Christian churches, do not merely represent the deity, but are in fact the physical presence of the deity. High Gods (*akua*, again) relate to human society through smaller, more accessible versions of themselves (*'aumakua*, or familiar spirits), and both independently can manifest as *kinolau* (Figure 6.3, left side). At the same time, through the transference of *mana*, deceased ancestors can be elevated to the level of familiar spirits (*'aumakua*), who in turn can manifest back into nature through *kinolau* or take a particular physical form (Figure 6.3, right side). Gods become nature, and humans become demigods, which in turn become nature. The circle between divinity, humanity, and nature is complete, and the boundaries among them permeable (Johnson 1987; Kamakau 1964).

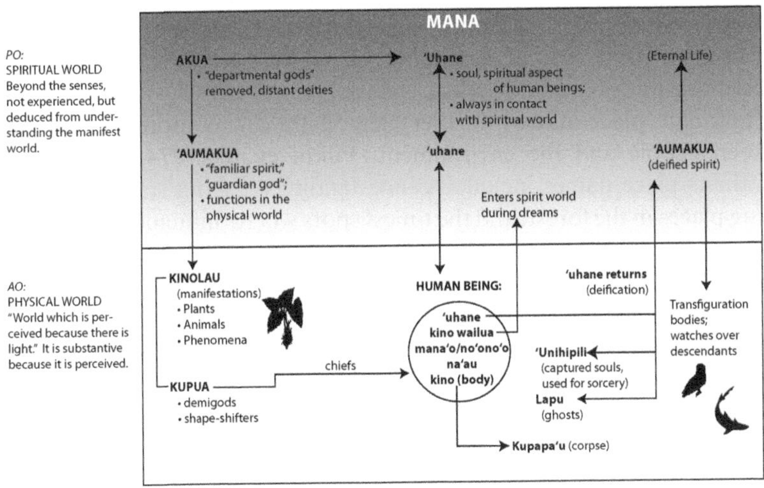

Figure 6.3 Hawaiian "spiritual ecology," showing network of relationship between the material world and the spiritual world in Hawaiian cosmology

Source: designed by Rubellite Kawena Johnson.

Place names embody and, to a degree, codify this fluidity between society, nature, and the spiritual world. There are two other ways in which words and names have a greater fluidity in Hawaiian geographical knowledge, and both of these are linked to the non-literate nature of the language and culture. The first is in the plurality of meanings for (particularly common) words. Since words exist only in the context of the moment when they are spoken, their meaning derives from that context. Thus there is what literate Anglophone culture considers a paucity of words. "Wai," meaning "fresh water," is also used for any liquid other than sea water, including bodily fluids. Since one is present when the word is spoken, one knows what the referent is. Gestures, vocal inflections, and facial expressions also contribute to the meaning. Thus what appears on paper to be a simple language is actually much more subtle and complex. Each word can be used for many different shades of meaning that are clear only in the context of speech. But these contexts do not translate into writing, which poses a text abstracted from its real-life situation. On the other hand, oral language can be weak in specific terms for abstract concepts that have no real-world context: a circle drawn on a piece of paper, was identified variously by non-literate individuals as a "rock" or a "basket"—it had to *be something* (Ong 1982).[4]

Second, many different names, though often quite similar in sound, might apply to a single place. Nogelmeier's (1985) study of the crater known commonly as "Punchbowl" produced a list of alternative, similar-sounding Hawaiian names for this one geographic feature, and despite his attempt to find the "most appropriate name," the fact remains that *all* of them were valid. Not fixed in print, any of a number of similar-sounding names—all with different meanings—would suffice.

Hawaiian place names further demonstrate the intimate relationship between people and the environment. Pukui et al. (1974, x) remark that Hawaiians place names include "canoe landings, fishing stations in the sea, resting places in the forest, and the tiniest spots where miraculous or interesting events are believed to have taken place." Kimura (1983, 178) and Kanahele (1986, 175–6) emphasize the evocative power of place names and the many roles they play in Hawaiian culture.

Place names intertwined with the culture not just as locational signs, but as poetry and humour. Luomala (1964), Pukui et al. (1974, 266–77), Handy and Handy (1972, 42–3), Kimura (1983) and Piʻianāiʻa (n.d.) discuss the richness of

4 A classic example of this is the missionaries' effort to translate the word "adultery" into Hawaiian. Hawaiian language has no term for the abstract concept "adultery," but has numerous terms for specific "adulterous" activities. Mr Andrews wrote that "In translating the seventh commandment, it was found they had about twenty ways of committing adultery, and of course as many specific names; and to select any one of them would be to forbid the crime in that one form and tacitly permit it in all the other cases" (*Missionary Herald* [*MH*], October 1836, 390–91).

place names in chants, songs, poems, and proverbs, and as mnemonic devices in genealogies. The versatility in the use of place names is especially found in the practice of *kaona*, or "hidden meanings," in the various forms of Hawaiian oral literature. Most Hawaiian poetry or song is composed of several layers of meaning, from the most obvious or literal, to sexual or romantic, to *kapu* or sacred subjects. Using *kaona* allows one to talk about certain people or subjects without addressing them directly, and as such "presupposes a knowledge of place, history, and personal relationships" (Pi'ianāi'a n.d., 8).[5]

Pukui, Elbert, and Mo'okini (1975, 258) break down the words used in Hawaiian place names as shown in Table 6.1. Aside from geographic terms such as words for water, hill, point, mountain, and so forth, there are a surprising number of plant names. Pukui et al. suggest that this can be construed as indicating agriculture as one of the primary interests of the early Hawaiians (Pukui et al. 1975, 260–61). Names of geographic entities, plants, animals, legendary supernaturals, and religious terms (implicit and explicit) reflect the "animist" nature of Hawaiian culture. People's names are rarely used. Pukui et al. have said that this reflects a Hawaiian perception of human beings as fleeting, transient. While there are stories of places being named for ancestors,[6] Hawaiians generally did not use people's names as place names, and cases in which they do so suggest a direct relationship with the land, rather than an honorific attachment of a label.

The Transformation

Following Cook's 1778 "discovery" of the islands, China-bound fur traders began stopping here as early as 1784. The location of this archipelago as the only reprovisioning point in the North Pacific ensured that all traders and explorers travelling between North America and East Asia (including Russia) stopped here. Local chiefs made use of trade with these ships to gain goods, weapons, and even military advisors to aid them in their internecine warfare. With the aid of Englishman John Young, Kamehameha succeeding in conquering all islands except Kaua'i, which he gained by treaty in 1807, thus uniting all the islands, for the first time, into what was now the Kingdom of Hawai'i.

Though Kamehameha himself remained true to his Hawaiian ways, the encroaching mercantilist economy, with its virtually instantaneous creation of

5 See Kimura (1983) for an excellent discussion of the role of language in Hawaiian culture.

6 See Kamakau (1976, 7). One Hawaiian tradition holds that Hawai'i-loa, discoverer of the islands, named the largest island "Hawai'i" after himself and the others after his children, as well as naming various land divisions after the eight navigators who sailed with him (Beckwith 1970, 364).

Table 6.1 Sources of Hawaiian place name words, by category

Category	Number
Geographical features	831
Inanimate nature	265
Words of size	203
Material culture	170
Plants and plant life	153
Actions	124
Other qualities (except colours)	115
Numbers and words of quantity	71
Colours	69
Birds	68
Body parts	65
Legendary supernaturals	53
People	44
Sea life	32
Mammals	23
Locatives	10

Source: after Pukui, Elbert and Mo'okini (1975, 258).

supply and demand, set the groundwork for the Islands' entrance into a ancient world capitalist order (Ralston 1984, 25–6). Large-scale exploitation of the environment for profit began with the sandalwood trade, 1810–1830, which opened the Islands to international commerce. The socio-political changes began with the1819 abolition of the *kapu* system, a combined religious and legal code which legitimised chiefly power as sacred. The previous consumption of goods and labour for religious concerns was now available for use by the *ali'i* in other activities, especially foreign trade. American Protestant missionaries arriving in 1820 encouraged social, cultural, and political forms conducive to capitalism, leading to the establishment in 1840 of a constitutional monarchy, and the redivision of lands—the 1848 *Mahele*.[7]

The result was that the previously fluid land-tenure system became fragmented into a fixed grid of privatized parcels. The structure of the *ahupua'a*, in which all had gathering rights in different resource zones, was dismantled. Commoners had to claim the land they used in order to maintain rights to it. Yet surveying techniques at the time were crude, and the Hawaiian

7 See Kame'eleihiwa (1992) for an important discussion on the causes of the *Mahele*.

sense of boundaries and markers were not "precise" in a Western sense. The result was disastrous. Many people received no land at all, and some who gained freehold tenure to cultivated plots lost former communal rights to grazing land or collecting areas (Ralston 1984, 31). Subsistence farming became difficult if not impossible. Wise (1965, 87–8) explains that Hawaiians were accustomed for generations to communal rights land and sea, and could not imagine life on another basis.

The *value* of land was thereby transformed. Plots could now be bought or sold, with the result that the Hawaiian people, lacking capital, were gradually dispossessed from their land. By the mid-1840s, taxes had to be paid in cash, forcing people from remote areas to seek employment in the port towns to earn money (Ralston 1984, 31). *Kuleana* (land claims by commoners) might be rented to a large plantation, only to have them disappear under a transformed landscape: where all the familiar landmarks had been, an expanse of rice or sugar cane would stretch out. "Ditches had been filled in, dikes had been levelled off, hedges had been cut down" (Wise 1965, 90).

The transformation of landholding and the gradual transference of land from Hawaiians to foreign plantation owners was accompanied by the gradual infiltration of foreigners into the Hawaiian Government, and a steady Westernizing of the governmental form. Over the course of the nineteenth century, plantations became the new powers in the Islands: the old system of *ali'i* had been replaced by a new ruling class of foreigners that, like the old, maintained control over most of the land and labour in the Islands (Herman 1996). These businessmen became so powerful that moves on the part of the Hawaiian monarch to curb them were met with by arms in the 1893 overthrow of Queen Lili'uokalani. The new Republic of Hawaii became, with annexation in 1898, the Territory of Hawaii. This progression, culminating in statehood in 1959, was accompanied by the near-complete transformation of land and the built environment. After 1898, with the Territory under control of the US federal government and its the strategic significance clearly understood, the Islands became a support system for US military personnel and operations. Save the sugar industry, federal spending became the largest source of income for the Islands. The combined resources of the local and federal government, US military and even academic Geography (Baldwin 1908; Barnes n.d.; Bryan 1915; Coulter and Serrao 1932; Freeman 1927; Guillemard 1908; Jarrett 1930; Jones 1938), came down on the Islands to lay bare the terrain, scrutinize it, map and chart it, and lock it thoroughly into the grid of geographic knowledge and control.

Names of the Fathers

The stop-and-start transformation of the Islands since Cook's visit has been pushed strongly by various *haole* forces, from early explorers and traders to missionaries-turned-plantation owners. As stated above, this transformation was accompanied by a steady increase of Westerners holding government positions. Similarly, the plantation industry grew and spread its tentacles to every major enterprise in the Islands: banking and finance, utilities, transportation, and more. These "captains of industry" and these legislators—often the same people—are commemorated in Honolulu's streets.[8]

This began with John Young, a British seaman recruited by Kamehameha as a military and cultural adviser, decades before there even were any "streets." Under the first constitutional monarchy in 1840, the House of Nobles consisted of traditional *ali'i* plus John Young (Young Street, Olohana Street—a nickname of his, from "all hands").[9] Eleven years later in 1851, Young was still there, plus three other haoles: Armstrong, Judd, and Wyllie (Armstrong Street, Judd Street, Judd Hillside, Wyllie Street). New Western names a decade later that now serve as street names include Bishop, McCully, Dowsett, Chamberlayne, and Green. By 1880, over half the House of Nobles was foreigners, including Cleghorn, Castle, Dominis, Isenberg, Martin, Mott-Smith, and Wilder—all of whom have at least one street to their names. By 1890, shortly before the overthrow of the monarchy, 23 out of 25 nobles were of foreign ancestry, as were 24 of 33 representatives and all eight ministers. Many Honolulu street names are found on these lists. Not only legislators, but other *haole* Fathers are "honoured" with street names. This bespeaks the power and importance ascribed their roles: four discoverers and early explorers, 11 missionaries, six business leaders and philanthropists, five political leaders, and six military figures (Pukui et al. 1975, 263). A cluster of streets is named for Roman Catholic bishops.[10]

This was a practice brought to the Islands from outside, particularly from the United States from whence most of these men came, and where it was not uncommon for entire towns to be named for their founding fathers. Streets themselves are a result of Western influence in the Islands, arising after 1820 and expanding markedly after the 1898 annexation. They mark a second phase of naming, a second overlay of toponyms that correspond with the "modern" period in the Islands. Though Hawaiian place names, as stated earlier, show

8 The extensive intermarriage of Hawaiian women with *haole* men produced a population of Hawaiians without Hawaiian family names but who nonetheless retain identity as Hawaiians. As some of these were prominent citizens, the increasing paucity of Hawaiian names in the rosters, and their invisibility in the "Blue Book" (discussed below), is not a thoroughly reliable indicator. But this loss of Hawaiian family names is another aspect of colonization and the Name of the Father.

9 Data for this section were taken from the rosters of the legislatures compiled by Lydecker (1918).

10 Another is named for ships of the Matson company. Pensacola Street is named for a US warship. Pukui et al. (1974, 263–4) have given a broader analysis of English names of streets and buildings, as well as persons (Hawaiian and non-Hawaiian) for whom "places" have been named.

little emphasis on family names, street names tell a different story. Yet among Hawaiians' names used as street names, only a few are obvious: Kamehameha Avenue (named after the king), Ka'ahumanu Street (his favourite wife), Kina'u Street (his daughter); Kalākaua Avenue and Lili'uokalani Street (the last two monarchs), and two for Prince Jonah Kūhiō Kalaniana'ole (Kūhiō Street, Kalaniana'ole Highway). While this too shows a commemorating of "fathers," it is by this time the transplantation of a Western practice onto Hawaiian culture and applied to a Western landscape form, much as the monarchs took on the regalia of their European counterparts.

More important, however, are the not-so-obvious Hawaiian names. The Armitage street map of 1949 lists and identifies 258 *haole* family names, but only 17 street names are identified on this map as Hawaiian family names, along with about 15 Portuguese, four Chinese, and two Japanese. Fifteen names on this list are Hawaiianized versions of *haole* names.[11] Yet research by Budnick and Wise (1989, 3) identifies 171 streets named for Hawaiian people—more than 120 named for original land awardees or grantees from the *Mahele*.[12] In other words, the Hawaiian names, from the *haole* point of view, had ceased to be Names of the Father—had ceased to have logos. This loss of subjectivity accompanied the transfer of land and power from under the Hawaiians, the loss of Hawaiian cultural context, and the abandonment of Hawaiian language during the Territorial period. Largely unintelligible to Hawaiians as well as non-Hawaiians, these names became words only, markers without reference.

The transplantation of Western family names onto the urban landscape thus becomes a record of the Western transformation of the Islands. More than merely "humanizing the landscape," these street names participate in the transforming of Hawaiian space into Western space, denoting—as did many travel texts of the time—these Western men as those who brought the Islands out of "barbarism" and fashioned them in their own image. It is a song sung repeatedly. Bates (1854, 136) was first to state that "The Honolulu of today is the creation of the foreigner, the result of his handywork." Such sentiments were echoed by Nordhoff (1874), Musick (1897), Davis and Armitage (1941) and even Michener (1963).

---

11 Pukui et al. (1974, 265) give the following breakdown of place names (including streets and buildings) of foreign, non-English origin: Portuguese (25); French, Belgian French, and Hawaiianised French (19); Chinese, Hawaiianised Chinese, and Portuguese Chinese (12); Japanese (12); German (4); Russian (2); and Spanish (1). There is also a smattering of other imported names.

12 Otherwise, their breakdown of Hawaiian street names in Honolulu reveals the same connection with nature discussed earlier: 407 for geographic and topographic locations (e.g., land sections, land divisions, mountains, valleys, streams, etc.); 189 for flora; 157 for stars, planets and galaxies; 92 for fish and birds; 27 for mythological gods and goddesses; and 23 for winds and rain (Budnick and Wise 1989, 3).

A layer of Western family names was imposed during the Hawaiian Kingdom, and served as one additional means by which the resident foreigners asserted their control over a territory not their own, but of which they desperately sought to gain control. This point is demonstrated by the end of this practice and the adoption of Hawaiian words as street names following annexation to the US. The conquest had been achieved at last. Now the methods of anti-conquest were needed to cover the tracks by alleging to respect and honour Hawaiian culture.

A Paper Trail

While existing Hawaiian place names were not overlaid or eradicated by a new code, annexation heralded their subjugation to the Western geographic grid of knowledge. This took place through the compilation of place name texts after annexation. A reading of these texts and their stated intentions, methods, and viewpoints, illuminates the imposition of logos. First, there is a focus on establishing authority, of declaring what is "officially" the name of a given place. Second, this authority comes from a non-Hawaiian (and mostly, from non Hawaiian-speakers) and is placed into the hands of some official (USor state) government agency. The lack of reference to Hawaiians themselves suggests that the very idea of a "Hawaiian" authority is anathema. Third, these Hawaiian names lose their cultural context and often their meaning in the process of their transposition into the authorized texts. Finally, the fluidity associated with Hawaiian names is lost by establishing an official, unambiguous designation.

From the very first Western text on Hawaiian place names—US Coast and Geodetic Survey (Treasury Department): "Hawaiian Place names," compiled by W.D. Alexander (1903)—the emphasis is on uniformity, fixity, and certainty. It is important for the West to *know*, for *truth* to be clarified and certified. The need to eliminate ambiguity and fix order into writing is explained in the preface, which states that this study was done because "the importance of securing uniformity in geographic nomenclature in all portions of the territory of the United States at the earliest possible date [was] being constantly felt in this office …" (Alexander 1903, preface). A brief biographical sketch of Professor Alexander, who compiled the text, is "inserted to show his peculiar fitness for this task, and it is believed that he is the most reliable living authority on the subject." Granting *Professor* Alexander, a non-Hawaiian, status as the "most reliable living authority" posits the supremacy of Western knowledge. Hawaiians are placed outside the discourse of knowledge regarding their own culture. The "results" of Dr Alexander's study then become property of the US Treasury Department, part of the larger "authorisation" of the new code.

"Hawaiian Place Names" by the Assistant Chief of Staff for Military Intelligence, Headquarters Hawaiian Department (Hayes 1929), is aimed at "attaining some degree of uniformity in the pronunciation" of Hawaiian place names. This list includes meanings of the names, though "inasmuch as the meanings are included only as a means of arousing interest in pronunciation, it is not believed necessary to include all the possible meanings of the various words." Hayes hopes that "this publication will … lead to contact between groups of army personnel interested in the subject and authorities of the Hawaiian language," which might further improve pronunciation. But it is dubious that these "authorities of the Hawaiian language" are Hawaiians themselves. The text is not concerned with contact between the military and native speakers. With selected meanings provided for curiosity only, the authority of meaning inherent in the place names is itself unimportant. Just pronounce them. Similarly, Jones and Addleman (1937, v) state of their dictionary that it "is in no way a scholarly treatise. It was not intended to be. The translated meaning of many of the place names will be questioned by Hawaiian scholars; they may even be amused …." The Hawaiian scholar (a *haole*) might be offended or amused, but how would a native Hawaiian react? They remain silent, invisible, outside of knowledge.

A different point comes from "A Gazetteer of the Territory of Hawaii," compiled by John Wesley Coulter (1935), PhD, of the University of Hawaii: that once a name is fixed on some "official source," that source becomes the new authority. Coulter's sources are maps[13] produced by US government agencies: the United States Geological Survey, the Territorial Survey, the survey of O'ahu by the United States Army, and the re-survey of O'ahu by the United States Geological Survey. Coulter states, "The names are listed exactly as they are spelled on the quadrangles and maps. No decisions have been made as to whether the names are correct" (Coulter 1935, 7–10). Yet correct or not, Coulter has produced a new authoritative directory. The identical procedure is employed in *Gazetteer (No. 4): Hawaiian Islands*, published by the Hydrographic office, United States Navy Department (1944). Data was compiled from the US Coast and Geodetic Survey (Department of Commerce), and "names were transcribed exactly as they appear on the source charts and maps" (Hydrographic Office 1944, iii).

Coulter's gazetteer demonstrates an active manipulation and assertion of authority, and the consequent pruning of the analogic Hawaiian landscape into a privatized, atomistic Western one. Echoing Waterman's difficulties in Puget Sound, Coulter (1935, 10) explains that

---

13 Harley (1988, 278) argues mapping reifies the conceptions of those who produce the maps and the relations of power which they represent. Though this study does not attempt a deconstructive history of cartography of Hawai'i, I affirm that control of naming is power, and mapping appropriates naming. See Fitzpatrick (1986) regarding the mapping of Hawai'i.

> The names of many geographical features in the islands known to the Hawaiians are not given in the gazetteer. There are in the office of the Surveyor of the Territory maps of parts of the Hawaiian island with *unimportant features identified by name*. The Hawaiians named outstanding cliffs, rocks, small streams and gulches, and some *trivial* land marks. Such names were thought *not to be of enough importance* to include in the work. Mr A.O. Burkland, during his supervision of the United States Geological Survey … chose the more important features to be named on the topographic sheets and the most authentic names. (Emphasis added)

A non-Hawaiian, US government official, through his selection of "importance" and "authenticity" weakens the cultural interface with the environment that Hawaiian place names served, and re-casts them into Western geographic discourse.

The paternalistic condescension towards Hawaiian people suggested here and characteristic during the Territorial period is important to understanding the disembodiment of Hawaiians from knowledge. Still another place names text exemplifies this attitude. T. Blake Clark, in an article on Honolulu's streets (1938, 5), painted the following picture of Hawaiians [*kanaka*]:

> Those kanakas were as busy as devils, pulling up fences, recklessly tearing down or setting matches to "homes" and "business houses" in the great new project of straightening Honolulu's streets. Like the cartoon firemen of today who rush into a man's house with axe drawn and leave his fine furniture a mangled wreck, these early Hawaiians took more delight in the destructive than in the constructive changes which they were busy making.

The "cartoon" depiction of the Hawaiians[14] reinforces the idea that Honolulu was built by and for Westerners, who constitute the progressive, productive members of society—in fact, who *are* the society. Clark places Hawaiians outside "society" again when he states that in one area of Honolulu, "there were few if any residences other than the grass houses of the Hawaiians"—a Hawaiian house apparently not qualifying as a residence (Clark 1938, 10). The social directory ("Blue Book") of that time is the clearest statement of "society" as *haole*. In this directory (Zillgitt and Snowden 1933), it is nearly impossible to locate a Hawaiian name amongst some 200 pages (see fn 17).

Clark describes the transformation of Honolulu from a dusty, disorganised assortment of structures to an organised urban centre with an orderly street

14 A different representational trope regarding Hawaiians during the Territorial period is a sort of recycled noble savagery, the "proud Hawaiian," described in very glowing terms (see, for example, Schnack 1915). But this picture invariably designates "ancient" Hawaiians, a noble past, a "they" not connected to Hawaiians of the present. This is part of the anti-conquest characteristic of the post-annexation period.

system. Pointing out that it had become practice to use Hawaiian words for street names,[15] he himself provides "good arguments against this practice:" Hawaiian names "sound alike" and "are often difficult to pronounce." Clark goes so far as to suggest using a simple numerically coded grid, such as that used in the Kaimukī section of Honolulu: "It would be simpler to extend such a uniform method of naming." Aside from this overt call for rectilinear order, which dates back to an 1832 missionary geography textbook,[16] Clark is again suggesting that Hawai'i is a place for English-speaking Americans, and that it is they who should encode this landscape.

The attempts to eschew ambiguity and authorize meaning can be almost comic. *Decisions on Names in Hawaii: Cumulative Decision List No. 5403*, by the United States Department of the Interior's Board on Geographic Names (US Board 1945), gives the reader the following instructions: "The underscored names are official for United States Government use. Where part of a decision is underscored, the use of the nonunderscored part is optional. Unapproved variant names and spellings ... are listed following the word 'Not' ... Former decisions no longer in force are listed without underscoring and marked 'Vacated'" (1954, 1).[17] Despite the attempt to establish a uniform code, the authors are clearly struggling with the variety and ambiguity characteristic of Hawaiian place names. The text eludes the question of Hawaiian authority altogether. It is a list of "decisions" on place names, determining what is "official" for government use. Whose decisions these are we do not know, except that it comes from the official source, the US DOI's Board on Geographic Names, and certifies what is "official" within the context of the US government.

Inasmuch as this insistence on fixed spellings and meanings is a component of print culture, the next technological-linguistic shift demands even further reduction of ambiguity. A Geographic Information Systems (GIS) print-out of Hawaiian place names and their locations in grid coordinates achieves this

15 First appointed in 1915, the seven-member Planning Commission (which, it seems, had the real power in naming streets) consisted of six *haoles* plus Prince Kūhiō Kalaniana'ole.

16 Woodbrige (1832). This geography [*hō'ike honua*] textbook was the first full length schoolbook produced by the Protestant mission, and can be seen as an attempt to rewrite the world into Euro-American capitalist terms for the young Hawaiian royalty (Daws 1968). It includes praise for the sensibility of the rectilinear organization of streets found in some American urban centres.

17 In 1965, Governor John A. Burns established the Hawaii State Board on Geographic Names in the Department of Planning and Economic Development. Place names as artifacts of Hawaiian environmental perception thus went into the hands of an organization with very different plans for the land. In 1968 the Board began a series of reports on "approved names." This function—and the Board—were transferred into the Department of Land and Natural Resources, where a new series of approved-name announcements began.

first by providing only one "approved" name, and second by the elimination of diacritical marks: use of the macron over long vowels (e.g., ē) and glottal stop (') which have arisen to flesh out the reductionist spelling of Hawaiian language produced by the missionaries.[18] These marks are now generally considered integral to Hawaiian language, and new street signs are required to have them. But such marks are not compatible with GIS. Nor is any ambiguity.

None of the above texts was authored or authorized by a native Hawaiian. Until Pukui et al.'s (1966) work, the authority to decide what is true, correct, and important laid in the hands of non Hawaiians, or at the very least, non native-speakers. The result is a non-Hawaiian sense of place, both within these texts, and through them, in the minds of those who rely on these texts as authoritative sources on Hawaiian place names.

The Language of Anti-Conquest

As Clark mentions, it became custom during the Territorial period to name streets with Hawaiian words.[19] This practice must be understood in the broader context of the political history of Hawaiian language. Literacy, introduced by the missionaries, had already weakened Hawaiian environmental discourse by creating more fixed meanings, by separating knowledge from direct experience, and by reducing the language into a literary by-product of its translation into English. But the final blow came with the replacement of Hawaiian altogether with English as the language of the Islands. The journal of Russian explorer Lisiansky foreshadowed this shift, presenting a vocabulary of Hawaiian words with the remark that "It is given more for curiosity than use, as there are several Europeans there, who may serve as interpreters; and, from the increasing civilization of the natives, the English language becomes better known to them every day" (Lisiansky 1814, 137).

By the mid-19th century, pro-annexationist discourse railed against the "almost universal rejection of the English language in the public schools, and the universal use of the Hawaiian in all clerical instruction of a public and

18 The glottal stop, or *okina* ('), is technically a diphthong: it constitutes a distinct sound whose inclusion or deletion can change the meaning of a word entirely. As such, it is now considered a *letter* in the Hawaiian alphabet. The macron (–), on the other hand, is more a guide to emphasis than a particular sound, but nonetheless can signify differences between words.

19 Ordinance 79-54 (City and County of Honolulu 1979) which made this practice official policy, states that "Street names selected shall consist of Hawaiian names, words, or phrases and shall be selected with a view to the appropriateness of the name to historic, cultural, scenic and topographical features of the area." Property owners along a street may petition for a change of name, but approval rests with the Director of Land Utilization.

private nature," and concluded that "the English language is the best medium, not only of commerce, but of civilization" (Bates 1854, 422–3).[20] With the influx of Westerners, a "universalizing" tongue such as English became "necessary" to cope with the internationalization of the Islands. The extent of this pressure to replace Hawaiian as the official language of the Kingdom is reflected in the *Penal Code of 1869*, Chapter XC, which states "Whenever there shall be found to exist any radical and irreconcilable difference between the English and Hawaiian version of any of the laws of the Kingdom … the English version shall be held binding." The same law was enacted by the Republic in their 1897 Penal Code.

With the impending death of Hawaiian sovereignty just prior to annexation, the alleged near-death of the Hawaiian language was seen as a positive indicator of progress by American travel writers: "The English language is almost universally taught" Musick wrote. "I was informed that there are but three Hawaiian schools now on the islands. The Hawaiian language, always narrow, is dying, and must in time yield to some more universal tongue" (Musick 1897, 442–7). Whitney was even more positive about the change: "There has always been manifest a native disposition to acquire familiarity with the English language" he wrote, stating that "at the wish of the natives themselves the language of the islands is English" (Whitney 1899, 248).

These remarks belie the health of Hawaiian language at the time of annexation. True, Kamehameha V had encouraged the teaching of English in Hawaiian schools. But there were nonetheless more than a hundred Hawaiian-language newspapers in print at the end of the 19th century. Hawaiians constituted one of the most, if not *the* most, literate populations of its time—in its own language (Kimura 1983, 189). Yet by this time, language also reflected class and power: Hawaiian was the language of low-order government service and the courts, low-order internal business, working-class jobs, and the subsistence life of the country districts; while English was the language of the high-paying, upper administration jobs and big business (Kimura 1983, 193).

It was the Republican government, after the overthrow of the Hawaiian monarchy, that brought an end to the widespread use of Hawaiian language. In the *Hawaii Session Laws of 1896* (Act 57, Section 30), the Department of Public Instruction is ordered to make "The English language … the medium and basis of instruction in all public and private schools" (though with provision for teaching another language, but not as the language of instruction), specifying further that "Any schools that shall not conform … shall not be recognized by the Department."

20 Originally all texts for teaching Hawaiians were in Hawaiian language. Missionaries felt that access to English language would also provide access to all the vices and immorality of Western culture. Conversely, they tried to keep their own children from learning Hawaiian, for exactly the same reason—that it would give them access to all the vices and immorality of a "heathen" culture.

Within two years after formal annexation, the last vestiges of institutional support for the Hawaiian language ended under the new American government.[21] The number of schools taught in Hawaiian dwindled to zero for lack of funding. For decades after the 1896 law, speaking Hawaiian was strictly forbidden anywhere on the school grounds, and physical punishment was often used on those who spoke it. In that psychological climate, especially during the WWII years and the drive to statehood, Hawaiians eager to prove their worth as equal citizens with the *haole* (and under social pressure to do so) abandoned the language that made them stand out as different. In that English-speaking, more racially loaded *American* environment, Hawaiian language had no value, and bad English was even worse.[22] To speak Hawaiian was to be a backward, ignorant person. Hawaiian language lingered only in certain pockets, such as particular families, hula schools, churches, and on the isolated island of Ni'ihau. Despite some grassroots efforts to perpetuate the language, by the 1980s the number of native speakers was estimated at less than two thousand.

This is the broader context of cultural colonization: the forcing of Hawaiian language to the brink of extinction. The decision to use Hawaiian words to name streets during the Territorial period, while at the same time closing the Hawaiian-language schools, is distinctly anti-conquest: it offers up a token of cultural respect at the same time that linguistic and cultural purging was being enacted. The Hawaiian language, like Hawaiian rule, was a thing of the past. Hawaiian place names became a commodity intended to develop a unique sense of place for the Islands, to differentiate them symbolically from the unifying Mainland cultural economy, and to create a local identity for the predominantly non-Hawaiian population. That virtually no one understands the place names attests to their anti-conquest positioning: without meaning, unable to speak. Their translated meanings (most of them often wrong) are presented, in texts such as the Armitage map, to portray a quaintness to the modern "Hawaiian" landscape.

---

21 Foreign language schools run by Asian (especially Japanese) immigrants to teach their children their ancestral languages became the main target of the Territorial government, which saw these schools as "if not distinctly anti-American … certainly un-American" (*Farrington vs. Tokushige* 1926, 4). In the early 1920s laws were enacted to regulate these schools, and aimed to see that "the Americanism of the students shall be promoted" (Department of Public Instruction 1922). Strict licensing and permits were required for schools and teachers, hours of instruction limited, and only texts issued by the Department of Public Education were to be used.

22 The rise of *pidgin*, or Hawaiian Creole English, from the plantation culture is itself a topic worthy of more discussion than is possible here. Local and mainland *haole* writers saw it as "bad English." Calling it "'da kine' plague," Michener (1963, 70–71) remarked that it is a "barbarous *lingua franca* derived from bad English, Hawaiian, Chinese, Portuguese and Japanese, all delivered in an incredible sing-song … it remains a damnable burden."

Not only county planners, but churches and developers have coined new Hawaiian place names.[23] Many such names, such as *Lanikai* [sea heaven]—a 1924 subdivision that should have been named *Kailani* [heavenly sea]—are inappropriately constructed by non-Hawaiian-speakers. At the same time, places charging admission prefer English names (Paradise Park, Polynesian Cultural Center), to hook into the minds of tourists who are already overwhelmed with all the "foreign" Hawaiian toponymy (Pukui et al. 1974, 243). Some popularized English names for Hawaiian places aim to capture the lost sense of the exotic, but do so in a way that has nothing to do with Hawaiian culture. "Sacred Falls" on O'ahu and "Seven Sacred Pools" in Hana, Maui are popular tourist sites where the unwary mainlander may seek to encounter a spiritual experience. But the Hawaiian names tell a different story: "Sacred Falls" is Ka-liu-wa'a, a name that likens the waterfall's appearance to a leak in the side of a canoe. The body of water known as Seven Sacred Pools is named Ohe'o (its meaning unknown), and was "sacred" only to the extent that it was *kapu* for menstruating women to bath in it.

The flood of Mainland Americans moving to the Islands since the late 1960s has resulted in the growth of subdivisions, the streets of which, by law, had to have Hawaiian names. Pukui et al. (1974) observe that "on the island of Hawai'i, streets in areas that are for sale along the Kona Coast have names tailored to the convenience of newly arrived visitors or of persons who have never been to Hawaii and are unable to master Hawaiian words ... Rarely is an effort made to search in *Indices of Awards* [from the *Mahele*] or in the State Archives for the ancient name of a place or for the names of the original Hawaiian owners, for fear such names might be hard to say or have unpoetic or risqué meanings" (1974, 243).

The greatest anti-conquest of all is the designation "The Aloha State." Vigorously employed by the tourism industry and stamped on Hawaiian license plates, this mix of English and Hawaiian terms reflects the American ideal of the Islands: it is a *state*—property of the US—and with a native culture that is gracious, warm, charming, welcoming. The appropriation of the term "aloha" to support tourism—the state's main industry—has been admirably discussed by Goss (1993). Seeing the phrase as implying "forgiveness for a forgotten history," Goss states (1993, 684) that the use of "*aloha*" is

> a means to reassure readers that, despite the historical oppression and the exploitative nature of the tourist-host relationship, the people will warmly welcome the visitor from the mainland United States. *Aloha* essentializes the Hawaiians as 'friendly natives', who greet tourists with the same guileless generosity and naive conviviality with which they are supposed to have

23 Pukui et al. (1974, 241–3) observe that "In only one domain, that of surfing sites, do the increments arise anonymously ...." More recently, names for skateboarding locales would be included here.

greeted Cook and Europeans who brought them, among other things, guns, disease, and economic exploitation.

The English-speaking population of the "Aloha State" appreciate an identity of "otherness" in their exotic new home with a place name code partially meaningful within Hawaiian culture itself, and partially meaningless. Seemingly honouring Hawaiian culture, this use of place names continues to serve the transformation of power and place that ensued with the discursive and economic re-ordering of the islands. But now, as Hawaiian language makes its resurgence, the backlash to the new politics of language points out that the cultural capital of Hawaiian language and place names is mutually exclusive with real Hawaiian power.

Re-conquest: The Adornment of the Land

Language emerges from the direct human interaction and relationship with the physical, social, and spiritual environment; it is, as Lopez (1986) puts it, a dialogue with the land. And as such, indigenous tongues speak a particular *language of place* that intimately and inextricably represents the existential experience of that place. It is a mode of environmental understanding that takes centuries to develop, and continues to evolve along with the advances in technology, diffusion of ideas, and changes in material culture.

I have suggested that literacy was an agent in breaking up an immediate relationship with the land and creating a new relationship mediated by written texts. But the loss of Hawaiian language goes much further, breaking up social and cultural relations as well, and imposing a language of an industrial society from someplace else. English language developed in a different climatic and geographical environment, and, especially as it modernized and Americanized before coming to dominate the Hawaiian Islands, does not have the sensitivity to the subtleties of light and colour, wind and rain, and emotive landscape that these islands present. The countless names for individual winds and rains and for conditions of the sea, the cultural link with the earth, the spiritual link with unseen but clearly felt forces—in all, a great body of geographic knowledge and an entire way of understanding that is indigenous and intertwined with this *place*—is lost with the peripheralization and near extinction of Hawaiian language. While Hawaiian place names preserve part of that geographic code, that code is meaningless to all but a few.

In 1966, the collaboration of Mary Kawena Pukui and Samuel H. Elbert provided the first authoritative work on Hawaiian place names that involved a person of Hawaiian ancestry. The considerably expanded Second Edition of *Place Names of Hawai'i* (Pukui et al. 1974) now serves as the authoritative text on Hawaiian place names, and is done in "dictionary" style with appended explanations that show Hawaiian place names as a geographic code, richly entwined with the culture. Sources of information include Hawaiian legends and mythology, archaeological studies, journals of early explorers, accounts of post-contact Hawaiian writers, and other literary sources, as well as the previous studies.

Differing in character from the previously discussed texts, this work may be said to reflect a more "Hawaiian" approach to place names. Aside from the usual range of geographic features and urban centres, this work includes locations such as surfing areas and Honolulu streets and buildings. Emphasising the vernacular as well as the "official" places of importance and their names, it has the additional quality of having been produced by native speakers and scholars of Hawaiian language. Budnick and Wise's *Honolulu Street Names* (1989) marks the completion of the turn-around. Subtitled "The Complete Guide to O'ahu Street Name Translations," this text focuses exclusively on *Hawaiian* names—Western family names and words are omitted—and includes an exhortation, in a foreword by Samuel Elbert, that "To pronounce Hawaiian street names as Hawaiians do is to respect Hawaiian culture, and to honor Hawaiians." But this text is not anti-conquest: it is re-conquest. Its elimination of Western street names from consideration or even mention, and its assertion that Hawaiian family names are on the land—thus highlighting the dispossession brought on by the Mahele—is part of the reclaiming of meaning engaged in the reclaiming of Hawaiian identity and land from the colonial past and present.

The relationship between control and order, as discussed earlier, predicts that the movement to genuinely revive Hawaiian language should meet resistance. This indeed is what occurs, and in doing so, points up the place name code as anti-conquest: it is "safe" for the colonizers as long as it remains silent—chatters, even sings, but does not speak. Exotic place names, and the beautiful (but incomprehensible) Hawaiian music that is an integral part of the tourism landscape of the "Aloha State," reassure the colonizer that through his control, peace and happiness exist—that the colonized state is a blissful one and the natives are happy. Any move to assert Hawaiian language as meaningful—of being logos—is political, and threatening. It threatens to yield the phallus—if not the land—back to the feminized Other.

At roughly the same time as the furore over Thurston Avenue, a related storm arose regarding the state-supported Hawaiian language immersion programme, *Pūnana Leo*.[24] This programme is designed to revive Hawaiian language as a mother tongue by placing young school children in a Hawaiian-

24 See Smyser (1991a, 1991b, 1992); Beamer (1991), Kamana and Wilson (1991), Keale (1992), and "Hawaiian immersion critic gets doused with objections" (1991). Hawaiian is one of two official languages in the State of Hawai'i, but as in New Zealand until recently, this official status did not translate into educational programmes. The *Pūnana Leo* ("Language Nest") programme began in 1984, based on the *Te Kohangareo* programme established in New Zealand/Aotearoa.

only teaching environment during the day, making Hawaiian the language of instruction. It was by no means an attempt to produce monolingual children; the overwhelming English-language social and cultural environment would ensure against that. Rather, this programme aimed to produce bilingual people fluent in Hawaiian as a *mother tongue* rather than as a second language.

In response, the contributing editor to the Honolulu *Star-Bulletin*, Bud Smyser, argued that raising children to speak Hawaiian as their first language would handicap them in regard to further education and employment. This socioeconomic argument itself retains the flavour of paternal colonialism. But Smyser's other fear was that teaching children in Hawaiian would breed separatism, similar to the situation of French in Canada. English creates the "melting pot" of American culture, he argued, adding that he himself would reject the admission of Puerto Rico to the Union as a Spanish-speaking state (1991a).

For refusing to allow Hawaiians this one small bit of rejuvenated culture, Smyser was strongly condemned for what many saw as advocating racism and cultural imperialism. But showing the efficacy of over 150 years of cultural imperialism, Hawaiians themselves fell on both sides of the argument. Billie Beamer (1991, A15) wrote, "This exclusive Hawaiian programme usurps the influential formative learning years ... History has taught us that due to the seclusion of 1,500 years, our one-dimensional stone-age ancestors were grievously overwhelmed and unable to cope with the flood of practices and ideas introduced by those from the exposed cultures." The language immersion programme, it was thought by some, would create a separate community within the Islands that possibly would have fewer opportunities because of inadequate English. But then, with Hawaiians as a group as the most socially and economically marginalized population in the Islands,[25] it was hard to see that there was anywhere to go but up.

Smyser's words barely masked his fear: fear that "our Hawai'i"—belonging to *us all* by virtue (in part) of English language—would no longer be ours, that allowing Hawaiians to reclaim their language was one step towards allowing them to reclaim the Islands themselves. Such is the fear of the colonizer: "Some of the advocates of immersion," he later stated, "are also advocates of Hawaiian sovereignty, and their tones are pretty strident" (Smyser 1991b, A14). The Thurston Avenue controversy, and the debate and comments that surrounded it, showed that indeed the politics of language, place names, and sovereignty are intertwined.

This linkage is further illuminated by the third relevant *resolution* to sail through this story: on 12 March 1997 the Honolulu City Council passed a

25 Constituting only twelve percent of the total population today, Hawaiians form 38% of the prison population, and proportionally have the highest percentages of homelessness and unemployment. They also have the highest proportion of health risks and the lowest life expectancy.

resolution that no longer required street names to be Hawaiian. Passing by a vote of 6:3, the one Hawaiian on the council voting in opposition to it, the bill was denounced before the Council by 13 persons testifying either as individuals or as representatives of Hawaiian organizations. No one testified in its favour. One person, recalling how her grandparents and great-grandparents were punished for speaking Hawaiian during class, remarked that eliminating the requirement for Hawaiian street names would send the wrong message and be a step backward ("Road-Naming ..." 1997). Charles Rose of the Association of Hawaiian Civic Clubs spoke of "the pain of being Hawaiian" when he went to Waikiki to find all signs in English and Japanese. At the same time, other Hawaiians pointed out that the legacy of intermarriage had left many Hawaiians without Hawaiian names. It should be possible, they argued, to name streets in honour of persons such as George Helm, Hawaiian activist and entertainer, who was lost at sea during a protest. *Star Bulletin* columnist Charles Meminger (1997) disagreed with the notion that attaching a Hawaiian name to something helps the culture:

> Every other cold, steel high-rise in Honolulu is named "Hale Something-or-other." Calling a gigantic empty overpriced condo building a "hale" [house] is really just a ploy by the developer to cash in on Hawaiian culture, not promote it ... Likewise, giving some little cul-de-sac in a sterile neighborhood of hurricane-bait cheapo townhouses a Hawaiian name doesn't really honor the Hawaiian culture. If anything, it's a cruel joke, considering how many Hawaiians have been waiting patiently for Hawaiian Homes lots for years. I'll bet a Hawaiian would rather get his promised lot and house on a street named Boardwalk or Park Place than live in a rented dump while everyone else moves into neighborhoods with lots of Hawaiian-named streets. In other words, arguing about giving streets strictly Hawaiian names only pays lip service to the real concerns of Hawaiians.

Meanwhile, the resurgence of Hawaiian language returns to the land in other ways. Hawaiian language classes at the University of Hawai'i at Mānoa are filled to capacity, having expanded from a couple of sections to more than 10, and with students wait-listed for admission. Hawaiian music, characterized primarily by use of Hawaiian language, has moved from being merely a charming backdrop to being an increasingly political vehicle for pro-sovereignty sentiments. When Hawaiian squatters took over Waimānalo Beach in 1994, their tent camps and taro patches were adorned with banners in Hawaiian language, or using Hawaiian words. The 1994 document returning the "target island" of Kaho'olawe to the state was written in Hawaiian, and the 1998 Aloha March on Washington, commemorating the Queen's 1894 visit to protest her overthrow, was accompanied by a traditional chant.

The 1976 voyage of the Hōkūle'a, a Polynesian voyaging canoe navigated from Hawai'i to Tahiti by traditional methods, and the subsequent construction of the even larger and more authentic Hawai'i-loa canoe (named for the legendary discoverer of the islands), have become foci for the regeneration of Hawaiian culture, providing examples of a technology which, for its time, was the most sophisticated in the world. Hula, once outlawed by the missionary-inspired government, then co-opted in a banal form for the tourism industry, has made a strong comeback among Hawaiian youth, and includes instruction in language and traditional culture. Hawaiian chanting, a nearly lost art that invokes the deities and the landscape, is slowly coming back into vogue. The Hawaiian language is fast becoming a calling-card for young Hawaiians seeking to recapture their own traditions and identity, while those of the older generation of Hawaiian activists sometimes regret with justifiable bitterness the loss of their mother tongue (Trask 1993).

The anti-conquest/re-conquest interface is captured in the Hawaiian-language motto of the State of Hawai'i: *Ua mau ke ea o ka 'āina i ka pono.* Not quite a "place name," but nonetheless the phrase by which Hawai'i represents itself, this sentence was spoken by King Kamehameha III after the British restored the sovereignty of the Islands—seized by Captain Paulet—to the Hawaiian monarchy. Admiral Thomas was sent to declare sovereignty restored, which he did at a place now known as Thomas Square. Missionary doctor-cum-politician Gerritt Judd translated the King's words as "The life of the land is perpetuated in righteousness." This remains the official and oft-quoted English version.

It is a lofty-sounded phrase. But what does it mean? I posed this question to students at the University of Hawai'i and received dumbfounded silence and glassy looks. The motto sounds good, it sounds *holy*. One looks at the sky while pondering its meaning. It is inscribed under the State Seal and is spouted by government officials. It is anti-conquest in the mouths of haoles and reconquest in the mouths of Hawaiians. But its meaning escapes into the ethers. Judd's translation hinges on the word *ea*, which he read as "life." But an alternative translation of *ea* is "sovereignty, rule, independence" (Pukui and Elbert 1986). With this simple change, the motto re-translated means "The sovereignty of the land has been restored, as it should be"—a phrase that makes perfect sense given the context in which it was uttered, and which is how it has always been understood by Hawaiian intelligentsia (Pi'ianāi'a 1987).[26] Few people in the Islands today are aware of this alternative meaning, and the traditional translation remains a banner for the state itself as well as for Hawaiian sovereignty groups. This misinterpretation *is perpetuated* since so few people can speak Hawaiian.

A modern Hawaiian proverb states, *O ka 'ōlelo Hawai'i ka wehi o ka 'āina*—"Hawaiian language is the adornment of the land." The loss of Hawaiian

---

26 "*Ua mau,*" at the beginning of the statement, suggests past tense, action completed—it *has been* restored—while "is perpetuated" would be "*e mau ana,*" indicating ongoing action. (Pi'ianāi'a n.d.).

language played a key role in dismantling Hawaiian geographic discourse, in the commodification of land and resources, and in the internationalisation of the Islands to the point of annexation. With the damage overwhelmingly done—an astoundingly American built-environment dominating the landscape—one wonders how much recapturing the language can do to revive traditional Hawaiian culture. But cultures change. While Hawaiian sovereignty movements are literally gaining ground, Hawaiian language is on the resurgence. Place names—the code of those who lost the struggle—are again becoming an intelligible vocabulary from which Hawaiian environmental discourse may re-emerge: a re-conquest, the adornment of the land.

References

Alexander, W.D. (1903), *Hawaiian Geographic Names*, US Coast and Geodetic Survey, Treasury Department Appendix No. 7—Report for 1902 (Washington: Government Printing Office).

Armitage, L. and Armitage, G.T. (1949), *Detailed Map and Guide of Honolulu and Oahu, with Map of the Hawaiian Islands* (Honolulu: Hawaiian Service).

Baker, R.J. (1950), *Sketches and Maps of Old Honolulu* (Honolulu: R.J. Baker).

Baldwin, C.W. (1908), *Geography of the Hawaiian Islands* (New York: American Book Company).

Barnes' Hawaiian Geography (n.d.), (New York: A.S. Barnes and Co).

Bates, G.W. (1854), *Sandwich Island Notes. By A Hāolé* (New York: Harper and Brothers, Publishers).

Beamer, B. (1991), "Program Could Produce Students Unprepared For Complex World," *Honolulu Star-Bulletin* 14 March.

Beckwith, M. (1970), *Hawaiian Mythology* (Honolulu: University of Hawaii Press).

Behdad, A. (1994), *Belated Travelers: Orientalism in the Age of Colonial Dissolution* (Durham, NC: Duke University Press).

Blaisdell, K. (1989), "'Hawaiian' vs. 'Kanaka Maoli' as Metaphors," *Hawaii Review* 13: 27,3): 77–9.

Blunt, A. and Rose, G. (eds) (1994), *Writing Women and Space: Colonial and Postcolonial Geographies* (New York: Guilford Press).

Bryan, W.A. (1915), *Natural History of Hawaii: Being an Account of the Hawaiian People, the Geology and Geography of the Islands, and the Native and Introduced Plants and Animals of the Group* (Honolulu: The Hawaiian Gazette Co. Ltd).

Budnick, R. and Wise, D.K. (1989), *Hawaiian Street Names* (Honolulu: Aloha Publishing).

Butler, J. (1993), *Bodies that Matter: on the Discoursive Limits of "Sex"* (London: Routledge).

Cheesman, P.R. and Cheesman, M.F. (1975), *Early America and the Polynesians* (Provo, UT: Promised Land Publications).

City and County of Honolulu (1979), Ordinance No. 79–54, Bill No. 46 Draft No. 3.

Cixous, H. (1981), "Castration or Decapitation?," *Signs: Journal of Women in Culture and Society* 7, 41–55.

Clark, T.B. (1938), "Honolulu's Streets," *Papers of the Hawaiian Historical Society* 20: 5–26.

Cocks, J. (1989), *The Oppositional Imagination: Feminism, Critique, and Political Theory* (London: Routledge).

Cosgrove, D.E. (1984), *Social Formation and Symbolic Landscape* (Totowa, NJ: Barnes and Noble).

—— and Daniels, S. (eds) (1988), *The Iconography of Landscape* (Cambridge: Cambridge University Press).

Coulter, J.W. (1935), *A Gazetteer of the Territory of Hawaii*, University of Hawaii Research Publications No. 11 (Honolulu: University of Hawaii).

—— and Serrao, A.G. (1932), "Manoa Valley, Honolulu: A Study in Economic and Social Geography," *The Bulletin of the Geographical Society of Philadelphia* 30: 2, 107–30.

Crocombe, R. (1990), "Naming and Claiming in the South Pacific," paper prepared for the South Pacific Place Names Conference, Wellington 5–7 November.

Davis, B. and Armitage, G.T. (1941), *Hawaii, USA* (New York: Frederick A. Stokes Co.).

Daws, G. (1968), *The Shoal of Time: A History of the Hawaiian Islands* (Honolulu: University of Hawaii Press).

Department of Public Instruction, Territory of Hawaii (1922), Regulations re Foreign Language Schools.

Dudley, M.K. (1986), "A Philosophical Analysis of Pre-European-Contact Hawaiian Thought," PhD disseration, University of Hawai'i.

—— (1990), *A Hawaiian Nation I: Man, Gods, and Nature* (Honolulu: Na Kane O Ka Malo Press).

Farrington vs. Tokushige. (1926), Petitioners' Brief, Supreme Court of the United States, October Term.

Fitzpatrick, G.L. (1986), *The Early Mapping of Hawaii, Volume I: Palapala'āina* (Honolulu: Editions Ltd).

Foucault, M. (1972), *The Archeology of Knowledge* (New York: Pantheon Books).

Freeman, O.W. (1927), *The Economic Geography of Hawaii* University of Hawaii, Research Publications 2 (Honolulu: University of Hawaii).

Goss, J.D. (1993), "Placing the Market and Marketing Place: Tourist Advertising of the Hawaiian Islands, 1972–92," *Environment and Planning D: Society and Space* 11, 663–88.

Grosz, E. (1990), *Jacques Lacan: A Feminist Introduction* (London: Routledge).

Guillemard, F.H.H. (1908), *Stanford's Compendium of Geography and Travel New Issue: Australasia. Vol II: Malaysia and the Pacific Archipelagoes*, 2nd edn, revised by A.H. Keane (London: Edward Stanford).

Hale, H. (1846), *Narrative of the United States Exploring Expedition During the Years 1838, 1839, 1840, 1841, 1842. Ethnography and Philology* (Philadelphia: Printed by C. Sherman. Reprinted 1970, Ridgewood, New Jersey: The Gregg Press).

Handy, E.S.C. and Handy, E.G. (1972), *Native Planters in Old Hawaii: Their Life, Lore, and Environment*, Bernice P. Bishop Museum Bulletin 233 (Honolulu: Bishop Museum Press).

—— et al. (1965) *Ancient Hawaiian Civilization* (Rutland: Charles E. Tuttle).

Harley, J.B (1988), "Maps, Knowledge, and Power," in Cosgrove, D. and Daniels, S. (eds), 277–312.

Hayes, C. (Assistant Chief of Staff for Military Intelligence) (1929), *Hawaiian Place Names* (Honolulu: Headquarters Hawaiian Department, Fort Shafter).

He Mau Manao Hamama I ka Poe Hawaii a me na Lahui Polinesia o ka Moana Pakipika (1922), Kakau ia e kekahi Lunakahiko o ka Ekalesia o Ieso Kristo o ka Poe Hoano o na La Hope nei (Honolulu: L.D.S. Hawaiian Mission).

Herman, R.D.K. (1988), "Mind and Earth: A Geography of Being," MA thesis, University of Hawai'i.

—— (1996), "The Dread Taboo, Human Sacrifice and Pearl Harbor," *The Contemporary Pacific* 8: 1, 81–126.

Honolulu Star-Bulletin (1991), "Hawaiian Immersion Critic Gets Doused With Objections," 5 March, A15.

Honolulu Star-Bulletin (1997), "Road-Naming Testimony Turns into a 2-Way Street," 13 March.

Hydrographic Office, United States Navy Department (1944), *Gazetteer No. 4: Hawaiian Islands* (Washington: United States Government Printing Office).

Jacobs, J.M. (1994), "Earth Honoring: Western Desires and Indigenous Knowledges," in Blunt, A. and Rose, G. (eds), 169–96.

Jarrett, L.H. (1930), "A Source Book in Hawaiian Geography," MA thesis, University of Hawaii.

Johnson, R.K. (1981), *Kumulipo: The Hawaiian Hymn of Creation* (Honolulu: Topgallant Publishing Co. Ltd).

—— (1983), "Native Hawaiian Religion," in *Native Hawaiian Study Commission: Report on the Culture, Needs, and Concerns of Native Hawaiians* (Washington: Native Hawaiian Study Commission).

—— (1987), personal communication, 5 March.

Jones, D. and Addleman, W.C. (1937), *Dictionary of Hawaiian Place Names*, 2nd edn 1942 (Honolulu: US Army Printing Plant).

Jones, S.B. (1938), "Geography and Politics in the Hawaiian Islands," *The Geographical Review* 27: 2, 193–213.

Kamakau, S.M. (1964), *Ka Po'e Kahiko: The People of Old*, trans. Mary Kawena Pukui (Honolulu: Bishop Museum Press).

—— (1976), *Na Hana a Ka Po'e Kahiko: The Works of the People of Old*, trans. Mary Kawena Pukui (Honolulu: Bishop Museum Press).

Kamana, K. and Wilson, W.H. (1991), "Immersion may save Hawaiian Tongue from Extinction," *Honolulu Star-Bulletin* 14 March.

Kame'eleihiwa, L. (1992), *Native Land and Foreign Desires* (Honolulu: Bishop Museum Press).

Kanahele, G.H.S. (1986), *Ku Kanaka Stand Tall: A Search for Hawaiian Values* (Honolulu: University of Hawaii Press and Waiaha Foundation).

Keale, M.K. Sr (1992), "Immersion Will Help Hawaiians Reclaim Culture," *Honolulu Star-Bulletin* 27 April.

Kimura, L.L. (1983), "The Hawaiian Language," in *Native Hawaiian Study Commission: Report on the Culture, Needs, and Concerns of Native Hawaiians* (Washington: Native Hawaiian Study Commission), 173–203.

Kirsch, P.V. (1985), *Feathered Gods and Fishhooks: An Introduction to Hawaiian Archaeology and Prehistory* (Honolulu: University of Hawai'i Press).

Lacan, J. (1977), *Ecrits* (New York: W.W. Norton).

Lindsey, J.H. (1947), "District and County Guide of the Territory of Hawaii," manuscript.

Lisianksy, U. (1814), *A Voyage Round the World in the Years 1803, 4,5, & 6* (London: Printed for John Booth, Duke Street, Portland Place; and Longman, Hurst, Rees, Orme, & Brown, Paternoster Row. Reprinted 1968, Ridgewood, NJ: The Gregg Press).

Lopez, B. (1986), *Arctic Dreams: Imagination and Desire in a Northern Landscape* (London: Picador/Pan Books).

Luomala, K. (1964), "Creative Processes in Hawaiian Use of Place Names in Chants," *Laographia* 22, 234–7.

Lydecker, R.C. (ed.) (1918), *Roster Legislatures of Hawaii 1841–1918* (Honolulu: Hawaiian Gazette).

Malo, D. (1951), *Hawaiian Antiquities*, Bishop Museum Special Publication 2, 2nd edn (Honolulu: Bishop Museum Press).

Menzies, A. (1920), *Hawaii Nei: 128 Years Ago*, ed. W.H. Wilson (Honolulu).

Michener, J.A. (1963), Background on the Fiftieth State. In "'Aloha' for the Fiftieth State," *New York Times Magazine*, 19 April 1959. Reprinted in *The New States: Alaska and Hawaii,* ed. W.P. Lineberry, 65–73, The Reference Shelf 35/ 5 (New York: H.W. Wilson).

Musick, J.R. 1897), *Hawaii … Our New Possessions* (New York: Funk and Wagnalls).

Nogelmeier, P. (1985), "Pūowaina," manuscript, Hawaiian Studies Institute, Honolulu, Kamehameha Schools/Bernice Pauahi Bishop Estate.

Nordhoff, C. (1874), *Northern California, Oregon, and the Sandwich Islands* (New York: Harper and Brothers).

Olivier, C. (1989), *Jocasta's Children: The Imprint of the Mother* (London: Routledge).

Ong, W.J. (1982), *Orality and Literacy: the Technologies of the Word* (New York: Methuen).

Penal Code of the Hawaiian Kingdom, compiled from the Penal Code of 1850 (1869) (Honolulu: Published by Authority).

Penal Laws of the Hawaiian Islands (1897), (Honolulu: Published by Authority).

Pi'ianāi'a, I.A. (n.d.), "The Expression of Place in Hawaiian Folk Songs," manuscript, University of Hawaii.

Pi'ianāi'a, A. (1987), personal communication, 6 February.

Portlock, N. (1789), *A Voyage Round the World; but More Particularly to the North-West Coast of America: Performed in 1785, 1786, 1787, and 1788, in the King George and the Queen Charlotte, Captains Portlock and Dixon* (London: Printed for J. Stockdale, and G. Goulding. Reprinted 1968, New York: De Capo Press).

Pratt, M.L. (1992), *Imperial Eyes: Travel Writing and Transculturation* (London: Routledge).

Pukui, M.K. and Elbert, S.H. (1986), *Hawaiian Dictionary: Revised and Enlarged Edition* (Honolulu: University of Hawaii Press).

Pukui, M.K. Elbert, S.H. and Mo'okini, E. (1974 [1966]), *Place Names of Hawaii*, 2nd edn (Honolulu: University of Hawaii Press).

Ralston, C. (1984), "Hawai'i, 1778–1854: Some Aspects of Maka'ainana Response to Rapid Cultural Change," *Journal of Pacific History* 19: 1, 21–40.

Sahlins, M. (1992), *Historical Ethnography, Vol. 1, Anahulu: The Anthropology of History in the Kingdom of Hawaii* ed. P.V. Kirsch (Chicago: University of Chicago Press).

Schnack, F.J.H. (1915), *The Aloha Guide: The Standard Handbook of Honolulu and the Hawaiian Islands* (Honolulu: privately published).

Shapiro, M.J. (1988), *The Politics of Representation: Writing Practices in Biography, Photography, and Policy Analysis* (Madison: University of Wisconsin Press).

Smyser, A.A. (1991a), "'Hawaiian Immersion' Idea May Divide Us," *Honolulu Star-Bulletin* 26 February, A14.

—— (1991b), "Questions Remain on Hawaiian Immersion," *Honolulu Star-Bulletin* 19 March, A14.

—— (1992), "Immersion May Fail a Careful Examination," *Honolulu Star-Bulletin* 24 March, A10.

Stewart, G.R. (1970), *American Place Names: A Concise and Selective Dictionary for the Continental United States of America* (New York: Oxford University Press).

Trask, H.K. (1993), *From a Native Daughter: Colonialism and Sovereignty in Hawai'i* (Monroe, ME: Common Courage Press).

Ulumaimalu, Nai'a (1998), "My Place in this World," *Holo I Mua, the Hawaiian Generations Magazine* 1: 2, 11–14.

United States Board on Geographic Names (1954), *Decisions on Names in Hawaii: Cumulative Decision List No. 5403* (Washington: Department of the Interior).

Valeri, V. (1985), *Kingship and Sacrifice: Ritual and Society in Ancient Hawaii* (Chicago: University of Chicago Press).

Vickers, M. (1993), "Council Hands Off Renaming Avenue to Folks in Makiki. Honlulu," *Honolulu Star-Bulletin* 11 March, A3.

Waite, D. (1993), "Thurston Ave. Caught in Overthrow Flap," *Honolulu Advertiser*, 10 February, A5.

Waterman. T.T. (1922), "The Geographical Names Used by the Indians of the Pacific Coast," *Geographical Review* 12: 2, 175–94.

Whitney, C. (1899), *Hawaiian-America* (New York: Harper and Brothers).

Wilkes, C. (1845), *Narrative of the United States Exploring Expedition During the Years 1838, 1839, 1840, 1841, 1842*, vols I and IV (Philadelphia: Lea and Blanchard. Reprinted 1970, Ridgewood, NJ: Gregg Press).

Wise, J.H. (1965), "The History of Land Ownership in Hawaii," in Handy, E.S.C. et al., 81–94.

Woodbridge, W.C. (1832), *He Hoikehonua: He Mea ia e Hoakaka'i I Ke Ano o Ka Honua Nei, a me Na Mea Maluna Iho* (Oahu: Mea Pai Palapala a Na Misionari).

Zillgitt, C. and Snowden, J.A. (1933), *The Blue Book of Hawaii Nei: a Society Directory of the Hawaiian Islands* (Honolulu: The Blue Book Publishing Co.).

Appendix

Glossary of Hawaiian terms:

Ahupua'a: land section usually extending from the uplands to sea
'Āina: land, as sustainer of life (*'āi* = to eat)
Akua: god, goddess, spirit; active consciousness, volition
Ali'i: person of the aristocratic or chiefly class
'Aumakua: familial or guardian spirit
Hale: house, building
Haole: white person, American, Englishman, Caucasian (formerly, any foreigner)
Hō'ike honua: Geography ("to show the earth")
Kaona: hidden meaning, concealed reference in poetry
Kapu: [taboo] sacred, forbidden; sacred law
Kinolau: "many bodies," physical manifestations of deities
Konohiki: headman of an *ahupua'a* land division, under a chief
Kuleana: land claims by commoners
Mahele: partition
Maka'āinana: commoner, populace (people that attend the land)
Mana: supernatural or divine power
Moku: large land sections; districts

Chapter 7
Irish Place Names: Post-colonial Locations

Catherine Nash

Introduction

In Ireland place names on official road signs appear in two forms, in the English language version which is the name most commonly used, and in the place name's original form in the Irish language. In Northern Ireland place names officially appear on road signs in English only. However, there are places in Northern Ireland where Irish versions of these names have been included in street signs, where local place names feature on roadside plaques and where place names in Ulster Scots are included on local road signs. These present-day patterns of naming reflect the long and contentious history of British colonization in Ireland, the significance of reinstating Irish place names after independence in Ireland and the contemporary politics of culture, identity and belonging in Northern Ireland. Official policies and popular approaches to place names are bound up with wider debates and conflicts about what counts as a collective past, the histories of settlement that are reflected in place names, the politicization of language, desires for cultural authenticity and arguments for cultural pluralism. In societies whose topomyies have been shaped by the power-laden relationships between different ethnic and linguistic groups, debates about place names are entangled with all the most difficult questions of cultural identity, ownership, rights and recognition; decisions about which place names are most appropriate or the recognition of different language versions reflect particular approaches to questions of cultural purity, cultural change and cultural diversity in society more widely.

This chapter explores the different ways place names in Ireland and Northern Ireland have been figured as part of a collective cultural heritage in relation to wider debates about identity and belonging. It builds on the work of those who have explored the powerfully symbolic and material assertion of meaning through place names in relation to capitalist modernization (Pred 1990, 1992), colonial settlement (Carter 1987), state formation (Cohen and Kliot 1992), nation building (Yeoh this volume), or official commemoration (Azaryahu 1996). The naming of places within European colonization has been read as a practice of cultural erasure and territorial appropriation (Brealey 1995). Getting rid of prominent colonial place names after independence asserts the value and validity of the culture denigrated under colonialism. However, the desire to recover, return to, and restore pre-colonial place names and by extension, pre-colonial culture more widely, has complex and contradictory implications. It challenges colonial cultural subordination, insists on the worth of those devalued cultural traditions and affirms a regenerated sense of collective cultural identity. Yet, the language of anti-colonial or national cultural purification that seeks to establish a fixed and singular collective heritage by erasing the evidence of colonization can leave little room for cultural pluralism. It can also imply that this reclaimed heritage is the rightful heritage of only those who qualify as being part of the native and national community of shared descent. Those who do not meet the criteria of being indigenous, also often strictly defined, neither fully belong in the nation nor can have any claim to share that national heritage. This is cultural nationalism at its most exclusive and exclusionary, but the obvious implications of its most extreme forms are implicit too in its more subtle versions. Official governmental policies and popular approaches to place names are entangled in these difficult questions of how to challenge the legacy of colonialism without recourse to a language of cultural purity and exclusive belonging.

This issue is central to societies dealing with the legacies of colonialism whose configurations of identity, cultural and belonging before and after national independence have been the subject of post-colonial studies. While this interdisciplinary and public critical engagement with colonialism highlights the issues common to societies that have been subject to European colonization, recent approaches have become more attentive to the specific dimensions of different post-colonial locations. This term, post-colonial locations, points both to the broad significance of questions of located identity and belonging—who belongs in a country and on what basis?—and the specific dimensions of these questions in particular places where ideas of being native, settler or immigrant have particular histories, meanings and implications. The post-colonial locations in question here are Ireland and Northern Ireland, places whose contemporary cultural and political forms have been profoundly shaped by British early modern colonialism, even if its discourses where shaped by Ireland's ambiguous place within the European and white "family" of nations. Considering questions of identity and belonging in Ireland and Northern Ireland through a post-colonial perspective points to the particular implications of ideas of being indigenous in these contexts. While the recognition of the indigenous rights and respect for indigenous cultures has been a deeply significant strand of social change and conflict in former European settler societies, the nationalist construction of true Irishness based on pre-colonial cultural purity in response to colonial denigration of the "native Irish," underlies the continued political and cultural tensions between those conventionally deemed "native" and "settler" in Northern Ireland—Catholic and Irish, or Protestant and British—and is both revived and revised in debates about the past and in response to social change in Ireland. This focus on the problematic nature of the category "native" in this context complicates accounts of place-name change which are critical of colonial naming practices but overlook the complex implications of discourses of cultural de-colonization.

This chapter explores the ways place names have been the focus of contrasting arguments about identity, culture and belonging in Ireland and Northern Ireland. It does so by firstly recounting the history of colonial place naming and anti-colonial and cultural nationalist approaches to place names in Ireland, and secondly by exploring a campaign to protect and celebrate local place names in Northern Ireland that began in the early 1970s and continues to insist on the significance of local place names as the shared heritage of all in Northern Ireland.

Irish Place Names: Colonial "Translation" and National De-colonization

The approach to Irish place names that was adopted as part of the mid-nineteenth-century British project to survey Ireland, by mapping the country and recording its place names and other features, is frequently referred to as a paradigmatic case of colonial cultural erosion. This project had been preceded by politically symbolic but less systematic practices of colonial naming. Early modern English interventions in the toponymy of Ireland were clearly part of the strategy of extending English authority in Ireland through plantation, military control and cultural repression. The most patent example of the ways in which colonial authority was inscribed in the land through naming was the use of the names Kingstown for Dún Laoghaire, Queenstown for Cóbh, King's County for Offaly and Queen's County for Laois. Their existence on maps of Ireland up until the early twentieth century was a legacy and reminder of early modern colonization. However, these changes were relatively minor compared to the toponymic transformation of the mid-nineteenth century. As part of the cartographic survey, place names in their Irish form were collected from local informants and converted to an English language version that was sometimes a direct translation, or substitution, but most often was an English form that bore some relation to the sound of the Irish version but not its meaning. These names became the officially sanctioned versions in colonial cartography and administration in Ireland.

There has been a sustained critique of this history of British colonial mapping and naming in Ireland over the last two decades, stimulated in part by Brian Friel's (1981) exploration of the collective cultural consequences of the colonial mapping and re-naming project for the local Irish speaking people subject to this transformation of their toponymic landscape in his play *Translations* of 1981. Mary Hamer (1989a, 1989b), for example, locates the decision to map Ireland in the political context following the act of Union of 1801 which dissolved the Protestant parliament in Dublin and returned authority over Ireland to Westminster. Along with the nineteenth-century pursuits of ethnology and philology, she argues, the cartographic survey could make Ireland knowable and governable and so secure the authority of the predominately Protestant landowners and British territorial claims over Ireland. "Masquerading as a process of systematic record," she argues, "the mapping of Ireland was a prolonged act of cultural displacement and textual processing, in which ancient place names and boundaries were incorporated and reinscribed" (1989a, 190).

The most contentious aspect of the project is the way in which place names were anglicized, replaced by English alternatives or simply mis-recorded as the field notebooks of the Ordnance Survey officers were filled with names transcribed from local people (Andrews 1975). As cartographic historian John Andrews argues, while translation has recently become "a metaphor symbolizing all the cultural mischief done by Englishmen in Ireland" (Andrews 1992, 11), translation was relatively rare. Irish names were altered, he argues, but mostly through the processes of "dictation" in which a non-Irish speaker recorded in English orthography a place name spoken by an Irish-speaker and substituted English words which partially matched the sound of the Irish place name elements but obviously not the meaning, converting for example, the place name Muine Beag, meaning "little thicket," for example, to Moneybeg with no literal meaning. Nevertheless, the process of effacing the collective narratives and local knowledges of folklore, mythology and history condensed in Irish place names and authorizing new largely meaningless derivative forms has been seen as a form of colonial cultural violence deeply tied to the late nineteenth-century decline of the Irish language. For the poet John Montague place names in Ireland are "shards of a lost tradition" which hint at shared social memory of a landscape whose collective meanings were part of a unifying, but now lost, repository of community knowledge—"The whole landscape a manuscript / We had lost the skill to read" (1982, 108).

It was this sense of cultural damage, or toponymic "degradation" as one early twentieth century critic put it ("An Irish C C," 1919), that informed the place name policies adopted by Irish cultural nationalists before and following the creation of the independent Irish Free State in 1921. Prior to independence the Gaelic League published two lists of the Gaelic versions of anglicized names in 1905 and 1911 (Laoide 1905, 1911), as part of its work to stem the decline of the Irish language. In his call for the de-anglicization of the Irish nation in 1892, the founder of the Gaelic League Douglas Hyde turned to the question of place names claiming that

> On the whole, our place names have been treated with about the same respect as if they were the names of a savage tribe which had never before been reduced to writing, and with about the same intelligence and contempt as vulgar English squatters treat the topographical nomenclature of the Red Indians ... I hope and trust that where it may be done without any great inconvenience a native Irish Government will be induced to provide

> for the restoration of our place names on something like a rational basis. (Ó Conaire 1986, 166–7)

After independence the work of the Gaelic League was eventually followed by official toponymic guides in 1937 and 1969 (Oifig an Phuist 1937, 1969). But even before independence the introduction of local government to Ireland in 1898 had allowed some urban local authorities to introduce place name policies in sympathy with Irish cultural nationalism. Dublin Corporation faced with "24 names of kings, queens and their families, 56 Lord Lieutenants, 96 nobles and other owners of property, various officials and celebrated persons" (Mac Mathúna 1992, 63) erected bilingual street nameplates in the first decade of the twentieth century, and the names Offaly, Laois and Dún Laoghaire were in use before they were officially sanctioned by the Free State (Ó Maolfabhail 1992, 18). In 1946 a place names commission, An Coimisiún Logainmneacha, was established by the Dublin government to advise on

> i. the investigation of the placenames of Ireland; ii. research in order to determine the correct original Irish forms of those names in so far as they can be established, and iii. the preparation of lists of those names, in their Irish forms, for publication and for official use. (Ó Maolfabhail 1992, 138)

A Placenames Branch was established by the Irish Ordnance Survey in 1955 to research and establish the correct Irish forms of geographical names. Though the first bilingual road signs in Ireland had been erected by the Automobile Association in 1937, the state assumed responsibility for road signs soon after and in 1952 it was decreed that all place names "shall be shown in Roman letters in Irish and English." In contrast, in Northern Ireland in keeping with the Unionist government's antipathy towards the Irish language the erection of bilingual signs was officially banned in 1947 and was still legally prohibited until 1995 (O'Reilly 1998).

In many ways the approach to place names adopted in Ireland after independence could be interpreted as an uncomplicated process of post-colonial cultural recovery. Yet, the idea of restoring pre-colonial place names, and pre-colonial culture more widely, enshrine an ideal of cultural purity that defines a collective heritage in the most narrow and exclusive terms. The existence of bilingual road signs in Ireland actually reflects a more complex post-colonial cultural politics. Though the drive of many strands of Irish cultural nationalism before and after independence was to rid Ireland of the corrupting cultural influence of colonialism and restore and protect a pure national Gaelic and, for some, Catholic culture, there has also been a parallel exploration of the meaning Irish nationhood and collective Irish identity that has been especially prominent over the last decades of the twentieth century and continues. This critical engagement with Irish national identity has addressed the challenge of asserting and celebrating cultural distinctiveness in ways which do not endorse a fixed, singular and exclusive account of that collective culture. How for example, can Catholicism, the Irish language and Gaelic traditions be envisaged as part of but not the strictly defining characteristics of Irishness and the only basis of belonging in Ireland? How can a commitment to holding on to cultural distinctiveness in the face of wider forces of change avoid a defensive idea of culture as ideally static and fixed? Arguments about place names in Ireland reflect these wider debates.

Rather than simply insist on the use of place names in their original Irish language forms, place name policy in Ireland effectively combines a commitment to the accuracy and authenticity of officially recognized Irish place names and a pragmatic approach to the widespread usage of their variously anglicized versions. For some at least, the existence of place names that reflect the colonialization of Ireland are part of the diverse heritage of the island rather than emblems of cultural corruption. At a seminar held in Dublin in 1992 to discuss place name policy organized by the Placenames Branch of the Irish Ordnance Survey and An Coimisiún Logainmneacha, the Irish historian Donnchadh Ó Corráin rejected the notion of somehow "ethnically cleansing" place names in Ireland. Renaming, he argued

> where it involves the removal of names that have become part of history, should not usually be undertaken. I do not for one moment suggest that the names of the chief governors of Ireland should be erased from the record or that there should be massive gaelicisation of non-Irish names. To my mind, such an undertaking is the work of the vandal—like that of the vandals who destroyed the Public Record of Ireland or dynamited Nelson's pillar in the name of republicanism of one kind or another. That kind of renaming would destroy the main strands of our history and corrupt the record of the past for ideology's sake. (Ó Corráin 1993, 37)

Citing evidence of "pre-Celtic, Gaelic, British, Old Norse, Norman French, Flemish and English names" (1993, 44) and the names introduced by the planters and "big-house" families, he suggests that "anyone who imagines so or may entertain any foolish ideas about purity of race is very quickly proved wrong by our placenames" (1993, 42). This rejection of a form of toponymic "cleansing" reflects recent moves to challenge the idea of national cultural purity. This has significant implications for the imagining of Irishness both in light of the complex history of settlement on the island and in relation to new patterns of immigration.

However, rejecting the idea of national racial or cultural purity in relation to place names does not mean relinquishing any claim that place names should bear some meaningful relationships to the places they name. Even before the massive building boom that has characterized the last ten years of economic prosperity, concerns were expressed about the names given to new housing estates in Ireland. In 1977 and 1986, the Southern Irish government issued circulars to local authorities, followed by guidelines in 1992, recommending that local authorities should try to ensure that these new names are historically linked to the area being developed and that traditional local names are used wherever possible. These recommendations responded to the increasing tendency of developers, keen to convey the status and prestige of their projects, to adopt English place name elements like Downs, Dene, Copse, Hurst or Spinney or arbitrarily use the names of trees in road and estate names like The Brambles or Ailesbury Oaks, in developments where, as Liam Mac Mathúna, an Irish language scholar, put it "cohorts of *Hadleighs*, *Hamptons*, *Westburys* and *Westminsters* team up with squads of *Closes, Copses, Downes* and *Mewses*" (Mac Mathúna 1992, 67–8; Ó Corráin 1992, 35). The issue is clearly to combine some sense of authentic connection with a place with a sense of historical complexity and cultural pluralism. In Ireland concerns about new place names reflect the rapid social changes that have accompanied recent economic modernization and globalization. In Northern Ireland, approaches to place names reflect attempts to move beyond exclusive and divisive models of culture and heritage as well as the continued significance of older categories of identity and difference.

Local Place Names, Identities and Belonging in Northern Ireland

In Northern Ireland the debates about culture, naming, place and language that Brian Friel's play prompted were also being stimulated from another direction. In 1973 the Post Office in Northern Ireland made the controversial decision to implement a system for the modernization of postal deliveries that had a particularly significant impact in rural areas. In order to use the newly-devised post code system, rural roads were named and houses allotted numbers along them. The person's name, house or farm name, townland name, and parish as basis of address were replaced by a house number, road name and nearest postal town. A campaign to challenge this system soon emerged whose focus was the implications of making townland names irrelevant in the new postal address system. Townlands are the smallest administrative unit of the land based on the traditional territorial division of the country into counties, baronies, parishes and townlands (Dallat 1991). The townland system is considered to be one of the "most distinguishing marks of Ireland and the term townland itself, while unknown in the homeland of the English language, is recognized in all parts of Ireland" (Ó Maolfabhail 1978, 3). As a subdivision of the civil parish, there are over 60,000 townlands in Ireland as a whole, ranging in size from 50 to 250 acres (Mac Aodha 1989; Ó Maolfabhail 1991, viii). Though the Post Office did not prohibit the use of townland names, it was felt that the scheme meant that their use was no longer necessary, recognized or legal. If they were to be redundant in addresses, it was feared that they would eventually be no longer used, spoken or remembered (Flanagan 1978). While reminiscent of the campaign by the Gaelic league to have Irish place names recognized by the Post Office at the beginning of the twentieth century, the campaign did not entail arguing for a return to the Irish forms of townland place names most of which are Gaelic in origin but exist now in English forms. The campaign did not seek to return them to their original forms but was an attempt to ensure that local names as they are currently known did not fall out of everyday use.

Resistance to the way the post code system was being implemented was mainly expressed through the seminars, lobbying, petitions and publicity organized by the Townlands Sub-committee of the Federation for Ulster Local Studies, an umbrella organization for many local history and community groups (Johnston 1977; Hume, 1989, 1990). In 1976 the Federation issued *The Post Office and Rural Addresses in Northern Ireland—A Federation Statement* which sought to increase public awareness of the damage being done through the inappropriate use of post codes and to channel resistance to the system (Canavan 1991, 56–7). More specifically, the Federation sought to revoke the power invested in the Post Office by the Local District Councils. On 13 October 1990, designated Townlands Day, a petition of almost ten thousand signatures in support of the Townlands Campaign was submitted to government (Hume 1991, 55).

The arguments for keeping the townland names alive were all about location, identity and cultural attachment, but the senses of identity and heritage expressed within the campaign did not fall into model of the "two communities" in Northern Ireland, one Catholic, Irish and nationalist and the other Protestant, British and Unionist each with completely separate senses of collective heritage. Instead, townland names were defined as part of the shared heritage of all people in Northern Ireland regardless of other differences of political affiliation or denomination. Saving these place names, it was argued, was "a task in which all Irishmen can unite regardless of political difference" (quoted in Dallat 1991, 9). The significance of this account of place names as a "cross-community" cultural heritage lies in its contrast with conventional accounts of belonging, identity and culture in Northern Ireland. Most of the townland names are derived from earlier names in the Irish language. Therefore these local place names were established centuries before the seventeenth-century colonial Plantation of Ulster. This means they could be claimed, along with the Irish language itself and Gaelic traditions, as the heritage of those who descend from "native" Gaelic stock.

Yet, while this model cultural ownership has a potent logic it is the product of the historical construction of collective identities in Ireland and their particular configuration in Ulster in the late nineteenth- and early twentieth century. The significance of Protestantism to the making of ideas of Britishness in British state formation had deep implications for the making of these collective identities in Ireland, for British antipathy to Catholicism helped shaped the coupling of anti-colonial Irish nationalism with Catholicism and Protestant senses of embattled difference in Ulster. The making of an Irish national identity defined in contrast to Protestant Britishness in turn informed the coupling of Protestantism, Unionism and Britishness in Ulster. Despite centuries of intermarriage between those initially descended from Gaelic inhabitants and Scottish and English settlers, the idea of two separate communities of descent in Ulster divided by differences of culture, religion and political persuasion developed in the late nineteenth century. For many people since, Protestant and Catholic alike, the Irish language and the pre-colonial Gaelic culture of Ireland is imagined as the heritage of those descended from the "native Irish." Irish place names and the Irish language itself have been used as emblems of cultural resistance and identity in a context in which expressions of Irishness were reviled by the Unionist dominated Northern Ireland government. While many argue that the Irish language is not the exclusive property of Catholic nationalists (Mac Póilin 1997), the erection of Irish street names and the re-naming of housing estates in west Belfast in the 1980s, along with the use of Irish in Republican murals and slogans, emphasized the link between the Irish language, Catholicism and Republicanism (O'Reilly 1998). These associations often reinforce the idea of the Irish language as an exclusive heritage and antagonist marker of difference. However, in spite of the politicization of the Irish language in Northern Ireland it was argued that the shared sense of the significance of townland names, whose origins mostly lie in earlier Irish versions, displaced or superceded political, class and cultural differences as people united against a shared threat.

This sense of common concern about the continued use and value of townland names was strongly linked to accounts of a shared rural culture and of the significance of local histories and local identities, both of which have been posited as alternatives to the model of the "two communities" and polarized perspectives on the past in Northern Ireland. The Townland Campaign was a feature of a much wider movement in Northern Ireland from the 1970s onwards to articulate the value of local history for providing enriching senses of collective identity and understanding (Nash 2005a). Arguments about the significance of shared rural traditions and the scale of the local as a focus of identification ran through this movement and the Townland Campaign. According to the campaign documentation, once evidence appeared of the decline and disuse of the townland name, protest was mobilized against the Post Office in a cross-community effort to save the nomenclature of the Irish landscape from a linear, street-based system designed for British towns and cities, and inappropriate in the country. Local place names, it was argued, were part of collective rural heritage that was shared across the conventional divisions of religion and political affiliation in Northern Ireland. For Pat Loughrey, the second great threat to the communal identity of rural Northern Ireland after mechanized agriculture was coming from the new post office address policy:

> The fact that country people do not live along roads was not allowed to impede the master plan. Linear addresses were imposed upon country communities who have always been known and had known themselves by spatial divisions [...] A generation of children are growing up who have never used their townland names in their addresses. A fundamental element in our identity is being lost.
>
> Townland names, like the landscape to which they relate, are precious records of the history, legends and mythology of their communities. For the countless generations who had no written literature, townland names became the index-cards upon which memories are stored. They are the pages of our literary landscape, the focus of our oral tradition. (Loughrey 1986, 211)

Those who argued for the significance of these local names to people's sense of belonging and identity did not argue that because they were original Gaelic Irish names that they really belong to Catholic nationalist people in Northern Ireland, or that they are the cultural heritage of the "native" population. While the Irish language has been caught up in arguments like this and used as a symbol of an exclusive nationalist culture, the argument put forward about these local names was that they belonged to all people in Northern Ireland.

Ideas of local attachments and identities were central also to these arguments about place names, identity and belonging in Northern Ireland. For the poet Seamus Heaney, writing in support of the campaign for the Federation for Ulster Local Studies, the local is precisely the scale of an identity that is located but un-antagonistic. Recalling the poet Patrick Kavanagh, Heaney writes of townland names as emblems of local knowledge and a "parochial imagination" that

> is never in any doubt about the social and artistic validity of its parish … Empowered within its own horizon, it looks out but does not necessarily look up to the metropolitan centres. Its impulses and possibilities abound within its boundaries but are not limited by them. It is self-sufficient but not self absorbed, capable of thought; undaunted, pristine, spontaneous; a corrective to the inflations of nationalism and the cringe of provincialism. (Heaney 1991, xi)

The townland name for Heaney

> connotes a totally uninsistent sense of difference, a freely espoused relation to an idiom and identity that are regional, authentic, uncoerced and acknowledged. It is a minimal but reliable shared possession, the kind of word that could be agreed upon in the beginning as a means to an end, the kind of word that could provide the right verbal foundation for talks about

> talks ... it is both gutsy and non-sectarian in an unself-congratulatory way. (Heaney 1991, xi)

Here townland names are pointers to authenticity and pluralism, to identity and dialogue. For Heaney, the Townland Campaign expressed the possibilities of recognizing shared concerns across conventional political divisions. Townland names stand as "unpolluted channels that remain open to the copiousness and multifariousness at the head of the past"; to lose them "is to hamper these bountiful recognitions and to attenuate the possibilities of a more informed and consequently more absolved future for everybody" (Heaney 1991, xii). Local place names were seen as part of the strong local identities that were common to all in Northern Ireland and indeed had the potential to be a positive alternative to conventional divisions of political and religious affiliation.

In addition to these arguments about townland place names as part of the shared rural heritage and local identities of all people in Northern Ireland, place names have also been the focus of efforts to challenge the conventional model of the "two communities" since they point not just to shared forms of cultural heritage across the "two communities" but are evidence of a more complex cultural heritage that cannot simply be contained in the model of "native Irish" and "Planter stock." Though most current townland names are based on original Irish versions, many place names also record the long histories of settlement of the island. For Tony Canavan, writing as the Development Officer for the Federation:

> townlands help tell the story of the settlement of Ireland from the Celtic peoples who established townlands, through the Norse, Normans, English, Scots and others who have settled in Ireland and left their mark on the landscape. Although predominantly Gaelic, townland names bear the distinctive influence of the different peoples who have become part of the Irish fabric. The continued use of townlands and their acceptance as their own by all sections of the community in Ulster makes them a unique and priceless element of our cultural heritage. [...] If we are seeking to heal the wounds that divide our society and to illustrate the richness of what we share, then townlands have a crucial role to play. For they are not only part of a past which we all share but are a living part of the present too. (Canavan 1991, 51–2)

This emphasis on cultural diversity has become a significant theme in attempts to address questions of identity and division in Northern Ireland and now runs through public discourse, government policy as well as community relations strategy where it was strongly articulated from the 1980s (Nash 2005b). Approaches to place names in Northern Ireland have both contributed to and reflect this development of a language of pluralism. Drawing on discourses of local history and cultural diversity within the government programme of "education for mutual understanding" (Canavan 1990; Dallat 1989, 26–7), townland names and maps have been enlisted in local history, inner city youth schemes and community reconciliation projects which emphasize the shared and diverse histories and cultures in the province (Doherty 1989). Place names have been explored in school projects in Northern Ireland in which the varied origins of place names in the region have been used to consider ideas of local history, identity and diversity (Bardon, 1991a, 1991b). Since 1987 the Northern Ireland Place-Name Project based in the Celtic Department, Queen's University, Belfast has worked to research the origin and meaning of place names in Northern Ireland (McKay 2004). Those involved recognize the cultural and historical significance of their efforts to reveal the complexities of place names and their linguistic, geographical, social and historical aspects" (Muhr 1990, 109; Muhr 1991, 25–6).

The Townlands Campaign has largely been a success in terms of changing governmental attitudes. Since 2001 it has been government policy in Northern Ireland to support and use townland names and this policy is approved by all the political parties (Dallat 2004). Though in most places the rural road names and house numbers that were established in implementing the post-code system still exist, postcodes can now be linked to local townland names, and their significance is emphasized both in the Ordnance Survey of Northern Ireland and in the new publicly accessible official postal address database *Pointer* which, working in conjunction with the Northern Ireland Place-Name Project, includes the original forms and meanings of modern place names. Work on celebrating and encouraging interest in local names and local histories still continues. The Northern Ireland Place-Name project launched a touring exhibition and booklet entitled Celebrating Ulster's Townlands in 1999 and runs a "Townlands in Art" competition for schools and adults groups to encourage interest in local place names and to foster strong senses of shared local attachment and belonging. In 2004 many of those involved in the original Townland Campaign came together in a conference organized by the Ulster Local History Trust in association with the Cavan-Monaghan Rural Development Co-operative Society to reiterate the significance of townland place names as a deeply potent symbols and sources of local history, identity and belonging (Duffy 2004; Johnston 2004; Loughrey 2004; Reid 2004) and as the focus of memory, family history and identification within the diaspora (Lambkin and Fitzgerald 2004).

However highlighting the ways townland place names were the inspiration for and focus of innovative and important explorations of constructive and potentially conciliatory approaches to identity and belonging is not to suggest that the Townland Campaign exemplifies the nature of identity politics in Northern Ireland. In part its significance lies in its contrast to the resilience of old categories of collective identity. Place names are also entangled in the emerging politics of culture in the post-conflict period in which ideas of culture and heritage are being mobilized in defensive and antagonistic as well as reflective and progressive ways. Local place names signs have been erected in Ulster Scots in parts of Antrim and Down as part of the wider turn to this tradition (Nic Craith 2001; Stapleton and Wilson 2004). This move is caught up in the wider politics of culture and language in Northern Ireland. Ulster Scots language and traditions can been seen as another strand of the shared cultural diversity of Ulster. They can also be taken to be the distinctive heritage of Protestants in Ulster and either incorporated into personal and collective reflections on alternative forms of belonging in Northern Ireland or incorporated into more antagonistic and competitive discourses of cultural rights in response to the apparent gains of Catholic nationalists. Combining language, history, culture and location, place names can be mobilized in the diverse ways in which configurations of identity and belonging are imagined and put to work. Place names have been incorporated into attempts to address division through ideas of shared and diverse traditions but place names also label a sectarian geography of residential segregation in Northern Ireland.

Conclusion

Approaches to place names are particularly telling indicators of wider concerns about and understanding of cultural purity, pluralism, heritage, authenticity and change. More specifically they demonstrate the implications of combining a critical approach to colonial cultural damage with an uncritical emphasis on cultural purity and authenticity in discourses and practices of decolonization. A simple narrative of colonial place naming and post-colonial re-naming fails to capture the complexity of post-colonial cultural politics. Yet avoiding a language of purity and fixity and exclusive national heritage in relation to place names, and culture more widely, does not mean simply discarding any claims to meaningful and appropriate place names. Similarly the problems of exclusive accounts of collective culture do not mean that ideas of belonging and heritage are necessarily regressive of divisive. Instead, as the examples from Ireland and Northern Ireland suggest, place names can be understood through alternative articulations of historical complexity and cultural diversity.

Acknowledgements

The chapter is an abridged and revised version of a journal article of the same name published in 1999 in the *Transactions of the Institute of British Geographers* 24: 4, 457–80. I am very grateful to Jani Vuolteenaho and Lawrence Berg for their invitation to include this shorter revised version here, and to Terhi Ainiala, Kaisa Rautio Helander and Jani Vuolteenaho who invited me to present it at the "Place names and identities in multicultural contexts" symposium held in Kárášjohka, Norway in 2006.

References

"An Irish C C," (1919), *Dysert-Diarmada; Or Irish Place-names their Beauty and their Degradation* (Dublin: M.H. Gill and Son).

Andrews, J.H. (1975), *A Paper Landscape: A History of the Ordnance Survey of Ireland* (Oxford: Clarendon Press).

—— (1992), "'More Suitable to the English Tongue:' The Cartography of Celtic Placenames," *Ulster Local Studies* xiv, 7–21.

Andrews, L. (1997), "The Very Dogs in Belfast Will Bark in Irish;" The Unionist Government and the Irish Language 1921–43" in Mac Póilin, A. (ed.), 49–94.

Azaryahu, M. (1996), "The Power of Commemorative Street Names," *Environment and Planning D: Society and Space* 14, 311–30.

Bardon, J. (1991a), *Placenames in the North of Ireland* (Belfast: Northern Ireland Centre for Learning Resources).

—— (1991b), *Investigating Placenames in Ulster: A Teacher's Guide* (Belfast: Northern Ireland Centre for Learning Resources).

Brealey, K.G. (1995), "Mapping Them 'Out:' Euro-Canadian Cartography and the Appropriation of the Nuxalk and Ts'ilhqot'in First Nations' Territories, 1793–1916," *Canadian Geographer* 39, 140–56.

Buckley, A.D. (ed.) (1998), *Symbols in Northern Ireland* (Belfast: Institute of Irish Studies, Queen's University of Belfast).

Canavan, T. (1990), "Local History—New Resources or New Responsibilities?," *Ulster Local Studies* 122, 58–61.

—— (1991), "Townlands in Ulster Today," in Canavan, T. (ed.) , 49–52.

—— (ed.) (1991), *Every Stoney Acre has a Name: A Celebration of the Townland in Ulster* (Belfast: Federation for Ulster Local Studies).

Carter, P. (1987), *The Road to Botany Bay: An Essay in Spatial History* (London: Faber and Faber).

Cohen, S.B. and Kliot, N. (1992), "Place-names in Israel's Ideological Struggle over the Administered Territories," *Annals of the Association of American Geographers* 82, 653–80.

Dallat, C. (1989), "Cultural Traditions in Northern Ireland," *Ulster Local Studies* 111, 26–7.

—— (1991), "Townlands—their Significance" in Canavan. T. (ed.), 3–10.

—— (2004), "Save our Townlands!," in Turner, B. (ed.), 39–42

Doherty, P.L. (1989), *Townland Maps* (Derry: Inner City Trust and Derry Youth and Community Workshop).

Duffy, P. (2004), "Townlands: territorial signatures of landholding and identity" in Turner, B. (ed.), 18–38.

Flanagan, D. (1978), "Placenames: A Matter of Identity," *Bulletin of the Ulster Place-Name Society* 1, 1–7.
Friel, B. (1981), "Translations," in *Selected Plays* (London: Faber and Faber).
Hamer, M. (1989a), "Putting Ireland on the Map," *Textual Practice* 3, 184–201.
—— (1989b), "The English Look of the Irish Map," *Circa* 46, 23–5.
Heaney, S. (1991), "Preface," in Canavan, T. (ed.), xi–xii.
Hill, M. and Barber, S. (eds) (1990), *Aspects of Irish Studies* (Belfast: Institute of Irish Studies, Queen's University of Belfast).
Hume, H. (1989), "Townlands Sub-Committee," *Ulster Local Studies* 111, 12.
—— (1990), "The Townlands Sub-Committee," *Ulster Local Studies* 121, 34–5.
—— (1991), "Epilogue," in Canavan, T. (ed.), 54–5.
Johnston, J. (1977), "The Townlands Sub-Committee," *Ulster Local Studies* 31, 29.
—— (2004), "Townlands and the Local Historian," in Turner, B. (ed.), 93–105.
Lambkin, B. and Fitzgerald, P. (2004), "Townland Diasporas," in Turner, B. (ed.), 88–92.
Laoide, S. (1905), *Post-sheanchas Containing the Provinces, Districts, Counties and Post-office Towns of Ireland, Béarla-Gaeilge/English-Irish* (Dublin: The Gaelic League).
—— (1911), *Post-sheanchas Containing the Provinces, Districts, Counties and Post-office Towns of Ireland, Gaeilge-Béarla/Irish-English* (Dublin: The Gaelic League).
Loughrey, P. (1986), "Communal Identity in Rural Northern Ireland," *Ulster Local Studies* 922, 205–11.
—— (2004), "The Heart's Townland," in Turner, B. (ed.), 9–17.
Mac Aodha, B.S. (1989), "The History and Nature of Irish Street Names," *Names* 37, 345–65.
Mac Mathúna, L. (1992), "Urban Placenames: Streets and Districts," in Ó Maolfabhail, A. (ed.), 53–70.
Mac Póilin, A. (1997), "Plus ça change: The Irish Language and Politics" in Mac Póilin, A. (ed.), 31–48.
—— (ed.) (1997), *The Irish Language in Northern Ireland* (Belfast: Ultach Trust).
McKay, P. (2004), "The Northern Ireland Place-Name Project," in Turner, B., 83–7.
Montague, J. (1982), *Selected Poems* (Oxford: Oxford University Press).
Muhr, K. (1990), "The Place-Name Research Project, Department of the Environment for Northern Ireland and the Celtic Department, Queen's University Belfast," in Hill, M. and Barber, S. (eds), 109–14.

—— (1991), "Studying Townland Names in Ulster: Suggestions for Local Investigation," in Canavan, T. (ed.), 25–9.

Nash, C. (2005a), "Local Histories in Northern Ireland," *History Workshop Journal* 60, 45–68.

—— (2005), "Equity, Diversity and Interdependence: Cultural Policy in Northern Ireland," *Antipode* 37: 2, 272–300.

Nic Craith. M. (2001), "Politicised Linguistic Consciousness: The Case of Ulster Scots," *Nations and Nationalism* 7: 1, 21–38.

Ó Conaire, B. (1986), *Douglas Hyde: Language Lore and Lyrics Essays and Lectures* (Dublin: Irish Academic Press).

Ó Corráin, D. (1992), "A Future for Irish Placenames" in Ó Maolfabhail, A. (ed.), 33–46.

Ó Maolfabhail, A. (1978), "Keep Townland Names Alive," *Ulster Local Studies* 32, 3–5.

—— (1991), "Introduction," *Liostaí Logainmneacha: Contae Lú/County Louth* (Baile Átha Cliath, Brainse Logainmneacha na Suribhéireachta Ordanaís).

—— (1992), "The Background and Present Role of the Placenames Branch of the Ordnance Survey" in Ó Maolfabhail, A. (ed.), 11–32.

—— (ed.) (1992), *The Placenames of Ireland in the Third Millennium: Proceedings of a Seminar 28th February 1992* (Dublin: Ordnance Survey for the Placenames Commission).

O'Reilly, C. (1998), "The Irish Language as Symbol: Visual Representations of Irish in Northern Ireland" in Buckley, A.D. (ed.), 43–62.

Oifig an Phuist (1937), *Eoluidhe an Phuist/Post Office Guide* (Dublin: Oifig an Phuist).

—— (1969), *Ainmneacha Gaeilge na mBailte Poist* (Dublin: Oifig an Phuist)

Pred, A. (1990), *Lost Words and Lost Worlds: Modernity and the Language of Everyday Life in Late Nineteenth-Century Stockholm* (Cambridge: Cambridge University Press).

—— (1992), "Languages of Everyday Practice and Resistance: Stockholm at the End of the Nineteenth Century" in Pred, A. and Watts, M.J. (eds), 118–54.

—— and Watts, M.J. (eds) (1992), *Reworking Modernity: Capitalism and Symbolic Dissent* (New Brunswick, NJ: Rutgers University Press).

Reid, B. (2004), "Labouring towards the Space to Belong: Identity, Locality and the Townland in Northern Ireland," in Turner, B. (ed.), 110–121.

Stapleton, K. and Wilson, J. (2004), "Ulster Scots Identity and Culture: The Missing Voices," *Identities: Global Studies in Culture and Power* 11: 4, 563–91.

Turner, B. (ed.) (2004), *The Heart's Townland: Marking Boundaries in Ulster* (Downpatrick: Ulster Local History Trust).

Yeoh, B. (1996), "Street-naming and Nation Building: Toponymic Inscriptions of Nationhood in Singapore," *Area* 28, 298–307.

Chapter 8
Proclaiming Place: Towards a Geography of Place Name Pronunciation

Robin Kearns and Lawrence Berg

Introduction

Place names are not only inscribed on maps and within the landscape on signs, they are also—and more frequently—spoken. In this chapter we argue that in certain times and places, the speaking of place names involves a proclamation of cultural politics. After critically reviewing a growing body of geographical work on place naming, we seek to extend this scholarship by exploring the role of pronunciation in constituting the socio-cultural landscape. In so doing, we aim to extend the scope of the cultural politics of place naming.

As Garth Andrew Myers (this volume, 86) argues, "Place-name study has largely languished in atheoretical caverns of geographical inquiry." This sentiment is echoed by Wilbur Zelinsky, perhaps the most sustained contributor to place name geography, who states that the study of names "… has barely advanced beyond its pioneering phase" (1997, 465). In addressing these concerns, we have previously argued that naming places is implicated in the ideological processes involved in forming what Benedict Anderson (1983) has called 'imagined communities' (Berg and Kearns this volume). In this chapter, which extends a growing body of work that critically examines place naming, we argue that even when widespread agreement on the name of a place prevails, *the way the name is pronounced reflects, and contributes to, the constitution of imagined communities.* In advancing this contention, we are in sympathy with Don Mitchell's (2000) contention that 'there is no such thing as culture.' It is our belief that while place names can be considered to be one of the material and symbolic artefacts of culture, from another vantage point they only become part of local cultural politics when spoken and thus inserted into the contested dynamics of place.

As evidence, we interpret selected media reports and published correspondence as data on the way place names are pronounced. We first review recent scholarship on place naming and situate this literature within the framework of cultural politics. We argue that there has historically been a largely atheoretical approach to toponymic classification and that this approach has effectively neglected struggles over the politics of place that often underlie naming processes. Only recently has a literature developed that critically examines questions of cultural politics and the power of place naming. Secondly, we focus attention on the issue of how names are communicated. In particular, we are interested in the way names are pronounced, a gap in understanding earlier identified by Zelinsky (1982). We follow recent explorations in cultural geography that have moved beyond the written and the visual to consider a wider sensory spectrum, including the influence of sound in constituting place (Smith 1994; Rodaway 1994). Section three examines how identity is narrated through vehicles such as pronunciation. Fourthly, we briefly sketch a background for our case study in terms of the (bi)cultural identity politics in New Zealand, a nation which shares with the United States and Canada "... a recency and transparency of the naming process" not encountered in European settings (Zelinsky 1982, 89). Section five profiles the issue of place name pronunciation in New Zealand and explores discourses reflecting conservative reactions to sympathetic pronunciations of place names in the print media. We end by reflecting on the ways that culture, place and identity are connected through variability in pronunciation, with sympathetic usages being associated with solidarity with indigenous struggle by Pakeha (white, of European ancestry), as well as with a reassertion of identity by Maori.

Naming and Power in the Landscape

We contend that the reinscription of place names on the map through historical research by toponomists has tended to overlook the struggles for place that underlie the actual socio-cultural processes involved in naming itself. One such struggle for place is the way in which names are pronounced. In this section, we address three questions:

i. why is place name study important?
ii. how has the literature dealt with this issue? and
iii. why is naming especially relevant in the case of New Zealand?

With respect to question one, geographers and others have collected and analysed the names of places to provide clues as to the historical and cultural heritage of places and regions. In *most* cases, such geographies have been descriptive, avoiding recognition that there is conflict over place naming. In her work on Britain, for instance, Margaret Gelling uses place names to "... provide vital evidence for dating and, indeed, for estimating the mixture of races" (1988, dust jacket). By analogy, therefore, names are seen by Gelling to be akin to the particulate matter within stratigraphic sequences that fascinate geomorphologists. The implication is that place names are evidence of innocent histories, rather than active contestations, a view reflected in an earlier volume by the same author titled *Place-names in the Landscape* (1984). It is our contention that rather than being entities *in* the landscape, names are a constitutive component of landscape itself. We believe Gelling's view of place names may be understandable in some states where there may be a relatively high level of consensus on issues of national identity. However, the linkage between place names and power is more readily apparent in a contested and postcolonial setting such as New Zealand.

In terms of question two, until recently the indisputable focus of much toponymic scholarship is the past and the view that a country's history is revealed in its place names (Rickard 1968). The place name literature has often avoided confronting the power/politics/place link in the issue of naming. However, there is a contemporary and growing literature that is beginning to engage in a more theoretically informed discussion of the cultural politics of place naming (all of the essays in this volume provide examples). Recently, for instance, there has been a recognition that place names mark "the spatiality of power relationships" (Myers this volume, 85). In another postcolonial setting, Douglas Herman links some toponymic studies to the predations of colonists:

> Like plants, animals, and people removed from their environment, these language specimens lost their significance as cultural economies of geographic meaning, becoming instead reduced to artefacts. (1999, 79)

While descriptively interesting work on place names among various linguistic groups has flourished among geographers (e.g. Jett 1997) and others (Callary 1997), until very recently "... toponomy ha[d] yet to be re-examined in light of human geography's recent philosophical and theoretical emphasis on power and language" (Myers 1996, 237). In an attempt to address this concern our recent analysis of the politics of naming places adopted a discourse-analytic approach to the contested naming of a number of localities on the Otago peninsula of New Zealand (Berg and Kearns this volume). In that paper we argued that the naming of places is an integral aspect of the relationship between place and the politics of identity, and that, to this extent, naming is a form of "norming." Given both the symbolism of names and their material qualities when concretized in the human landscape, we contended that they (potentially) provide normality and legitimacy to those who dominate the politics of place representation.

A number of recent studies by other geographers have lent support to this perspective on place names. Indeed, "... rather than treating toponyms as independent clues on environmental change, there is increasing interest in reading them within the wider social context" (Yeoh this volume, 72). In Israel, for instance, Saul Cohen and Nurit Kliot (1992) illustrate how the nation state selects place names for the administered territories in order to reinforce Zionist ideology (also see Azaryahu 1992). Maoz Azaryahu (1986, 1988, 1990, 1992, 1996) has undertaken perhaps the most sustained analysis of the politics of street naming, focussing most often on Berlin at various periods.

Like other forms of place naming, the naming of streets provides a means by which those in authority can insert "the official version of the past into the semiosphere" (Azaryahu 1990, 33). Elsewhere, Paul Carter (1987) observes analogous processes in the colonial Australian landscape in which naming "... produces a form of 'linguistic settlement' that *produces* places" (Berg and Kearns this volume, 19). Such intentionality is frequently gendered, with the colonial New Zealand landscape strewn with the memory of (British) male military figures such as Nelson, Marlborough and Wellington. Ron Crocombe (1991) also notes the gendered character of naming places in Pacific Island societies, arguing that the naming process is an attribute of masculinity, power and control, a theme more thoroughly interrogated by Herman (this volume) who examines the phallogocentrism of naming in Hawai'i.

Garth Myers (this volume) makes the point that Cohen and Kliot's work is silent on the possibility of Palestinian *resistance* to Israeli naming. His work on naming in Zanzibar considers ways in which groups constructed as Other create alternative toponomies and thereby resist official naming practices. Myers's invocation of the spatialities of power and language is a theme adopted by Brenda Yeoh (this volume) in her analysis of street-naming practices in Singapore. This greater attention to geographies of resistance (Pile and Keith 1997) seems to us to be one of the more promising directions in place name geographies, but one that can be developed beyond the conventional interest in the naming process itself. We contend that resistance to naming can occur on at least two levels: the creation and deployment of alternative names (see Myers this volume) and the use of alternative pronunciations for established names, a possibility we explore in the next section of the chapter. "Resistance" also needs to be thought not only in terms of the politics of recognition invoked by marginal groups, but also in terms of the resistances enacted by members of hegemonic groups in response to such politics (Morin and Berg 2001).

Our third question is why the matter of place names is especially relevant to the case of New Zealand. In a similar fashion to elsewhere in the western world, place name studies in Aotearoa/New Zealand have tended to consider Maori and Pakeha toponymies separately and descriptively. The most prominent compiler of Maori place names was George Graham (1926) who "... assiduously collected traditions and information remembered by the elders of the Tamaki tribes" (Simmons 1980, 11). More recent collectors have moved beyond the simple listing of names to categorizations and rudimentary statistical analyses (Griffiths 1988, 1990), a trend also evident among some geographers (see Zelinsky 1992, 1993). This type of work treats names as artefacts, something described as a "'pins on the map' method" and an "antiquated approach" (Fair 1997, 467). Perhaps reflecting the 'two worlds' (Salmond 1991) of Maori and Pakeha experience, Rickard's survey of *Historic Place Names of New Zealand* begins with the rather contradictory statement that "I have tried as far as possible to ... provide connecting links. Maori names have not been dealt with" (1968, i). In summary, New Zealand scholarship on place names has involved classifying items with little integration between Maori and Pakeha interpretations. With this in mind, and given the everyday character of contested cultural politics of identity in New Zealand, it seems apposite to study the production of identity inherent in the pronunciation of place names.

(Pro)claiming Place Names

Before proceeding further, we now consider the processes involved in the otherwise taken-for-granted process of (place)naming. As Zelinsky (1992, 171) remarks, the significance of place naming lies in an *initial* or *local* decision with "... conscious deliberation and an encounter with something novel." Zelinsky's comment refers to the origins, rather than the ongoing use of, a name. We contend that, through the *performance* of speaking, naming is appropriated into the frequently contested contemporary context.

People enunciate names, and these acts have an inherent flexibility that surpasses the more static character of the written word. According to Entrikin (1991, 55), "in ... everyday conversation ... place names have a semantic depth that extends beyond the concern with simple reference to location ...." This comment provides some leads as to how we might conceptualize the role of speech in articulating the politics of place naming. Speech is performed in time and place, structured within everyday settings. In the language of Anthony Giddens (1984), speech is thus part of the structuration of everyday life.

Locating place name pronunciation in the banal routines of everyday life offers an approach that can be interestingly entwined with Butler's (1990, 1993, 1995) notion of performativity. Performativity is defined not as a singular 'act," but as "the reiterative and citational practice by which discourse produces the effects that it names" (Butler 1993, 2). Drawing on this sense of the performative, we can see that the specific pronunciation of names can be more than a descriptive act. Instead, the performative utternace of names must be seen as a constitutive act—one that instantiates specific meanings and implicit politics in everyday interactions between people and place (names).

The act of pronouncing names also involves *strategy:* the conscious choice of *how* a name is pronounced. Our principal interest in this chapter is in the ways that pronunciation can be deployed to assert a politics of representation. John Robb implicitly suggests this possibility by saying that "... toponomy is a creative process, but not just in the original acts of naming" (1996, 170). While he specifically refers to the selectivity employed by collectors and classifiers of names, we can build on this remark to signal the creativity employed by speakers in their choice of *how* to (pro)claim a place. In developing this possibility, we can draw on Giddens's (1984) distinction between speech and language.

Speech, he argues, occurs within the particularities of time and place. In this sense, it can be seen as an embodied and emplaced activity: speech literally takes place. Language, on the other hand, is less place-bound than speech. It is not the product of any one speaker, but instead is both a product of, and resource for, a multitude of speakers. Giddens's conceptualization offers an important pointer to the significance of speech in the (re)production of place. For place naming analyses have tended to focus exclusively on the written representation of names, relying on a model of naming as "language." Written place names are seen as records of the naming process that endure beyond the specificities of time and place. In writing, names become formalized within the cartographies and bureaucracies of the state. In this way, place names are mapped into official discourse and they become part of what Lefebvre (1991) refers to as *representations of space,* a form of spatialized bureaucratic rationality that colonizes everyday life (also see Simonsen 1996).

We contend that pronunciation can potentially be a form of resistance. First, this is because speaking, in almost all circumstances, is a social act. As Mitchell (2000 151) says "Resistance can't really be private. What would be the point?" Place names are publicly pronounced, and there is thus scope to not only *identify* a point on the map, but also to *make* a point through (metaphorically) mapping out one's politics of place in speech.

To develop this contention, we can build on the ideas of Michel de Certeau (1984) who writes of walking as an example of the way the banal undertakings of everyday life can be acts of resistance " ... to domineering institutions that seek to order our lives" (Mitchell 2000, 151). Like walking, speaking in vernacular idiom may also be foreign to the more formalized language of public agencies, creating "shadows and ambiguities within them" (1984, 101). For de Certeau, everyday practices such as speaking are *tactics* that are small-scale, diverse and "an art of the weak" (1984, 37). Tactics can be contrasted with *strategies* that are the domain of the powerful and among other things, serve as the base from which relations can be managed (de Certeau, 1984). Like de Certeau's example of walking, so too talking and, in particular, acts of pronunciation create narratives that are, or can be, subversive to official policies and narratives. Tactics begin "below radar," but may, as in the case of Maori pronunciation in New Zealand, transform the strategies of the state itself. Mitchell (2000, 153) takes issue with de Certeau in suggesting that resistance can only be achieved when a collective *social* movement (rather than a collection of *individual* movements) occurs. Power, according to Mitchell, is made possible by visibility. Thus social movements must engage in strategic resistance that may, beyond individual tactics, involve challenging and transforming institutional strategy (e.g. broadcasting policy) around the banal activities of everyday life (e.g. pronouncing place names).

Our foregoing argument suggests that a focus on the written in place name geography ignores an important subset of the cultural politics of place, namely the everyday (re)production and contestation of place naming through the spoken word. As Garth Myers (this volume) states, "There is simply no way to appreciate the performance aspects of names without hearing them called, played with as tactics of empowerment, or used as vehicles of derision." Here we are calling for an analysis of the *spaces of representation* (Lefebvre 1991), of the banal geographies of everyday life rather than of the official geographies inscribed in archives, maps, road signs and state documents which, to follow Le Febvre, constitute an example of the *representation of spaces*. Although in Aotearoa/New Zealand, the distinction between official discourse and everyday life is sometimes unclear, we wish to focus our analysis on a set of processes qualitatively different to those involving bureaucratic naming processes and discussed by Berg and Kearns (this volume) and Stokes (1994). We wish to move, metaphorically, from the map to the moment and read the resistance within speech as a creative and assertive act that can operate independently from (and contradictorily to) the forces of hegemony and dominant discourse. We contend that within performance, there is room for resistance and the creation of alternative realities. Thus, in the following sections we sketch our research on place-name pronunciation in Aotearoa. In so doing, we aim to advance a theoretical understanding of how not only place name geographies, but also the geographies of pronouncing place names, point to the multiple ways that space is spoken and places are produced.

Naming and Narrating Identity

The importance of speech and language in place-making and place naming cannot be underestimated. According to Allan Pred (1989) "no worlds exist without words or other symbols" and "all the forms of social life that constitute the world are enabled and sustained by language." A useful account of how we constitute the world and narrate identities has been developed by Graham Dawson (1994) who proposes that we engage in the act of composing stories (narratives) through which we tell ourselves and others about the day's events, our relationships, and our evolving concept of Self. In other words, the Self is formed in relation to the world and through this is a psychological process of narrating identity. Every day, taken-for-granted narratives about place—and, by extension place-names—are composed in ways that interact with hegemonic readings and understandings of the politics of location.

The pronunciation and composition of place-names involves alternatively a process of learning or selection, whereby the subject either mimics 'received' practices or actively chooses different phonetic emphases and intonations to produce an alternative coherent whole. If the former, the composer might highlight some phonetic aspects at the expense of others, and interpret the significance of these pronunciations within the context of hegemonic socio-spatial relations. In so doing, a person constructs not only a place-name, but also a position for the Self within the politics of naming places. This is a complex psychological process that allows the Self—sometimes working unknowingly within and sometimes working knowingly against hegemonic discourses—to make sense of the events related to naming and claiming places so that their troubling, disturbing aspects may be managed, repressed, contained, or similarly worked through (after Dawson 1994). In this way, narratives of place in turn and inevitably become bound up with the storyteller's own—perhaps implicit—fantasies about desired and feared outcomes in terms of nationhood. More specifically, the pronunciations deployed within Aotearoa/New Zealand potentially signal a politics of inclusion or exclusion (Sibley 1997) in which Maori are included as Treaty partners, or excluded as the unsettling Other. Importantly, the story that gets told is always the one preferred amongst other possible versions. Further, the story will involve a striving, not only for a formally satisfying narrative or a coherent version of events, but also for a version of the Self that can be lived with in relative psychic comfort (Dawson 1994, 23).

Place narrative is also an inevitably social process, inasmuch as there is frequently an audience in mind (be it "real" or "imagined") and the speaker is searching for certain responses from listeners. We position ourselves in "society" through speech, accepting or rejecting certain shared values. Such "choices" are inevitably constituted in articulation with hegemonic social relations and discourses of identity, whether the individuals making them know it or not. Resistance to Maori pronunciations can be both political (part of a collective viewpoint) as well as psychological (in which the listener prefers not to have her/his psychic comfort disturbed). These two possibilities are, we contend, closely connected, with an aversion to both linguistic change and political issues often experienced. While some conservative speakers passively recite learnt and largely reflexive pronunciations, other speakers are more actively searching for certain kinds of responses, or seeking to endorse politicized positions, when they pronounce places.

Drawing upon the multiple meanings of "composition," Dawson extends the implications of narrating identity to everyday life. Narration is a deeply embedded cultural practice—all of us engage in the act of composing stories (narratives) that we tell ourselves and others about the day's events, our relationships and our evolving concept of self. Everyday, taken for granted narratives of place—and thus place names—are composed in ways that interact with hegemonic readings and understandings of the politics of location.

As narrative constructions, place names are heavily influenced by dominant, or hegemonic, cultural forms. These forms constitute the set of values and discursive practices that define what it means to be "in place." Drawing on the thinking of John Eyles (1985), place name pronunciation speaks to the recursive relationship between sense of place and our place-in-the-world. Yet the demands placed on individuals by these hegemonic cultural norms are not uniform. In fact, in their articulation with different (and sometimes contradictory) aspects of social life, they are more often than not fractured and contradictory. The achievement of a unified place identity is thus a virtual impossibility. Nonetheless, there does exist a hegemonic discourse of community that defines a narrow set of racialized, gendered, sexualized, able-bodied and classed possibilities for people who wish to fit in with the dominant understandings of place and the politics of identity. Narration of identity as a general process allows people to locate themselves imaginatively within complex social worlds. As imagined forms, place names are thus 'made up' through hegemonic processes; they also materialize in the social world as structured forms with real effects (see Berg 1999).

Placing Names in the Identity Politics of Aotearoa

A brief overview of the (bi)cultural politics of place in Aotearoa/New Zealand is an important precursor to considering contemporary issues of pronunciation. The current relationship between Maori and Pakeha arises out of Britain's sustained colonization of New Zealand in the 1800s and the active opposition to this process by some Maori. The British Crown and (some) Maori chiefs signed the Treaty of Waitangi in 1840, guaranteeing *te tino rangitiratanga*, or Maori sovereignty, and *mana whenua*, or control of land (Belich 1986; Orange 1987; Walker 1990, 1992). According to Nairn and McCreanor (1991) a "collective forgetting" of the Treaty soon beset British settlers, allowing for a British Act of Parliament to establish 'responsible' settler government without Maori representation. British desire to acquire Maori land drove them to armed conflict in order to establish their dominance (Belich 1986). While the British troops were never able to inflict any decisive military defeat, they managed to grind Maori resources down to the point where they were unable to organize effective and sustained resistance to Pakeha colonization (Nairn and McCreanor 1991). More than 1.6 million acres of tribal lands were confiscated following the New Zealand Wars, ostensibly as punishment for Maori "rebellion" against the Crown (see e.g. Belich 1986; Orange 1987; Walker 1990). A more successful instrument for divesting Maori of their lands in the North Island involved the Native Land Court system, implemented in the 1860s. Between 1840 and 1911, the proportion of New Zealand land controlled by Maori dropped from 100 percent to just 11 percent (Pawson 1992; also see Orange 1987; Walker 1990). Maori population decline was equally dramatic, dropping from about 80,000 in 1840 to just 41,993 individuals in 1891. Conversely, the European settler population increased from a few hundred individuals in 1840 to more than 770,000 in 1901 (Pool 1991).

Some Maori have struggled against the state (and against international capital) for recognition of their rights since the early colonial era. In more recent times, this struggle has been reflected in a resurgence of Maori cultural forms, and increasing calls for self-determination (see, e.g. Awatere 1984; Walker 1990). The state and Pakeha New Zealanders have responded with a number of concessions (Barber 1989), along with a few significant ones such as formation of the Waitangi Tribunal. The Tribunal is a standing commission of inquiry created in 1975 to examine Maori Treaty grievances. It was initially empowered to examine only claims dating from 1975, but the *Treaty of Waitangi Amendment Act, 1985* allows it to adjudicate claims dating from 1840 (Stokes 1992). Maori have won a number of significant claims put to the Waitangi Tribunal, and this has led to government recognition of land rights, allocation of greater fishing resources to Maori, and the establishment of numerous statutory bodies to deal with Maori concerns (Walker 1990, 1992). These victories were, however, tempered by the New Right monetarist policies implemented during the period 1884–1999 by both Labour and National governments (Le Heron and Pawson 1996; Kelsey 1995).

The recent politics of restitution and assertion of Maori identity remains unpopular with a major fraction of the Pakeha population, and this unpopularity is mapped into the politics of language as a whole in New Zealand. Contemporaneous with language politics and the renegotiation of actual ownership of resources, there is contest over the *symbolic* ownership of place. At key sites, Maori have asserted an appropriation of the landscape. For instance, at Mangakiekie/One Tree Hill, a prominent vantage point in the heart of Auckland where repeated assaults on the eponymous lone pine tree (an exotic species that symbolized colonial power relations) led to its demise (Kearns and Collins 2000). In sum, our argument derives from within a context in which Maori are successfully claiming large tracts of land and have had their language and economic rights upheld for the first time in 150 years. Cumulatively, this situation presents a threatening and uncertain environment for members of the status quo. Thus, pronunciation of Maori place names in public situations such as national radio and television touches on an acute question for many New Zealanders*: whose place is this?*

We claim that rather than there being scientifically correct pronunciations, there might be politically correct versions of pronunciation which are supportive of the aspirations of Maori people. In this way, we can recognize a diversity of pronunciations that signify support for decolonizing Aotearoa/New Zealand. This process can therefore be seen as a political strategy that recognizes and affirms diversity and difference in Aotearoa—an approach that appears to threaten the stability and "harmony" of the hegemonic identity politics drawn upon by many Pakeha New Zealanders.

Our rather stark representation of the "divide" between Maori and Pakeha must be tempered somewhat with a recognition of the hybridity and creolization (Bhabha 1994; Young 1985) of these groups within Aotearoa/New Zealand. Indeed, debate over the nature of categories *Maori* and *Pakeha* has led to changes in census definitions (Gould 1992). However, in general, the spatialized politics of Aotearoa/New Zealand have remained locked in vantage points born out of colonialism. Thus, both epistemological spaces and material places have been conceptualized with an overriding binary discourse of "Maori" and "Pakeha"—a dualism formed by a specific type of Orientalism that developed within British colonialism in Aotearoa/New Zealand (McClean et al. 1997). Assimilationist and integrationist ideas have, since the 1970s, been joined by a discourse of bi-culturalism which—as with "multiculturalism" in other national settings (Kobayashi 1993)—has tended to reinforce the hierarchies of the colonial Maori/Pakeha binary. We argue that this underlying identity logic of biculturalism needs to be re-thought as "both/and," rather than its current "either/or." Thus, it is important to remember that Maori *and* Pakeha can be found on both sides of the colonial/anti-colonial divide. Nonetheless, while some Pakeha actively support Maori people in their quest for recognition of their rights, there still remains a large proportion of Pakeha for whom Maori rights are a significant source of anxiety. The choice by Pakeha—especially broadcasters—to pronounce place names using traditional phonetic inflections, sympathetic to Maori speech, is therefore one source of contemporary anxiety among some fellow European New Zealanders.

Pronunciation of Maori Place Names

> It is time, or rather well past the time, that Tauiwi [non-indigenous New Zealanders] took a leaf out of its own book and made the effort to pronounce Maori names and place names properly. (Te Rauroha Mokai Tamehana, letter to the *New Zealand Herald*, 29 June 1994)

> Oh-ah-maru [Oamaru], as we're calling it these days. (Brian Edwards, New Zealand National Radio, 10 August 1996)

We see the politics of pronunciation as derivative of a wider politics of naming that we have previously explored with respect to the South Island of New Zealand (Berg and Kearns this volume), but is equally prevalent in North Island localities. For instance, the re-naming of the electoral seat of Whanganui (formerly spelled "Wanganui") led to a "staggering proportion" of re-enrolment cards being returned with the "h" struck out and obscenities written on them (*New Zealand Herald*, 5 August 1996, 3). Actual and proposed name changes are not without their Pakeha defenders. A correspondent to the *New Zealand Herald* writes of other readers seeming to be "increasingly hysterical at the suggestion that the volcanoes of Auckland resume their original Maori names. Some correspondents decry this as racist and warn of ethnic minorities running amok naming mountains (and molehills) at random" (PM, letter to the Editor, *New Zealand Herald*, 15 March 1994).

We now survey the underpinnings of pronunciation within Aotearoa/New Zealand. At the outset, it is important to signal a fundamental difference within the constituent cultures of New Zealand: for Maori, speaking is traditionally privileged and writing is secondary. The opposite tends to be the case with western European cultures, and this issue can be witnessed in the fact that we are publishing this study whereas many Maori might be more likely to debate the topic of our chapter on a *marae* (communal meeting place) or on radio. We also acknowledge the difficulties for anyone attempting to pronounce Maori words, whether they are native speakers or not. Complex compound words, such as "Kahukuranui"—which could be pronounced "KAhu-KUranui" or "KAhukura-NUi"—present problems for even the most fluent Maori speaker if they have not heard the word pronounced before. As Bruce Biggs (1986, 126) points out, "it appears that there is no way to deduce from their apparent structure the pronunciation of compound words. We must … listen to speakers who know how to pronounce them." Biggs goes on to endorse Hugh Young's (1985) exhortation to undertake fieldwork to map out geographies of pronunciation. Biggs implicitly suggests that there might be ways to identify a singular correct pronunciation for Maori names. Indeed, this may well be pragmatically useful, for there are no common forms of emphasis used in the written form of place names to guide the novice. However, we are not so much interested in identifying true or correct pronunciations, as we are in identifying forms of pronunciation that articulate particular positions in the cultural politics of place naming (but see Appendix for some indicative pronunciation guidance).

According to prominent Maori educator Irihapeti Ramsden, "It is possible to estimate the level of political consciousness of people by the way they use and apply names" (personal communication, 14 April 1999). Here Ramsden, who developed "cultural safety" as a codified approach to bicultural relations (see Dyck and Kearns 1995), is suggesting that a measure of political awareness can be gleaned from: (a) what names are used; and (b) how they are pronounced. Her position implies that pronunciation *becomes* political rather than being a political statement *per se.* In other words, pronunciation is based on access to knowledge that *enables* making a political statement. It is indeed possible that some speakers (including, perhaps, Maori) are not necessarily aware of the statement that they are making through pronunciation as they have only had access to more hegemonic (Europeanized) ways of speaking.

We should also note that as an outcome of the concessions to Maoridom made in the latter part of the twentieth century, *te reo* (Maori language) is now more widely taught and spoken than in the recent past. Reflective of this trend, is the increasing use of the unofficial alternative name for New Zealand (Aotearoa). However, while Maori is an official national language (through the Maori Language Act 1987), its status is to be found more in principle, than in practice in contemporary New Zealand. Notwithstanding the existence of the Maori Language Commission which was set up in 1987

"… to promote the use of Maori as a living language and as an ordinary means of communication" (Maori Language Commission 2002), government agencies and most public organizations conduct the vast proportion of their business in English. Most organizations—including universities—have names in both languages, yet in everyday life they are known by their English-language titles. Increasingly, however, government employees are taking Maori language classes and are expected to know how to conduct themselves at ritual events such as a *powhiri* (welcome) on a *marae.* In schools, while *te reo* may not be universally taught as with French in Canada, it does at least tend to be introduced to children regardless of ethnicity through songs and exposure through temporary immersion on school *marae* visits. At the level of public ceremony, most political events such as visits by dignitaries involve a *powhiri.* While Maori elders are often relied upon to do most of the speaking, it is not uncommon for public figures to speak, and authentically pronounce, Maori words and phrases. At a typical university graduation or sporting event, for instance, verses of the national anthem may be sung consecutively in English and Maori to no particular consternation. However, and tellingly, it is on the rare occasion when the anthem is sung completely in Maori (as occurred at an All Blacks rugby test in 2000) or when someone wishes to defend themselves in court in *te reo* that there are cries of "foul play" aired in the print and broadcast media.

The discursive politics we identify often gets played out explicitly in the 'Letters to the Editor' pages of local newspapers and also through talkback radio. In these settings, it is rare to find support for sympathetic pronunciations of Maori place-names in the public arena (e.g. radio, television). On the other hand, the few who do support the "proper" pronunciation of local place-names present eloquent arguments. A personal narrative gleaned from a correspondent to New Zealand's major daily newspaper illustrates the significant connections between naming, place and the politics of identity:

> My time at school was spent having the English language rammed into my head, mind and being and I as a Maori, was prohibited from speaking my own language. When I first came to Tamaki Makaurau [Auckland] in the late 1950s I had to change my name to accommodate the needs of my employer and workmates for the simple reason they could not pronounce my name nor did they show any interest in knowing how to … In January this year I dropped the name I had adopted these past 40 years and have regained, retained and am now using my Maori name again, which is my *mana* [social power, charisma], my pride, my *iwi* [tribal relationship], which I shoved aside for the sake of the thankless *tauiwi* [newcomers, white New Zealanders] … As for me, now I have regained my heritage and culture, I know where I am going; I know where I am from. I am not going anywhere, this is my *turangawaewae* [place to stand, home] and I no longer have to try to impress

> Pakehas. But I can try to educate them in the proper pronunciation of Maori words. (TMTR, letter to the Editor, *New Zealand Herald*, 29 June 1994)

This lengthy quote highlights the significant relationship between place and identity in Maori cosmologies. In *tikanga Maori* (or Maori ways), people identify themselves first through sets of relationships to place and other people, such as the names of their sacred mountain, local waters (river or lake), tribal canoe, significant ancestors, *marae* and local community (Yoon 1986). In the context of some specific ritualized encounters, it would be rude to identify oneself by name at the outset. Instead, one's identity is traditionally established by reference to relationships *with* places and ancestors. We do not mention these points to suggest that there is an essentialized Maori identity that is stuck in the past. Clearly such is not the case, as Maori cosmologies are socially constructed and enacted through social practices that evolve and change over time. We do wish to suggest, though, that for many Maori, identity is inextricably defined by, and through, place names.

Place names have been passed down through an oral tradition that gives life to place and prioritizes proclamation over inscription. Thus the recording of place names on maps and signposts is, inevitably, associated with the colonial imprint on the landscape. The pre-eminent site for the practice of the oral tradition has been the *marae,* a set of buildings and spaces where communities meet according to particular rituals of encounter and interaction (Tauroa and Tauroa 1986). *Marae* are Maori spaces where interaction is defined in terms of indigenous Maori culture(s) and are therefore important sites for placing speech and speaking place.

It is one thing, as the foregoing correspondent argues, to expect "proper pronunciation" of names, but not all names occupy the same spaces in terms of peoples' sense of significance or ownership. The names of individual people, while significant in themselves (especially to the individual), are not necessarily invested with as much widespread social significance as place names. By way of illustration, we draw on an editorial from the *New Zealand Herald* that refers to the actions of a controversial Member of Parliament:

> Everyone has the right to decree the pronunciation of his [sic] name, as Mr Tau Henare (Toe Henna/re) [in original] did in Parliament this week. But he is probably destined for disappointment when he instructs fellow MPs upon the correct pronunciation of Taupo, Tauranga and the like. Broadcasters have been intoning "Toe-po" [Taupo], "Toe-runga" [Tauranga] for years and, thankfully, the popular pronunciation remains. Sometimes, it simply sounds better. (24 June 1994, 8).

Contested Pronunciations

By way of indicative evidence, we now present selected results of an informal survey of materials drawn from New Zealand newspapers through the years 1994–1999, searching for any reference to the issue of place names and their pronunciation. In this choice of evidence we are dealing with a paradox: we are not writing about speech itself, but rather we are writing about writing about speech. To supplement these data on writing about the 'speaking' of places, we also gained the viewpoints, through correspondence and personal interview, of representatives of New Zealand's state broadcasting corporation.

We begin with a letter to *The New Zealand Herald*, which illustrates the uncertainties experienced by some Pakeha in New Zealand:

> Sir—Taupo, that lovely lake and its township, known to generations as Towpo has become variously Tawpo, Tarpo and Tupo on television … Please give us back our Towpo and forget the silly correct speak. (RAB, letter to the Editor, *New Zealand Herald*, 25 April 1996)

The writer draws upon conservative notions of tradition as well as unitary concepts of people and place in order to contest what they see as the fragmentation and permissiveness of contemporary political correctness. They clearly want to re-invoke a singular unitary New Zealand. Moreover, their vision of the "correct" pronunciation of place is one that accords with the traditions of the colonizers rather than the colonized, indicating the significant material implications of the politics of pronunciation at play here. Another correspondent writes that:

> … the trend for newsreaders and weather announcers is to forgo the colloquial enunciation of place names such as Taupo and Tauranga … please let us continue to apply common usage rules to favourite place names. (AJL, letter to the Editor, *New Zealand Herald*, 11 November 1998)

In a third example, a correspondent suggests that sympathetic pronunciations nevertheless depart from any authenticity:

> Without exceptional vocal skills it just isn't possible to speak two languages in one sentence. Even professional broadcasters end up pronouncing Taupo as Toepaw, which is further from "correct" pronunciation than the cricket-commentary approximation. (PC, letter to the Editor, *New Zealand Herald*, 1 January 1999)

The foregoing correspondence refers to, and affirms, the significant role of television in propagating specific pronunciations of people and place in Aotearoa/New Zealand. The state broadcasting corporation, Television New

Zealand Ltd, has claimed that there was no official date at which 'proper' pronunciation of Maori names became policy, rather "there has been gradual change to the pronunciation standards" (A. Harley, personal communication, 1993). Nonetheless, the policy of adopting more sympathetic pronunciations of Maori names can be seen to stem from the late 1980s, and can be linked to a series of significant social policy changes in New Zealand (Kelsey 1995; Le Heron and Pawson 1996).

Of course, television itself, is not a pronouncer of names; rather, it is specific *an*nouncers who become *pro*nouncers, and in so doing contribute to the politics of pronunciation. As the last of the foregoing correspondents signalled, there is a fundamental difficulty in speaking "two languages in one sentence." Of course, this is hyperbole, for English language is near-universally used in public life in New Zealand, but the writer's point is that there cannot be fence-sitting when it comes to pronunciation: choice, by design or default, constitutes an implied political sympathy to the ears of hearers.

Nightly weather presenters on television perform a particularly potent role in (re)defining the policies of pronunciation because the subject matter of their pronouncements is the variable climatic conditions across a range of familiar places. More so than actors and newsreaders, their routinized script requires them to recite lists of place-names that are beamed into households nation-wide. The shift to a purportedly more authentic articulation of place has not gone without comment. Such change has been associated with a particularly urban masculinity, such that nationally known (and Pakeha) weather presenter, Jim Hickey, has been labelled as "rugged but sensitive" for his Maori pronunciation which reflects an "urban chic" (Grant 1998). As a *New Zealand Herald* editorial acknowledges:

> Maori names of people and places have taken a terrible pounding from Pakeha pronunciation over the years. So much so that, even today, a semi-correct reference to Te Kauwhata or Horowhenua is liable to flummox many a telephone operator. But the standard seems to be improving, thanks no doubt to the assiduous efforts of broadcasters among others. (24 June 1994, p8)

Television "personalities" therefore inevitably speak with a ratings-driven authority and, in so doing, contribute to the authoring of place. Their pronunciations may be revered or reviled, but nevertheless, they contribute significantly to the discursive politics of place.

A hegemonic understanding of the pronunciation of place, one that in Aotearoa is usually less than sympathetic to Maori aspirations, tends to dominate the "Letters to the Editor" sections of local newspapers. Three further examples will suffice. In the first instance, the writer invokes rhetoric in opposition to a supposed political correctness as an argument against sympathetic pronunciation:

> Sir,—We certainly hear some weird pronunciations of Maori words by our television presenters—Rottorua [Rotorua], Towpow [Taupo], Teemaru [Timaru] and Tauranga with a "g." Let us have a bit of sense with this political correctness. (RB, letter to the Editor, *New Zealand Herald*, 30 May, 1996, p8)

Later that year the same correspondent invokes British colonization as a rationale for the Anglicization of the New Zealand landscape. She raises concerns for:

> ... the visitors and tourists to this country who have much difficulty in getting their tongues around the vast number of Maori place names we already have here. In fact they sometimes might have difficulty, owing to the lack of anglicised place names there are compared with Maori, remembering that the British colonised this country. So fair's fair. (RB, letter to the Editor, *New Zealand Herald*, 3 December 1996, 8)

A third and final example sees a correspondent revisit the concern for tourists

> Overseas visitors, particularly Americans, are becoming increasingly frustrated by the strangulated, unintelligible pronunciation by radio and television people of some of our Maori names. (LF, letter to the Editor, *New Zealand Herald*, 5 February 1999)

The concern for visitors seems especially ironic, given that tourists are conventionally eager to embrace the exotic elements of the visited Other. Thus, in this case, we suspect correspondents are deflecting their hegemonic understandings away from their own (Pakeha) positioning in society. We can read into these reactions concern for economic as well as cultural security, as the issue of pronunciation is potentially seen as undermining the comfort of tourists. It is also noteworthy that the correspondent refers to "our" Maori names. This reference reveals that the appropriation of Maori land and resources throughout the colonial period casts a shadow into post-colonial Aotearoa/New Zealand such that place names—Maori or European—are considered the property of Pakeha. This proprietary view extends beyond the material process of placing names on maps and landscapes and into the public politics of pronunciation.

Conclusion

> Mispronunciation becomes normalised and finally people try correct pronunciation. The oral map grid which covered this place for a thousand

> years becomes seriously blurred as words lose meaning. Tragically, mispronunciation accelerates language destruction. (Irihapeti Ramsden, *National Radio*, 11 April 1999)

The foregoing comment reflects a recognition by a prominent Maori educator of the power of place-name pronunciation in (re)defining the geographies of everyday life. We began by arguing that within and beyond academic geography toponymists have tended towards an explicitly atheoretical approach to understanding the geographies of naming places. However, critical geographers have begun to focus on the ambivalent geographies of power inscribed in the landscape through naming places. A number of geographers have thus focused their studies on the various ways that the process of putting names on maps involves people positioning their identities. Herman (this volume, 126) suggests that "... the politics of language, of place names, and of sovereignty are intertwined" and Azaryahu (1990, 1992, 1996) argues that names play a fundamental role in the spatial organization and semiotic construction of places. Names are thus used by the powerful to insert their hegemonic version of history into the landscape (Berg and Kearns this volume). Our choice of gendered language is purposeful, as this process was part of a masculinist colonial project of naming and claiming the landscape.

While an advance on earlier descriptive studies, there has been a tendency for recent critical studies of toponomy to focus on the *representations of space* (Lefebvre 1991) found in the official discourse of maps and street names. While this focus is important, it nevertheless poses a problem because the analysis of official discourse does not offer much insight into how official naming policies are resisted through everyday practices. Myers's (1996) analysis of the role of "nicknames" for places in Zanzibar—particularly in his use of Scott's (1990) discussion of the "arts of resistance" through performance and manipulation—pointed to a new direction for the analysis of the identity politics of naming places. As Myers (this volume) stresses, fieldwork is a key to coming to terms with the performative (Butler 1990, 1993, 1995) aspect of naming. While on the one hand, we advocate fieldwork to support our contentions, on the other hand we follow Nast (1994) in claiming that we are always and already in the field. Further, our positioning within contemporary society and culture of Aotearoa/New Zealand through the 1990s equipped us to be in the field without any special processes of entry or exit (Berg and Kearns 1998). Our decision to use written evidence was thus made for both pragmatic and conceptual reasons: recording and transcribing pronunciations would be a costly and laborious task outside the scope of this exploratory project. Notwithstanding, we acknowledge the limited nature of our "data" and believe that further empirical work in this field will be a worthwhile activity.

We have argued that speech acts are social processes and therefore it is in the context of talk that we (re)constitute community and pronounce the significance of place. Speaking names is a process of narration that is a deeply embedded cultural practice. Narration itself is a complex psychic process of constructing and positioning the Self in relation to Others. Everyday, taken-for-granted narratives about place—and thus place names—are composed in ways that interact with hegemonic readings and understandings of the politics of location. Spoken names are therefore part of the everyday *spaces of representation* (Lefebvre 1991). Thus, the way place names are pronounced reflects, and contributes to, the constitution of imagined communities and spaces of social life.

Our case material was drawn from a nation undergoing major changes in identity. Within the social landscapes of Aotearoa/New Zealand, there is not only the reassertion of Maori cultural and economic rights, but also—at least in Auckland—there is an ascendant Pacific identity as well as recent moves towards closer national links with Asia in lieu of Europe. The issue of the pronunciation of place names provides a vantage point from which to examine this shifting link between culture, place and identity. Within New Zealand where, outside the major urban centres, Maori place names abound, pronunciation debates provide an indication of the transformation of the practice of everyday life (de Certeau 1984). We have shown that in either deference to, or support for, cultural rights broadcasters have adopted Maori place name pronunciation. In turn this strategy on the part of public figures has provoked reaction from the members of the public. Thus, to use the language of de Certeau (1984), the *tactic* of resistance on the part of most Maori and some Pakeha (using Maori pronunciation) has been adopted as a *strategy* by the state broadcaster. This is turn is invoking counter-resistance by those averse to change in their personal speech habits and/or sympathetic with the (neo)colonial project.

In the course of our preliminary review of evidence, we noted dismissals of sympathetic pronunciations of Maori place names as 'politically correct." It is noteworthy that this observation is not particular to New Zealand. Indeed, in his survey of Hawaiian experience, Herman similarly notes that proposals for renaming away from the colonial legacy are "... immediately assaulted as revisionism and political correctness" (1999, 77). The apparent need to denigrate particular pronunciations of place in favour of others highlights the significance of such an ostensibly banal aspect of human geography. If place matters, then obviously saying the name of particular places "properly" matters very much. We acknowledge that choice of pronunciation might be based simply on a lack of awareness of alternatives, rather than ethnic group membership or a sympathetic politics. Indeed, while our interpretation has stressed the potential for mispronunciations to be neocolonial speech acts, we recognize the opacity of speech, and the possibility that the issue of pronunciation might also be read as reflecting the diversity inherent in an increasingly hybrid culture in New Zealand. Nonetheless, the very fact that a significant proportion of the populace might be unaware of the various place name pronunciation "choices" is indicative of the power of hegemonic social relations that underpin and reinforce neocolonialism. The key question remains as to who gets to map out proper pronunciation, and how and why such topographies of naming might be contested.

References

Anderson, B. (1983), *Imagined Communities* (London: Verso).

Awatere, D. (1984), *Maori Sovereignty* (Auckland: Broadsheet Publications).

Azaryahu, M. (1986), "Street Names and Political Identity: The Case of East Berlin," *Journal of Contemporary History* 21, 581–604.

—— (1988), "What is to be Remembered: The Struggle over Street Names in Berlin, 1921–1930," *Tel Aviver Jahrbuch für deutsche Geschichte*, XVII, 241–58.

—— (1990), "Renaming the Past: Changes in 'City-text' in Germany and Austria, 1945–1947," *History and Memory* 2, 32–53.

—— (1992), "The Purge of Bismarck and Saladin: The Renaming of Streets in East Berlin and Haifa, a Comparative Study in Culture Planning," *Poetics Today* 13, 351–67.

—— (1996), "The Power of Commemorative Street Names," *Environment and Planning D: Society and Space* 14, 311–30.

Barber, K. (1989), "New Zealand 'Race Relations Policy," 1970–1988," *Sites: A Journal for South Pacific Cultural Studies* 18, 5–16.

Belich J. (1986), *The New Zealand Wars and the Victorian Interpretation of Racial Conflict* (Auckland: Auckland University Press).

Berg, L.D.(1999), "A (white) man of his times? Sir George Grey and the narration of masculinity in Victorian New Zealand," in R. Law, H. Grant and J. Dolan (eds), *Masculinities in Aotearoa/New Zealand* (Palmerston North: The Dunmore Press).

Berg, L.D. and Kearns, R.A. (1996), "Naming as Norming: 'Race," Gender, and the Identity Politics of Naming Places in Aotearoa/New Zealand," *Environment and Planning D: Society and Space* 14, 99–122.

—— (1998), "America Unlimited," *Environment and Planning D: Society and Space* 16, 128–32.

Bhabha, H.K. (1994), *The Location of Culture* (London: Routledge).

Biggs, B. (1986), "Maori Spelling," *Te Reo* 29, 119–29.

Butler, J. (1990), *Gender Trouble Feminism and the Subversion of Identity* (New York: Routledge).

—— (1993), *Bodies That Matter. On the Discursive Limits of 'Sex'* (New York: Routledge).

—— (1995), "Burning Acts Injurious Speech," in A. Parker and E.R. Sedgwick (eds), *Performativity and Performance*, (London: Routledge), 197–227.

Callary, E. (1997), "The Geography of Personal Name Forms," *The Professional Geographer* 49, 494–500.

Carter, P. (1987), *The Road to Botany Bay: An Essay in Spatial History* (London: Faber and Faber).

Cohen, S. and Kliot, N. (1992), "Place-names in Israel's Ideological Struggle over the Administered Territories," *Annals of the Association of American Geographers* 82, 653–80.

Crocombe, R. (1991), "Naming and Claiming in the South Pacific," *Journal of the Pacific Society* 50, 1–19.

Dawson, G. (1994), *Soldier Heroes: British Adventure, Empire and the Imagining of Masculinities* (London: Routledge).

De Certeau, M. (1984), *The Practice of Everyday Life* (trans. S Randell) (Berkeley: University of California Press).

Dyck, I and Kearns, R. (1995), "Transforming the Relations of Research: Towards Culturally Safe Geographies of Health and Healing," *Health and Place* 1, 137–47.

English Lingusitics Department, Chemnitz University of Technology (2001), *Thje Varieties of English: Exploring New Zealand English*, <www-user.tu.chemnitz.de/~tofu/nz_maori.htm>, accessed 12 March 2002.

Entrikin, N. (1991), *The Betweenness of Place: Towards a Geography of Modernity* (Baltimore, MD: Johns Hopkins Press).

Eyles, J. (1985), *Senses of Place* (Warrington: Silverbrook Press).

Fair, S.W. (1997), "Inupiat Naming and Community History: The *Tapqaq* and *saniniq* Coasts near Shishmaref, Alaska," *Professional Geographer* 49, 466–80.

Gelling, M. (1984), *Place-names in the Landscape* (London: J.M. Dent and Sons Ltd).

—— (1988), *Signposts to the Past: Place-names and the History of England* (Chichester: Phillimore).

Giddens, A. (1984), *The Constitution of Society* (Cambridge: Polity Press).

Gould, J.D. (1992), "'Maori' in the Population Census, 1971–1991," *New Zealand Population Review* 18, 35–68.

Graham, G. (1926), *Maori Place Names*, MS 120 (Auckland: Auckland Institute and Museum Library).

Grant, F. (1998), "It's a Great Time to be a Weatherman Mate," *New Zealand Herald*, June 4, E2.

Griffiths, G.J. (1988), *40 Common Elements in Maori Placenames* (Dunedin: Otago Heritage Books).

—— (1990), Names *and Places in Southern New Zealand* (Dunedin: Otago Heritage Books).

Harley, A. (1993), letter to R. Kearns from A. Harley, Chief Assistant to the Director of News and Current Affairs, TV New Zealand.

Herman, R.D.K. (1999), "The Aloha State: Place Names and the Anti-conquest of Hawai'i," *Annals of the Association of American Geographers* 89: 1, 76–102.

Jett, S.C. (1997), "Place-naming, Environment, and Perception among the Canyon de Chelly Navajo of Arizona," Professional *Geographer* 49, 481–93.

Kearns, R.A. and Collins, D.C.A. (2000), "Maungakiekie/One Tree Hill: Contesting the Iconography of an Auckland Landscape," *Australian-Canadian Studies* 18: 173–88.

Kelsey, J. (1995), *The New Zealand Experiment: A World Model for Structural Adjustment?* (Auckland: Auckland University Press).

Kiwiweb (undated), <www.kiwiweb.co.nz/generic/iwi/maori-pronunciation.htm>, accessed 12 March 2002.

Kobayashi, A. (1993), "Multiculturalism: Representing a Canadian Institution," in J. Duncan and D. Ley (eds) *Place/Culture/Representation* (London: Routledge), 205–31.

Lefebvre, H. (1991), *The Production of Space* (trans. D Nicholson-Smith) (Cambridge, MA: Blackwell).

Le Heron, R. and Pawson, E. (1996), *Changing Places: New Zealand in the 1990s* (Auckland: Longmans).

Maori Language Commission/Te Tauawhiri (2002), <www.tetaurawhiri.govt.nz/english/ accessed>, accessed 12 March 2002.

McClean, R., Berg, L.D. and Roche, M.M (1997), "Responsible Geographies: Co-creating Knowledges in Aotearoa," *New Zealand Geographer* 53: 2, 9–15.

Mitchell, D. (2000), *Cultural Geography: An Introduction* (Oxford: Blackwell).

Morin, K.M. and Berg, L.D. (2001), "Gendering Resistance: British Colonial Narratives of Wartime New Zealand," *The Journal of Historical Geography* 27: 2, 196–222.

Myers, G.A. (1996), "Naming and Placing the Other: Power and the Urban Landscape in Zanzibar," *Tijdschrift voor Economische en Sociale Geografie* 87, 237–46.

Nairn, R.G. and McCreanor, T.N. (1991), "Race Talk and Common Sense: Patterns in Pakeha Discourse on Maori/Pakeha Relations in New Zealand," *Journal of Language and Social Psychology* 10, 245–62.

Nash, C. (1999), "Irish Placenames: Postcolonial Locations," *Transactions of the Institute of British Geographers* 24, 457–80.

Nast, H.J. (1994), "Opening Remarks on 'Women in the Field'," *The Professional Geographer* 46, 54–66.

Orange, C. (1987), *The Treaty of Waitangi* (Wellington: Allen and Unwin).

Pawson, E. (1992), "Two New Zealands: Maori and European," in K.J. Anderson and F. Gale (eds), *Inventing Places: Studies in Cultural Geography* (Cheshire, Melbourne: Longman), 15–33.

Pile, S. and Keith, M. (1997), *Geographies of Resistance* (London: Routledge).
Pool, I. (1991), *Te Iwi Maori: A New Zealand Population, Past, Present and Projected* (Auckland: Auckland University Press).
Pred, A. (1989), "The Locally Spoken Word and Local Struggles," *Environment and Planning D: Society and Space* 7, 211–33.
Rickard, L.S. (1968), *Historic Place Names of New Zealand* (Auckland: Minerva).
Robb, J.G. (1996), "Toponomy in Lowland Scotland: Depictions of Linguistic Heritage," Scottish *Geographical Magazine* 112, 169–76.
Rodaway, P. (1994), *Sensuous Geographies: Body, Sense and Place* (London: Routledge).
Salmond, A. (1991), *Two Worlds: First Meetings Between Maori and European 1642 and 1772* (Auckland: Viking).
Scott, J. (1990), *Domination and the Arts of Resistance: Hidden Transcripts* (Newhaven, CT: Yale University Press).
Sibley, D. (1997), *Geographies of Exclusion* (London: Routledge).
Simmons, D.R. (ed.) (1980), *Maori Place Names of Auckland: Their Meaning and History* (Auckland: Auckland Institute and Museum).
Simonsen, K. (1996), "What Kind of Space in What Kind of Social Theory?," *Progress in Human Geography* 20, 494–512.
Smith, S.J. (1994), "Soundscape," *Area* 26, 232–40.
Stokes, E. (1992), "The Treaty of Waitangi and the Waitangi Tribunal: Maori Claims in New Zealand," *Applied Geography* 12, 176–91.
— (1994), "Geographic Naming in New Zealand: The Role of the New Zealand Geographic Board," *New Zealand Geographer* 50, 51–4.
Tauroa, H. and Tauroa, P. (1986), *Te Marae: A Guide to Customs and Protocol* (Auckland: Reed Methuen).
Walker, R. (1990), *Ka Whawhai Tonu Matou: Struggle Without End* (Auckland: Penguin).
—— (1992), "Maori People since 1950," in G.W. Rice (ed.) *The Oxford History of New Zealand*, 2nd edn (Auckland: Oxford University Press), 498–519.
Yeoh, B.S.A. (1996), "Street-naming and Nation-building: Toponymic Inscriptions of Nationhood in Singapore," *Area* 28, 298–307.
Yoon, H-k. (1986), Maori *Mind, Maori Land* (Berne: Peter Lang).
Young, H. (1985), *A Guidebook to Nga Ingoa o Aotearoa: An Oral Dictionary of Maori Place Names Part 1 Te Upoko o te Ika* (Wellington: Replay Radio Division, Radio New Zealand).
Young, R. (1995), *Colonial Desire: Hybridity in Theory, Culture and Race* (London: Routledge).
Zelinski, W. (1982), "By their Names You Shall Know Them: A Toponymic Approach to the American Land and Ethos," *New York Folklore* 8, 85–96.

—— (1992), "On the Naming of Places and Kindred Things," in S.J. Bronner (ed.), *Creativity and Tradition in Folklore: New Directions* (Logan, Utah University Press).
—— (1993), "Parsing Greater Washington's Namescape," *Names* 41: 4, 344–60.
—— (1997), "Along the Frontiers of Name Geography," *Professional Geographer* 49, 465–66.

Appendix: Characteristics of Maori Pronunciation

Unlike English, the rules for Maori pronunciation are simple, systematic, and apply to all Maori words. Understanding how to make the correct Maori sounds (that are represented by English alphabetical symbols) is very helpful in pronouncing Maori words correctly. The Maori alphabet consists of consonants H, K, M, N, P, R, T, W, NG and WH and of vowels A, E, I, O and U.

- A is pronounced 'ah' as in are or bar (e.g. in names like Wanganui).
- E is pronounced 'eh' as in egg or bed.
- I is pronounced 'ee' as in ski.
- O is pronounced 'or' as in core.
- U is pronounced 'oo' as in glue (e.g. in names like Motueka).

The consonants are pronounced the same as in English except that:

- R is spoken with a slight rolling of the tongue;
- NG is pronounced with a soft 'n' and soft 'g' as in singing;
- WH is pronounced like an F (as in the names Horowhenua and Whenuapai).

Other points to note:

- consonants never run together but will always be separated by at least one vowel;
- pronounce every letter. The letter h should not be unsounded in words like kahikatea, Waitohi or Ohakune. All syllables should be preserved in names such as O-ta-hu-hu or Pa-ra-pa-ra-u-mu;
- often a long vowel sound will be represented by use of a macron (a line at the top of the vowel). Alternatively a double vowel will be used, e.g. the word 'Maori' may be written 'Maaori' to indicate the long 'a' sound;
- 2 or more vowels will often run together and these vowels should be vocalized individually but made to run together so a smooth sound is produced, eg. E-I when run together smoothly almost sounds like "ay" (e.g. Whangarei), A-U almost sounds like "oh" (e.g. Whau Valley);
- there is no 'S' in the Maori language so it is not correct to add an 'S' at the end of a Maori word to make it a plural.

Sources: Kiwiweb (undated), The English Linguistics Department, Chemnitz University of Technology, 2001.

Chapter 9
Street Names as Memorial Arenas: The Reputational Politics of Commemorating Martin Luther King Jr in a Georgia County

Derek H. Alderman

Commemorative street naming is an important vehicle for bringing the past into the present, helping weave history into the geographic fabric of everyday life. Named streets, like any place of memory, can become embroiled in the politics of defining what (or who) is historically significant or worthy of public remembrance. I am interested in street names as "memorial arenas," public spaces for actively interpreting the legacies of historical figures and, in many instances, debating the meaning and relative importance of memorializing these figures. Specifically, my work focuses on one of America's most widespread yet under-analyzed commemorative practices—the naming of streets after slain civil rights leader Martin Luther King Jr. As of 2003, at least 730 cities in the United States had attached King's name to streets and of the 50 states, only 11 (or 22 per cent) had no streets named after him (Alderman 2006). Although a nationwide movement, street naming for King is proposed by local activists—usually African American—and decided by city and county leaders as they face the task of selecting a road that is most fit or appropriate for remembering him (Tilove 2003; Dwyer and Alderman 2008). As Roger Stump (1988, 215) observed in a seminal article, roads named for Martin Luther King Jr are "public symbols of community values, attitudes, and beliefs, revealing the character of both the figure commemorated and the community that has honored him."

King's commemoration is part of a larger movement on the part of black Americans to address the exclusion of their experiences from the national historical consciousness. Although the past contributions of African Americans certainly do not begin or end with King, he has become the most widely identified symbol of the civil rights movement and black heritage in general. King's status rose when the federal government established a holiday to honour him in the early 1980s, and a 1999 Gallup public opinion poll revealed the national prominence of his reputation—34 per cent of surveyed Americans named Martin Luther King Jr as the most admired person of the century, placing him second only to Mother Teresa (Newport 1999). Yet, to say that King is a highly admired figure is not to say that his commemoration does not generate public debate. Indeed, the attaching of his name to streets and roads often evokes great controversy not only between blacks and whites but also within the African American community. In part, this is because people—despite their general respect for the civil rights leader—interpret and personally connect with his historical legacy in different and sometimes competing ways.

The objective of this chapter is to explore the reputational politics of naming streets for Martin Luther King Jr. "Reputational politics" is an approach that focuses on the socially constructed and contested nature of commemorating historical figures and the discursive rivalries that underlie the memorialization of these figures. By way of case study, I visit Bulloch County, Georgia and examine the failed 1994 attempt by African Americans to have a new perimeter highway named for King. Black activists there struggled with military veterans, who sought to use street naming to publicly recognize their own sacrifices and achievements. Through newspaper editorials, speeches at local government meetings, and a name-suggestion campaign, outspoken veterans were successful in representing King's historical legacy as less important and less racially inclusive than their own commemorative cause. While this debate exposed traditional racial divisions within the county, it also displayed an interesting counterintuitive pattern—because of divided loyalties, many black men went on record in support of naming the perimeter highway for veterans rather than for King.

Discourse analysis is used to examine how participants in the Bulloch County struggle viewed and represented the commemoration of Martin Luther King Jr in multiple and competing ways, facing three important issues as they engaged in the reputational politics of commemorating King. The first was the *legitimacy of commemoration* or the politics of constructing the historical and political worth of remembering one figure or cause over another. The second was the *resonance of commemoration* or the politics of making a commemorated figure universally relevant or resonant to the various social groups that constitute the public. The third dimension was the *hybridity of commemoration*, which refers to the political complications that can arise when social actors, specifically African American activists, have sometimes overlapping commemorative interests. Before presenting the case study, it is necessary to provide some basic background on street naming as a memorial arena, the notion of reputational politics, and the politics of commemorating King in America.

Street Names as Memorial Arenas

The naming of streets after historical figures and events is an important part of modern political culture; it not only provides spatial and semiotic orientation to the city but, also serves to naturalize or legitimatize a selective vision of the past (Azaryahu 1996a; Azaryahu 1997). The seemingly ordinary and practical nature of street names makes the past "tangible and intimately familiar" (Azaryahu 1996a, 321). Named streets are powerful memorial spaces because they inscribe a commemorative message into many facets of daily urban life such as through road maps, phone book listings, the sending and receiving of mail, the giving of directions, advertising billboards, and of course, road signs themselves. Signs contribute significantly to the creation of a sense of place (Weightman 1988).

Street names are unique and potentially politicized memorials because of their geographic connectivity, their ability to touch diverse neighborhoods and social groups who may or may not identify with the person or event being memorialized (Alderman 2006). Street naming is also controversial because, with the commemoration of one vision of the past, there is often a de-commemoration of another. The naming of streets is an uneven negotiated process open to a multiplicity of meanings and interests (Yeoh this volume). And while powerful elites often control the commemorative messages communicated through street names, the naming process can also serve as a form of resistance as marginalized groups seek to redefine how the public interprets the past (Gonzáles Faraco and Murphy 1997). Street naming thus can become a contest about who has the power to determine how the landscape is represented and whose history will be told on and through the landscape.

America's street-name landscape has long served as a place for social groups, including minorities, to gain public recognition of their historical achievements (Chacón 1995; Seligman 2001). Recently, however, the United States has witnessed a flurry of commemorative street-naming campaigns led by racial and ethnic groups. Joseph Tilden Rhea (1997, 5) suggested that the movement to recognize the role of minorities in American history is not simply "because of a general drift toward cultural pluralism." Rather, it is the result of direct political action by minority activists, who seek a greater identity within society by challenging white-dominated interpretations of American history and creating new representations of their past within cultural landscapes.

Despite the ever-increasing importance of the street name as a platform for elevating minority heritages, little attention has been paid to the role of African Americans in commemorative street naming and place naming in general. There is ample evidence that African Americans view the naming of streets and other public places as a means of asserting their historical value and legitimacy within the country. Although Martin Luther King Jr is the figure most frequently commemorated, communities have identified streets with other notable black Americans such as Rosa Parks, Harriet Tubman,

Malcolm X, and Thurgood Marshall. As Melvin Dixon (1994, 20) so keenly observed: "Not only do these [street] names celebrate and commemorate great figures in black culture, they provoke our active participation in that history. What was important yesterday becomes a landmark today."

On the Importance of Reputational Politics

A critical analysis of King's commemoration requires not only recognizing the cultural importance of street names but also understanding the politics that surround the remembrance of historical figures. Heroes and other notable historical figures play an important role in public remembrance of the past. Gary Fine (1999, 226) discussed this very point when he wrote: "By shaping images of their leaders, social groups create social mnemonics to help audiences define events within a moral context." The historical image of a person—whether a hero such as Abraham Lincoln, a villain such as Benedict Arnold, or enigmatic personalities such as John Brown and Joseph McCarthy—is a social product open to multiple and competing constructions and interpretations (Ducharme and Fine 1995; Gaynes 1995; Schwartz 1997). There can be any number of different discourses or common ways of thinking and talking about a person and his or her contribution to society.

The struggle to define an individual's memorial legacy is what Fine has called "reputational politics." The historical reputation of a person is not simply made by the individual in question but also is used and controlled by social actors and groups who seek to advance their own commemorative agenda and divert the agendas of other parties. Custodial agents or "reputational entrepreneurs" carry out the shaping and control of historical reputations (Fine 1996). Fine (1996, 1161–2) also recognized that the "control of history may be contentious, and the claims of one group may be countered by another that wishes to interpret the same ... person through a different lens." In the politics of constructing and asserting the historical importance of commemorating one figure over another, reputational entrepreneurs engage in "discursive rivalries," the trading back and forth of statements and claims about the commemorative legitimacy and meaning of their respective heroes' reputations (Fine 1996, 1160). As this chapter illustrates, discursive rivalries over the meaning of historical figures do not simply appear out of thin air, but often accompany, revolve around, and participate in the production of memorial spaces and places, which give the past a tangibility and visibility.

The commemoration of fallen heroes, martyrs, and great leaders has been the most common motive for designating a physical site as a "sacred" memorial space (Foote 1997). While memorial spaces affirm the importance of certain historical figures, they also offer insight into the social tensions that often underlie the remembrance of these legacies and reputations, particularly when shaping the contours of national identity is at stake (Johnson 1994; Azaryahu 1996b; Edensor 1997). Memorializing Martin Luther King Jr began after his death in 1968, but he did not receive "widespread popular acclaim immediately after death" (Foote 1997, 75). Indeed, it took over two decades to convert the Lorraine Motel, the site of his assassination, into a national civil rights centre and museum. It was not until 1976 that Atlanta, King's hometown, placed his name on a street. The majority of King streets have been named in the years following the passage of the Martin Luther King Jr federal holiday, which brought official legitimacy to his historical reputation. The transformation of King into a national hero and the creation of memorial spaces in his honour have occurred only after many years of opposition and debate.

The Politics of Remembering King in America

As with other major figures in American history, public memory of King is the result of ongoing negotiation and debate within society. In King's case, many of these debates have revolved around the designation of his birthday as a holiday, both at the federal and state level (Alozie 1995; *Boston Globe* 1999). Before being passed in 1983, the federal holiday was the centre of heated discussion in the US Senate with opponents accusing King of being a communist and having extramarital affairs (*The Economist* 1983). But even before the holiday proposal ever made it to the floor of the Senate it had traveled a long and arduous journey through a variety of political struggles. Passage of the holiday came 15 years after Representative John Conyers (D-Michigan) first proposed the idea in legislation and 13 years after Congress received an unprecedented 6 million signatures in support. Perhaps the most interesting of these commemorative struggles took place within the African American community between members of the Southern Christian Leadership Conference (SCLC) and the civil rights leader's wife, Coretta Scott King. While the SCLC sought to honour their recently fallen leader through increased social activism and protest, Mrs King placed more emphasis on establishing the King Center in Atlanta, Georgia and favouring calls to establish a holiday in her husband's memory. The two parties even differed on when best to commemorate the civil rights leader—while the SCLC focused on 4 April, the date of King's death, the King family preferred 15 January, the date of his birth (Daynes 1997).

While King's annual commemoration is more accepted now than in the past, it remains a contested terrain in many communities. Ceasar McDowell has described the still controversial nature of the King holiday: Martin Luther King "may have spoken the common language of human dignity. But to many white people, he remains primarily a black man, and his birthday a black holiday, for black people" (quoted in *Boston Globe* 1998, A1). African Americans in Memphis were outraged by a McDonald's restaurant calendar that labelled 16 January 1989 as "National Nothing Day" (*Chicago Defender* 1989). In May of 2000, Jesse Jackson visited Wallingford, Connecticut, to mark a new state law forcing the town to observe the King holiday. According to news reports, Wallingford had been the only town in the state that still kept its offices open on the holiday. During his visit, Jackson confronted a group of white supremacists opposed to King's commemoration, citing the group as proof that there is "unfinished business" in the struggle for civil rights (*Jet* 2000, 34). While more of the country's businesses are recognizing the holiday, only 26 per cent give a paid day off to their employees (*NPR Morning Edition* 1998). Nor do African Americans all agree on how to observe King's birthday. For example, in 1998, a rift emerged in Houston's black community as two factions struggled over which one should organize and lead a parade to mark the MLK holiday. African American leaders also disagreed about the extent to which the civil rights leader's memory should be "commercialized" (*Houston Chronicle* 1998, A15).

Controversy over the legacy of Martin Luther King Jr has not been confined to the making and observing of holidays, however. In Memphis, Tennessee, black activist Jacqueline Smith has spent over 20 years protesting the conversion of the Lorraine Motel—the site of the King assassination—into the National Civil Rights Museum. Her protest has been literally street politics in that she has lived, eaten, and slept on the sidewalk across the street from the museum (Jones 2000). She advertises the museum as the "National Civil Wrong Museum," distributes protest literature, and provides museum visitors with an alternative vision of how to commemorate King. According to Smith, "The best monument to Dr Martin Luther King Jr would be a centre at the Lorraine offering housing, job training, free community college, health clinic or other services for the poor" (interview with author 23 December 1995). While Smith's protests take place outside the civil rights museum in Memphis, more subtle yet no less important struggles occur within civil rights museums and memorials. Owen Dwyer (2000) noted a tension in civil rights memorials over the commemoration of local, grassroots participants versus charismatic, national leaders such as King. In the case of the civil rights museum in Savannah, Georgia, local mobilization efforts are given much greater attention than King and the larger national movement. Arguably, this reflects the fact that African Americans in Savannah were able to carry out several successful protests in the 1960s without significant outside leadership. In some instances, they actually opposed King's direct involvement in these protests (Tuck 2001). Black Americans connect with King's historical reputation in many and sometimes contradictory ways, particularly when it involves weighing his reputation against other commemorative interests that are of equal or greater importance.

Like holidays, museums, and other memorials dedicated to King's memory, street names serve as important arenas for debating how King is best remembered. In Americus, Georgia, city officials did not rename a portion of US Highway 19 until black community leaders planned a boycott of city businesses. The controversy in Americus was made worse by the comments of an assistant fire chief, who said that he did not oppose naming half of the street for King if the other half was named for James Earl Ray, the man convicted of assassinating the civil rights leader (*Atlanta Journal-Constitution* 1992). In Dade City, Florida, vandals painted the name "General Robert E. Lee" over nine Martin Luther King Jr Boulevard signs, an incident symptomatic of the South's ongoing struggles over identity and memory. In a single year, almost 100 street signs with King's name in Hillsborough County, Florida were either spray painted, shot at, or pulled completely from their poles (*St. Petersburg Times* 1990). These controversies are not confined to the American South. African Americans in Milwaukee protested and marched against the city's decision to restrict the naming of Martin Luther King Jr Drive to the boundaries of the black community, a common complaint among black activists across the nation (Kenny 1995). In 1990, a debate erupted in Portland, Oregon when the city council voted to rename Union Avenue after King. Over two dozen people picketed and heckled the street-naming ceremony and more than 50,000 people signed a petition opposing the name change, although King's memorial was eventually allowed to stand (*Seattle Times* 1990).

Despite the growing frequency and controversial nature of commemorating King, scant attention has been devoted to examining how individual communities debate and struggle over the street-naming issue. This case study of Bulloch County provides an opportunity to fill this void and analyze some of the issues that shape the "reputational politics" of defining the historical importance and social relevance of King's memory.

Commemorating King in Bulloch County, Georgia

Although the historical significance of King is certainly not limited to one state, Georgia serves as a useful context within which to study his commemoration. Georgia has the largest number of places with a street named after him (Alderman 2006). Further, since King was born in Atlanta, which is now the location of a national historic site that recounts his life and work, he recently was recognized as "Georgian of the Century" (Gulliver 2000). Despite these strong historical connections, African Americans across Georgia have faced significant controversy and opposition when attempting to attach King's name and memory to streets and roads (*Atlanta Journal-Constitution* 1993; *Atlanta Journal-Constitution* 1999). Perhaps no location illustrates this fact better than Bulloch County, where black activists carried out two unsuccessful street-naming campaigns before succeeding in late 2002. The first of these two campaigns is the focus of this discussion. By examining where street naming has failed rather than succeeded, we can perhaps gain greater insight into the challenges that confront African Americans as they engage in the politics of redefining public representation of the past, particularly King's meaning and importance within contemporary society.

Bulloch County is located approximately 70 miles west of the Atlantic Ocean. It lies between the two population centres of Macon, in central Georgia, and Savannah, on the coast. Statesboro is the county seat, the location of Georgia Southern University, and the setting for many of the political struggles described here. According to 2000 US Census data, Statesboro has a population of 22,698 while Bulloch County's population is 55,983. Although not constituting a majority, African Americans have a significant presence in both the city (with 41 per cent of the population) and the county (29 per cent). Bulloch is an important case study in the sense that it affords us the opportunity to analyze the "reputational politics" of commemorating Martin Luther King Jr within a predominantly rural county. According to Bruce Williams and Bonnie Thornton Dill (1995), persistent inequality characterizes African Americans in the rural South, and the exclusion of southern blacks is most pronounced at the county level. At least in the case of Georgia, there is strong historical evidence that there are stark differences between black activism in rural and urban areas (Tuck 2001).

On 1 February 1994, a special committee of the local chapter of the National Association for the Advancement of Colored People (NAACP) requested that the Bulloch County Board of Commissioners name the then-unfinished perimeter road after Martin Luther King Jr Although some suggestions had been made in passing, this constituted the first formal request to name the highway. Because the road would run through both the city (Statesboro) and the county, a joint perimeter-naming committee was organized. Then, at a county commission meeting a month later, a representative of the local American Legion officially requested that the road be named "Veterans Memorial Parkway" in honour of area military veterans (Minutes of the Bulloch County Board of Commissioners 1994a; *Statesboro Herald* 1994a). Recognizing the politically contested nature of naming the perimeter road, the joint city-county naming committee asked citizens to mail in name suggestions. From these suggestions, the committee would recommend two names (a primary and a secondary suggestion) to be considered by both the city and county governments. Yet, as the committee pointed out and as was eventually realized, a high number of suggestions would not guarantee that one particular road name would be favoured over another (*Statesboro Herald* 1994b; *Statesboro Herald* 1994c).

Bulloch County residents mailed in 2,196 suggestions, offered more than 100 different names. Of the suggestions submitted, 1,680 (or 77 per cent) wanted the perimeter named "Veterans Memorial." Over 18 per cent of submissions suggested naming the road after Paul Nessmith, a farmer and former state representative currently living in the county. (Nessmith would later be eliminated from consideration because he was not deceased.) Only 72 (or 3.3 per cent) of suggestions called for naming the perimeter for Martin Luther

King Jr Behind King in the number of suggestions was Bulloch Memorial Parkway with seven (or 0.32 per cent) of suggestions. There appeared to be concerted support for having the perimeter's name serve as a memorial in some way since the memorializing of veterans and King combined assumed almost 80 per cent of submitted suggestions. An additional 12 suggestions requested that the word "Memorial" be included in the road's name more generally, as in the cases of Bulloch Memorial Parkway, Memorial Scenic Parkway, and Memorial Parkway (*Statesboro Herald* 1994f).

At first glance, the low number of suggestions submitted for naming the perimeter for King would indicate indifference on the part of the African American community. Although blacks accounted for more than 25 per cent of the county's population at this time, little more than 3 per cent of submissions called for the perimeter to be identified with King. But the low number may reflect, instead, the manner in which the suggestion campaign was advertised. From all indications, the call for road name suggestions was announced only a few times in the local newspaper. Rather than interpreting these mail-in results as showing weak support for King, it may be more useful to see them as showing very strong, mass-organized support for memorializing veterans. In fact, the vast majority of the suggestions for "Veterans Memorial" came in as form letters. As will be discussed later, King's weak following in the suggestion campaign may also be indicative of the extent to which African Americans battled a divided loyalty between commemorating King versus military veterans. Nevertheless, the low quantitative support for King should not divert our attention away from the qualitative value of analyzing the efforts and comments of individual "reputational entrepreneurs" and the "discursive rivalry" that took place between African Americans and veterans over defining King's historical legacy and the appropriateness of naming a road for him.

After the 1 June 1994, deadline for the submission of suggestions, the perimeter-naming committee members reduced the 100 suggested road names down to five. These suggestions were placed on a line ballot and ranked one through five in order of committee members' preference. The committee then submitted the two highest-ranking names to the city and county governments for a vote. The committee preferences from highest to lowest were Bulloch Memorial Parkway, Veterans Memorial Parkway, University Parkway, Bulloch Heritage Parkway and Martin Luther King Jr Parkway. Preference for Memorial Parkway, University Parkway, and Bulloch Heritage Parkway perhaps reflected attempts to sidestep the controversial decision of choosing a memorial either to veterans or King. Despite the numerical results of the original mail-in suggestion campaign, the naming committee chose to recommend the names Bulloch Memorial and Veterans Memorial Parkway to the Statesboro City Council and the Bulloch County Board of Commissioners (Minutes of the Bulloch County Board of Commissioners 1994b; *Statesboro Herald* 1994g). Thus, despite a disagreement between city and county officials and last-minute lobbying on the part of African American leaders, the new perimeter highway was eventually designated Veterans Memorial Parkway (*Statesboro Herald* 1994h).

Newspaper and government archives, interviews with several participants in the struggle, and written suggestions by citizens to the perimeter-naming committee provide a means of identifying multiple and competing discourses about King as an historical figure, his reputation or legacy, and the resonance of this legacy to contemporary social life. As suggested by Berg and Kearns (this volume), discourse analysis is useful in exposing how common-sense beliefs are represented and made socially important through public dialogue about place naming. In Bulloch County debates, at least three factors shaped the "reputational politics" of naming a street for Martin Luther King: the legitimacy of commemoration, the resonance of commemoration, and the hybridity of commemoration.

The Legitimacy of Commemoration

Dydia DeLyser (1999) has discussed how authenticity is not an inherent condition or quality but a notion that has different meanings to different people in various social and spatial contexts. Historical legitimacy—while often represented as a universal and objective standard—is similarly open to competing constructions. In the case of naming a street for King in Bulloch County, the first and perhaps most obvious barrier was the belief held by opponents that memorializing Martin Luther King Jr was less important or less legitimate than other competing historical figures or commemorative causes. Specifically, supporters of naming the perimeter for King found themselves in a struggle with the local chapter of the American Legion. While carried out on many fronts, local newspaper editorials were a common platform for struggles to determine the primacy of commemorating King versus veterans.

After the initiation of the mail-in suggestion campaign, school board member Charles Bonds (an African American) attempted to establish the legitimacy of naming the perimeter after King. In a 22 May 1994, letter to the editor, he reminded readers that King had been awarded a Nobel Peace Prize in 1964 for his non-violent campaign against racism. In building a reputational account of King's legacy of equality and justice, Bonds also pointed to the moral authority and public image that Bulloch County stood to gain (or lose) in deciding whether or not to commemorate the civil rights leader:

> Isn't it ironic that we, the citizens of Bulloch County, have minimized and viewed as mediocre what Dr. King has done for our country in fostering equality? His fight for equality and justice has not only led to racial equality but human equality. South Africans, blacks, coloreds, whites, Indians, and others have come to grips with the recognition of the importance of equality

> and justice … Can't we dedicate a minuscule portion of our county's constructions in memory of Georgia's and the world's greatest citizen? … Name the new bypass "The Martin Luther King Jr Boulevard" and let visitors and newcomers to our fine community know that we value equality and justice for all who may drive on the named highway. (*Statesboro Herald* 1994d, A4)

In a letter to the editor a week later, local veteran Edgar Godfrey reacted to Bond's comments, calling into question the legitimacy of King's dedication to peace rather than equality: "If his [Bond's] criteria is to honor this road with the name of someone who has been designated as a 'peacemaker,' let us honor the truly great peacemakers of our locality—the veterans who gave their lives or endured great physical hardship to preserve our continued peace and freedom. As a former combat infantryman, I salute the veterans of all races and propose that the new perimeter road be named 'Veterans Memorial Parkway' in their honor" (*Statesboro Herald* 1994e, A4).

What is perhaps most evident in Godfrey's statement is the implication that King is somehow not "truly great" or somehow less worthy of commemoration than war veterans. Ray Hendrix, leader of Bulloch County's American Legion chapter articulated this belief much more firmly, when he lobbied the city council to name the perimeter for veterans. Hendrix proved to be an especially powerful "reputational entrepreneur" in not only legitimizing the cause of veterans but also delegitimizing King's commemoration. For example, Hendrix was quoted as saying: "I respect Dr King's accomplishments but the soldiers and sailors who fought for America's freedom helped make it possible for Martin Luther King to be a great man" (*Statesboro Herald* 1994h, A1). In Hendrix's statement is an attempt to represent the achievements of King as not only subordinate to, but also dependent on, the historical legacy of veterans.

In the reputational politics of representing King's commemoration as less important, Hendrix further connected the legitimacy of his memorial cause to the idea that area veterans had long been overlooked in terms of public recognition: "We've [veterans] supported everything in this county. Veterans of Bulloch County do volunteer work, support scouting, support other community programs and nothing has ever been dedicated on behalf of the veterans. Veterans of this county include all citizens, represent all people, and naming the road Veterans' Memorial Parkway is the least you can do" (Minutes of the Bulloch County Board of Commissioners 1994b, n.p.).

As seen here, the politics of memory engage not only marginalized social groups such as African Americans contesting the hegemonic order for power to reconstruct collective memories of the past. The politics of memory also involve competition with other groups, such as veterans, who also perceive themselves as "subordinate" to the prevailing commemorative powers. Indeed, several mail-in suggestions submitted to the perimeter-naming committee expressed the sentiment that honouring veterans represented a justifiable cause. As these two supporters of "Veterans Memorial Parkway" wrote: "I am writing in support of this idea [naming Perimeter for veterans], as we know there are many veterans who served our country that have never gotten the recognition they deserve. This would certainly be a fitting tribute to everyone who has served in the uniform of our country" and "As you know, we have thousands of veterans living and working in the Statesboro/Bulloch County area. Many of them served our country at a very unpopular time … I believe this would be a fitting way to finally say 'Thank You'" (Suggestions Submitted to Perimeter Naming Committee 1994, n.p.).

Running throughout these statements is a belief that naming the perimeter could correct an unequal relationship between veterans and the larger community. But supporters of Martin Luther King Jr also used this type of argument. Although writing in very subtle terms, the author of the following mail-in suggestion defined the legitimacy of naming a road after King in terms of its ability to inscribe a new vision of race relations into the landscape: "I want to convey my personal suggestion to name the road: Martin Luther King Parkway. I agree that this naming would affirm an important segment of our Statesboro and county population and would be a healing and unifying act" (Suggestions Submitted to Perimeter Naming Committee 1994, n.p.). The implication here is that attaching King's name to the perimeter highway will "unify" and "heal" the rift between African Americans and the larger city and county population. In order for street naming to facilitate such healing, people must personally identify with the social group and their commemorative cause. Building public identification with the past is not simply a matter of defining the legitimacy of a commemorative movement. As argued in the next section, it is also a matter of resonance.

The Resonance of Commemoration

In struggling to name streets, African American activists confront the barrier that not everyone feels that King's commemoration has resonance or relevance in their lives. Resonance, as Michael Schudson (1989) has pointed out, is a key factor in shaping the ultimate power of a cultural object. The extent to which the reputation of a figure resonates with the public determines, in large measure, the cultural influence of that historical representation and the population who will have their identity defined by this commemoration. Events in Bulloch County illustrate how variation in public identification with King affected the struggle to rename a street after him.

The importance of resonance to the commemoration was evident in comments presented earlier. As veterans Godfrey and Hendrix suggested, each of their respective suggestions would represent a wider population than if the road were named for King. Godfrey, for example, stated that he wanted to name the perimeter in such a way as to honour "the veterans of both races."

Hendrix, in advocating "Veterans Memorial Parkway" stated that veterans "represent all people." These comments imply that naming the perimeter after King is a symbol that resonates only with African Americans. In contrast, naming the perimeter in honour of veterans, they insist, is a biracial or multicultural symbolic project. The idea that the commemorative cause of veterans cuts across race and other lines of identity was articulated very well in an anonymous suggestion to the perimeter-naming committee: "To be fair and impartial to all peoples, I suggest we go with the name of 'Veterans Memorial Parkway.' This name would be all inclusive of males, females, all races, color, creeds, and all wars and skirmishes. With this name no one could claim they have been left out or not considered, for all families have, or have had a veteran in their family at some time" (Suggestions Submitted to Perimeter Naming Committee 1994, n.p.).

Thus, "Veterans Memorial Parkway" was represented as a "one commemoration-fits-all" type of naming, one that would unify rather than divide the city and county's population. Of course, the implication, then, is that naming the perimeter after King would not offer such unity. The widespread nature of this belief was further substantiated when I interviewed a Statesboro city planner, who also serves as a pastor in one of the county's African American churches:

> *Alderman*: Why did the city council vote to name the perimeter for veterans rather than for MLK?
> *City Planner*: Members [of the city council] felt veterans represented a larger community than King. However, Dr. King stood for everyone, fighting for the poor. Poverty knows no race. (Interview with author, 18 May 1997)

The reputational politics of commemorating King through street naming are shaped by a prevailing assumption that his historical relevance is limited to the black community and that streets named for him represent only African Americans. As illustrated in the city planner's comments, however, African Americans counter by emphasizing the universal importance of King's legacy. In his own case, he builds a class-based reputational account of the civil rights leader, reminding us how issues of poverty and economic inequality dominated the last years of King's life and career. For blacks in Bulloch County as well as across the country, the politics of commemorating King are about building an image of the civil rights leader that resonates across racial boundaries. Another anonymous letter sent to the perimeter-naming committee put forth the idea that King's importance was not limited to one race or even one country: "I think that Martin Luther King had profound effects on *all of America*. He was an inspirer and friend whose love helped to change the *attitudes of the world*. He has enlightened so many lives with his on-living dream. I think the road should be named MLK Drive. Having the Drive named such would express the thanks he deserves [emphasis in original]" (Suggestions Submitted to Perimeter Naming Committee 1994, n.p.).

In the struggle to define the public resonance of commemorating veterans over King, even timing played an important role. The deadline for submitting perimeter-name suggestions to the Bulloch County government ended on 1 June 1994, only a few days after Memorial Day. More importantly, however, 6 June 1994, marked the 50th anniversary of "D-Day." The local newspaper carried veterans-related articles in the weeks preceding and coinciding with the D-Day anniversary. One of the anonymous suggestions submitted to the perimeter-naming committee on 30 May 1994, reflected how the naming of the perimeter was considered in close relation to the veterans' holiday: "As we have participated in our local observation of Memorial Day and seen it celebrated nationwide, what better tribute we can pay the memory of those who have served and died than to name this road in their honour? *It seems a right time* [emphasis added]" (Suggestions Submitted to Perimeter Naming Committee 1994, n.p.).

It is quite possible that the Memorial Day celebration and the anticipation of the D-Day anniversary expanded the resonance of honouring veterans due to the power of holidays to focus briefly, but intensely, public attention and identification. In this instance, street naming has to be examined within the larger commemorative genre that it shares with holidays. Of course, African Americans have used this same relationship to their own commemorative advantage. For instance, the original request by African Americans to name the perimeter after King took place on 1 February 1994, only two weeks after the celebration of the King federal holiday.

The Hybridity of Commemoration

Berg and Kearns (this volume) have suggested that the positions and identities of cultural actors involved in place naming are often ambiguous, complexly intertwined, and sometimes contradictory. There can be hybridity and interdependence in people's interests, thus leading to some rather unexpected political formations (Rose 1994). The politics of naming Bulloch County's perimeter road cannot be understood without considering the multi-positionality of African Americans within the debate to memorialize Martin Luther King Jr versus military veterans. King's memory, while certainly important, was not the only commemorative cause that African Americans had an interest in pursuing.

In many of my informal conversations with African American leaders in Bulloch County, the right of veterans to be commemorated through street naming was rarely criticized or refuted, although many asserted the greater necessity of memorializing King. Some African Americans found themselves in a difficult ideological position both as supporters of honouring both King and veterans. According to a local American Legion leader, many black men signed letters of support for naming the perimeter for veterans (interview with author 13 April 1998). This issue of commemorative hybridity was brought into sharper focus during an interview with an African American pastor who had spoken in support of naming the perimeter highway for Martin Luther King Jr at county commission and city council meetings. He described the difficulty in "placing" himself within the debate over commemorating veterans versus King:

> *Alderman*: How did you feel about the perimeter being named for veterans instead of King?
> *African American Pastor*: I wanted King's name on the perimeter road but I couldn't feel bad about honoring veterans ... When they named the perimeter for veterans, I was touched. I am a veteran and I have people [family] who were killed in war ... We walk around this country free because somebody died. When you come up with something for veterans, there are not many people who are going to fight that, even though you want the other part [King's commemoration] ... We should honor those who sacrificed so much. (interview with author 18 May 1997)
>
> *Alderman*: When you see Veterans Memorial Parkway, do you identify with it?
> *African American Pastor*: You believe it! And not just me. A lot of black veterans identify with it. You are not going to find a young man who served [in the military] who doesn't. (Interview with author 18 May 1997)

These comments illustrate how the politics of naming streets after King were complicated by the fact that some reputational entrepreneurs found themselves caught in the middle of two commemorative campaigns with which they identified and which they perhaps supported.

In the case of the pastor I interviewed, the reputational politics of naming the perimeter became so hybrid and intertwined that he saw similarities in both commemorative causes, drawing strong connections between King's quest for justice and the role of veterans in fighting for freedom. On this point he said: "Dr King wanted justice and he gave his life for it. That is what the soldier was fighting for. If that man Hitler had won, we wouldn't have any justice. That's why I went [to war]. I went because I wanted my children to live in a free country. And that's what he [King] was doing. It's the same thing. Rather than being a war in a foreign country, it was some one working for peace right here in the US" (interview with author 18 May 1997). The pastor did not recognize the potential contradictions of comparing King's efforts to achieve peace—based on a philosophy of non-violence and passive resistance—with the efforts of veterans to win peace through violent warfare. Nor does he mention King's controversial opposition to the Vietnam War. Instead, he creates a reputational account that finds commonality in King's death for freedom and the personal sacrifice of military veterans, making an analogy between the fight for civil rights in the US to the fight in Europe to topple Hitler. Regardless of whether these comments are representative of other African Americans or not, they provide keen insight into how one black veteran interpreted and represented the meaning of King's legacy and reputation in the face of choosing between two commemorations near and dear to his heart. From a political standpoint, the representation of King as another soldier in the war against injustice and oppression proposes problems for the extent to which African Americans can represent the civil rights leader's legacy as more legitimate and more resonant than the memory of military veterans. It is perhaps important to recognize that the reputational politics of memorializing King, or any historical figure for that matter, is a path-dependent process, in which commemorative images of the past are constructed and realized dialogically in the context of other memorials (Olick 1999).

Concluding Remarks

Street names serve as important arenas for social actors and groups to actively interpret the past, define historical reputations, and debate the relative value of remembering one commemorative cause over another. Discursive rivalries over the meaning and reputation of historic figures such as King give insight into the conditions under which commemorative street naming occurs. King streets also demonstrate how the American toponymic landscape has become an important site of social struggle as minorities, particularly African Americans, seek to have their historical contributions and struggles, which are often not single or monolithic, seen and heard publicly.

References

Alderman, D.H. (2006), "Naming Streets after Martin Luther King, Jr: No Easy Road," in R. Schein (ed.), *Landscape and Race in the United States* (New York: Routledge Press), 213–36.

Alozie, N.O. (1995), "Political Intolerance and White Opposition to a Martin Luther King Holiday in Arizona," *Social Science Journal* 32(1), 1–16.

Atlanta Journal-Constitution (1992), "Failure to Name Street to Honor MLK May Bring Boycott," 28 November, B9.

—— (1993), "Back Streets Get King Name," 12 September, Fl and F4.

—— (1999), "The Many Signs of MLK," 17 January, A1.

Azaryahu, M. (1996a), "The Power of Commemorative Street Names," *Environment and Planning D: Society and Space* 14(3), 311–30.

—— (1996b), "The Spontaneous Formation of Memorial Space: The Case of *Kikar Rabin*, Tel Aviv," *Area* 28(4), 501–13.

—— (1997), "German Reunification and the Politics of Street Names: The Case of East Berlin," *Political Geography* 16(6), 479–93.

Berg, L.D. and Kearns, R.A. (1996), "Naming as Norming: 'Race,' Gender, and the Identity Politics of Naming Places in Aotearoa/New Zealand," *Environment and Planning D: Society and Space* 14(1), 99–122.

The Boston Globe (1998), "Many Measures of King's Legacy; Day Viewed as American Paradox," 19 January, A1.

—— (1999), "New Hampshire Governor Signs MLK Holiday into Law," 8 June, B8.

Chacón, R. (1995), "Cesar Chavez Boulevard: Efforts to Suppress a Commemoration and Chicano Political Empowerment in Fresno, CA," *Latino Studies Journal* 6(2), 73–93.

Chicago Defender (1989), "Outrage Expressed Over King Labeling," 18 January, 3.

Daynes, G. (1997), *Making Villians, Making Heroes: Joseph R. McCarthy, Martin Luther King Jr and the Politics of American Memory* (New York: Garland Publishing, Inc.).

DeLyser, D. (1999), "Authenticity on the Ground: Engaging the Past in a California Ghost Town," *Annals of the Association of American Geographers* 89(4), 602–32.

Dixon, M. (1994), "The Black Writer's Use of Memory," in G. Fabre and R. Meally (eds), *History and Memory in African American Culture* (New York: Oxford University Press), 18–27.

Ducharme, L.J. and Fine, G.A. (1995), "The Construction of Nonpersonhood and Demonization: Commemorating the Traitorous Reputation of Benedict Arnold," *Social Forces* 73(4), 1309–31.

Dwyer, O.J. (2000), "Interpreting the Civil Rights Movement: Place, Memory, and Conflict," *Professional Geographer* 52(4), 660–71.

—— and Alderman, D.H. (2008), *Civil Rights Memorials and the Geography of Memory* (Athens, GA: University of Georgia Press).

The Economist (1983), "Honoured But Still Controversial," 22 October, 23–4.

Edensor, T. (1997), "National Identity and the Politics of Memory: Remembering Bruce and Wallace in Symbolic Space," *Environment and Planning D: Society and Space* 15(2), 175–94.

Fine, G.A. (1996), "Reputational Entrepreneurs and the Memory of Incompetence: Melting Supporters, Partisan Warriors, and Images of President Harding," *American Journal of Sociology* 101(5), 1159–93.

—— (1999), "John Brown's Body: Elites, Heroic Embodiment, and the Legitimation of Political Violence," *Social Problems* 46(2), 225–49.

Foote, K. (1997), *Shadowed Ground: America's Landscapes of Violence and Tragedy* (Austin, TX: University of Texas Press).

González Faraco, J.C. and Murphy, M.D. (1997), "Street Names and Political Regimes in an Andalusian Town," *Ethnology* 36(2), 123–48.

Gulliver, H. (2000), "King Took America to the Mountaintop," *Georgia Trend* 15(5), 11–19.

Houston Chronicle (1998), "City Council Won't Touch Parade Feud: Two MLK Events Set, But Only One Has Permit," 7 January, A15.

Jet (2000), "Newsmakers," 15 May, 34.

Johnson, N.C. (1994), "Sculpting Heroic Histories: Celebrating the Centenary of the 1789 Rebellion in Ireland," *Transactions of the Institute of British Geographers* 19(1), 78–93.

Jones, J.P. (2000), "The Street Politics of Jackie Smith," in G. Bridge and S. Watson (eds), *The Blackwell Companion to the City* (Oxford: Blackwell, 2000), 448–59.

Kenny, J. (1995), "Making Milwaukee Famous: Cultural Capital, Urban Image, and the Politics of Place," *Urban Geography* 16(5), 440–58.

Minutes of the Bulloch County Board of Commissioners (1994a), 1 February.

—— (1994b), 21 June.

National Public Radio Morning Edition (1998), "Martin Luther King Holiday," 19 January: transcript #98011903–210 (available through Lexis/Nexis).

Newport, F. (1999), "Mother Teresa Voted by American People as Most Admired Person of the Century," *Gallup News Service* 31 December, available online at http://www.gallup.com/poll/releases/ pr991231b.asp.

Olick, J.K. (1999), "Genre Memories and Memory Genres: A Dialogical Analysis of May 8, 1945 Commemorations in the Federal Republic of Germany," *American Sociological Review* 64(3), 381–402.

Rhea, J.T. (1997), *Race Pride and the American Identity* (Cambridge, MA: Harvard University Press).

Rose, G. (1994), "The Cultural Politics of Place: Local Representation and Oppositional Discourse in Two Films," *Transactions of the Institute of British Geographers* 19(1), 46–60.

Schudson, M. (1989), "How Culture Works," *Theory and Society* 18(2), 153–80.

Schwartz, B. (1997), "Memory as a Cultural System: Abraham Lincoln in World War I," *International Journal of Sociology and Social Policy* 17(6), 22–58.

Seattle Times (1990), "Street's Name Switch Riles Portland Residents, Fierce Public Backlash to Avenue Named after Martin Luther King Jr," 4 March, D5.

Seligman, A. (2001), "The Street Formerly Known as Crawford," *Chicago History* 29(3), 36–51.

Statesboro Herald (1994a), "County Asked to Withdraw Endorsement," 2 March, A11.

—— (1994b), "Main Street Makes Request to Council," 18 May, A1 and A11.

—— (1994c), "Perimeter Road Name Ideas Rolling In," May 20, A1 and A6.

—— (1994d), "Recognize Accomplishments of King by Naming Road for Him," 22 May, A4.

—— (1994e), "Name Road for Veterans," 29 May, A4.

—— (1994f), "Perimeter Road Naming Deadline Nears," 6 June, A1.

—— (1994g), "City, County to Settle Debate over Naming Perimeter Road," 21 June, A1 and A3.

—— (1994h), "City and County Differ on Name of New Road," 22 June, A1 and A8.

St Petersburg Times (1990), "King's Fight Still in the Streets: Re-Naming of Roads Incites Controversy," 23 April, B1.

Stump, R.W. (1988), "Toponymic Commemoration of National Figures: The Cases of Kennedy and King," *Names* 36(3/4), 203–16.

Suggestions Submitted to Perimeter Naming Committee (1994), County Courthouse Annex, Bulloch County, Georgia, 1 April–1 June.

Tilove, J. (2003), *Along Martin Luther King: Travels on Black America's Main Street* (New York: Random House).

Tuck, S.G.N. (2001), *Beyond Atlanta: The Struggles for Racial Equality in Georgia, 1940–1980* (Athens, GA: University of Georgia Press).

Weightman, B.A. (1988), "Sign Geography," *Journal of Cultural Geography* 9(1), 53–70.

Williams, B.B. and Dill, B.T. (1995), "African Americans in the Rural South: The Persistence of Race and Poverty," in E. Castle (ed.), *The American Country Side: Rural People and Places* (Lawrence, KS: University Press of Kansas), 339–51.

Yeoh, B.S. (1996), "Street-Naming and Nation-Building: Toponymic Inscriptions of Nationhood in Singapore," *Area* 28(3), 298–307.

Chapter 10
Indexing the Great Ledger of the Community: Urban House Numbering, City Directories, and the Production of Spatial Legibility

Reuben Rose-Redwood

Introduction

On 1 March 1768, King Louis XV declared that all cities and towns in the kingdom of France—with the exception of Paris—were required to affix numbers to houses in their jurisdiction, largely for military purposes. It was a common, yet unpopular, practice during this period for soldiers to be housed in civilian quarters with individual families in France. The chief aim of the 1768 house numbering decree was to make it easier to keep track of troops when they were dispersed throughout a given city or town. Being the centre of royal power, Paris was exempt from this rule because soldiers lived in official barracks rather than among civilians. Interestingly, evidence suggests that one of the earliest recorded cases of the numbering of houses actually occurred in Paris on the Pont Notre Dame in 1463. It was not until 1779, however, that a concerted effort was made to number the houses of Paris when the publisher of a city guide and directory, Marin Kreenfelt, hired a team of men to inscribe numbers upon houses along the Parisian streets. Kreenfelt's plan was met with considerable resistance from one of the magistrates of the Paris Parlement, François-Louis Joly de Fleury, who swiftly ordered the Lieutenant General of Police to put an end to Kreenfelt's scheme. Yet, by the end of the eighteenth century, the city authorities relied extensively on house numbers as the geographic foundation of taxing, governing, and policing the population (Pronteau 1966; Miles 1972; Garrioch 1994, 2002; Hunt 1995).

Other cities across Europe also began numbering houses during the second half of the eighteenth century. There is some evidence of numbered houses along several streets in London as early as 1708, but it was only after the Court of Common Council passed a law in 1765 that the city was officially required to number houses and place street signs at the corners of each intersection (Miles 1972; Joyce 2003). The first systematic house numbering plan for Copenhagen was established in 1771 in response to a petition by the city's first directory publisher. Geneva began its first extensive house numbering campaign in 1782, and the numbering of houses was introduced in the city of Berlin in the 1970s. Other cities soon followed the general trend of adopting the practice of house numbering as part of what, in retrospect, can be seen as a pivotal historical moment in the rationalization of urban space (Lynch 1960; Tscherny 1992; Garrioch 1994; Azaryahu 1996; Shaw and Coles 1997; Tantner 2007).

Despite the central role that house numbering has played in the production of urban space, the subject has drawn surprisingly little attention among geographers, historians, and other scholars until quite recently. Daniel Smail has suggested that the general topic of street addressing remains one of the "least studied" areas of historical inquiry. As Smail (2000, 37) contends:

> [t]he oversight is understandable, for the address has none of the innate appeal of race, gender, madness, or sexuality. But there are good reasons for studying the historical construction of the address ... The historical invention of the address as a nesting set of geopolitical terms was arguably one of the major geographic contributions to modern national identity, the other being the better-known creation of an idea of nation as a geographical entity with borders ... So important has the address become that few of us nowadays ... pause to consider the modern association between identity and address and the ways in which having an address frames us as citizen-subjects embedded within a series of geopolitical entities culminating in the nation-state.

In his own work, Smail examines the practice of geographic addressing in Marseille prior to the widespread adoption of house numbering during the eighteenth century. He explores how various strategies of spatial identification were developed by non-state actors, such as public notaries, during the late medieval period.

A classic historical study of house numbering is Jeanne Pronteau's *Les Numérotages des Maisons de Paris du XVe Siècle a Nos Jours* (1966), which has yet to be translated into English. David Garrioch (2002, 238) has brought some of Pronteau's ideas into circulation among English-speaking scholars by arguing that house numbering in Paris "was partly a reflection of an official desire to classify, to make urban space more uniform. But it was also a response to new ways of moving round the city" (also, see Garrioch 1994). Various other scholars have noted the importance of house numbering but have not made this topic the central focus of their investigations (Azaryahu 1996; Henkin 1998; Joyce 2003). One exception is a fascinating, yet little-known, study by Joel Morbito (1965), which considers the "physical and social implications" of street naming and house numbering in the United States from a modernist perspective. More recently, Anton Tantner (2007) has explored the history and politics of house numbering in Austria during the 1770s. Additionally, Christopher Thale's (1986, 2007) work has brought the question of house numbering to the attention of historical geographers. In a recent article in the *Journal of Historical Geography*, Thale provides a brief overview of urban house numbering in the United States and offers an insightful analysis of the spatial politics of house numbering reform in twentieth-century Milwaukee. Thale (2007, 126) critiques the traditional assumption that "[house] numbering promoted an unproblematic public interest," and he convincingly illustrates how the renumbering of a city's houses can be a politically contentious issue.

Drawing upon the work of scholars such as Maoz Azaryahu, Thale (2007, 126–7) highlights the symbolic importance of house numbers and argues that they constitute "a text written on every building, linking each to the whole and conveying such definitions daily to all." As part of what Azaryahu (1996, 324) calls a "city-text," house numbers are seen as an element of "the toponymical grid that makes the city geographically intelligible." This description of the city-as-text is not a mere superimposition of semiotic terminology upon the historical record. Rather, this chapter maintains that the early proponents of house numbering in the United States—city directory publishers and later city engineers—were themselves entranced with imagining the city as a "text." This modernist conception of the city-as-text was based not upon the discursive analysis of textual narratives, as we find in the so-called "new" cultural geography (e.g., Duncan 1990). Instead, the main focus related to what I shall call the "physical typography" of the city-text. The cityscape was conceived of as a "text" that required adequate "page numbers," an alphabetized "index," and a coherent "page layout." It is my contention that this typographical conception of the "layout" of the city-text is key to understanding the modernist project of constructing legible urban spaces.

This typographical framework continues to inform the spatial rationalities that underpin the current practice of digital geo-coding, which relies extensively on the maintenance of comprehensive street address systems (Monmonier 2002; Curry et al. 2004). The encoding of geographic space often entails the creation of a *spatial regime of inscriptions* that is literally inscribed into the spaces it thereby constructs. Once spatialized, a system of inscriptions enables the construction of a network of "circulating references" (Latour 1999) that establishes the basic contours of a "regime of truth" (Foucault 1980). House numbering systems are a prime example of such a coordinated regime of spatial inscriptions and are, therefore, central to the making of a "geo-coded world" (Curry 1999a, 1999b; Pickles 2004; Rose-Redwood 2006).

More than a simple way-finding device, the house number has historically been utilized as a technology of spatial individualization and remains an integral part of the broader project of envisioning the city as a "legible" social space (Scott 1998; Smail 2000; Joyce 2003). Here I examine how the spatial practice of house numbering directly linked the production of abstract space together with the formation of place-specific knowledges of governmentality in nineteenth-century US cities and towns (Lefebvre 1991; Foucault 1991, 2007). Previous scholars have noted that city directory publishers played an important role in the early history of urban house numbering in the United States, yet the precise strategies that directory publishers employed to spatially individuate the urban population, as well as the tactics of resistance that they encountered, have not been examined in any detail in the US context.

This chapter illustrates how city directory publishers were among the leading proponents of numbering houses at a time when many local governments had yet to firmly commit to systematic house numbering as an essential responsibility of the local state. Supported in large part by a city's business community through sales and later advertisements, the directory industry sought to facilitate commercial transactions by improving the "value of time" through the rationalization of urban space. Additionally, by linking the name of the "head of household" to a designated street address, the city directory also reinforced existing gender hierarchies while claiming to provide an objective, albeit imperfect, "mirror" for the local community. Although not directly part of the state apparatus, the directory industry's involvement in rationalizing urban space—both physically and textually—was an important component of what governmentality scholars call the "technologies of government" that individuated the population as a means of rendering social life "governable" (Dean 1999; Rose 1999).

In order to understand how the numbering of houses was embedded within a broader project of geographical knowledge production, it is necessary to examine the political economy of city directory publishing in nineteenth-century America as well as the power dynamics of gathering and compiling spatialized knowledges. Many residents resisted attempts by directory canvassers to acquire their name and address, especially in the South during the Civil War. The rapid spatial changes that American cities experienced during the nineteenth century complicated the process of imposing a static numbering system upon an urban landscape in the midst of constant transformation. This spatial contradiction was eventually "resolved" through the decimalization of house numbering plans, the establishment of Cartesian quadrant systems, and the implementation of numbering by street frontage.

By the turn of the twentieth century, house numbering had been integrated into the workings of municipal government, and the influence of directory publishers was surpassed by that of city engineers. It was at this time that the discourse on street and house numbering was institutionalized among city engineers and managers. Attempts were made to "scientifically" design streetscapes according to the principles of spatial legibility. The comparison of the layout of a city with the typographic design of a book was again employed as a means of envisioning the city as a text. The logic of legibility was pervasive in the increasingly technical literature on house and property numbering, and it is this very literature that now informs those calling for a renewed focus on creating comprehensive street address systems as the basis of geo-coding the world. The chief aim of this chapter is to rethink the entire notion of viewing the city as a "text" by historicizing the textualization of urban space as a crucial strategy in the production of spatial legibility.

"An Index to the Great Ledger of the Community" and "the Soul of Commercial Transactions"

Cadastral mapping and property numbering were widespread practices in the colonial settlements of North America (Kain and Baigent 1992), but the use of house numbering as a "technology of government" did not arise until the latter eighteenth century. When the British captured Manhattan during the American Revolution, one account suggests that they began numbering houses across the Island to establish military control (*Trow's New York City Directory* 1878, vii). Prior to the war, very few streets in Manhattan had street signs posted at intersections, and the population relied chiefly on shop signs, descriptive designations, and local familiarity when navigating through the city's streets (Hoffmann 1937; Isaacs 1940; Henkin 1998). In 1793, New York's Common Council eventually passed its first systematic house numbering law, which followed Philadelphia's lead in the emerging trend of separating odd and even numbers on opposite sides of each street.

Philadelphia had adopted the odd/even system in 1790, when the U.S. Marshal Clement Biddle conducted the first federal census and renumbered the city's houses (Morbito 1965; Alotta 1990). A rudimentary system of house numbering had previously been devised in 1785 by the editor of the first Philadelphia city directory, John Macpherson, after he had unsuccessfully attempted to convince the local authorities to number all of the houses in the city (MacPherson 1785). Macpherson's numbering plan followed the English practice of consecutively numbering houses along one side of the street and then proceeding down the other side continuing the same sequence (Alotta 1990; also, see Isaacs 1940). The English method was quickly deemed inadequate, since it resulted in the juxtaposition of high and low house numbers directly across from each other. After renumbering Philadelphia's houses using the odd/even method, Biddle then published his own city directory (Biddle 1791). This new plan had its own problems, however. As Robert Alotta (1990, 119) has pointed out, "[i]f a vacant lot existed, Biddle skipped it and numbered the next house," which led to various difficulties as new houses were later built. These spatial contradictions were not adequately addressed until the mid-nineteenth century.

As the case of Philadelphia illustrates, the introduction of house numbering into American urban life was directly tied to the development of city directory publishing in the United States. London is generally credited with having published the first city directory in 1677. However, it was only during the eighteenth century that city directories were published annually in London, and by the nineteenth century the directory industry was global in its reach

(Williams 1913; Shaw and Coles 1995, 1997, 2002; Coles and Shaw, 1997). Philadelphia (1785) and New York (1786) were the first American cities to publish directories. Four other US urban centres—Boston (1789), Charleston (1790), Baltimore (1796), and Hartford (1799)—had published city directories before the end of the eighteenth century. The number of cities publishing their first directory continued to rise until the trend eventually peaked in the 1850s and fell precipitously with the onset of the Civil War (Figure 10.1). The same overall pattern is evident when examining the total number of city directories published in a given year (Figure 10.2). The total number peaked in 1859 at over 90 and then fell below 30 in 1862. The number of directories published annually stabilized after the war, remaining between 60 and 70 each year from the 1870s until the end of the century (*City Directories of the United States* 1983).

Many of the early directory publishers were either real estate brokers or book publishers (Burton 1956). In some instances, those publishing a city directory had previously been employed by the government to conduct a census and then compiled a directory with the information collected. Business directories provided details for locating a city's leading commercial establishments, whereas city directories consisted primarily of an alphabetized listing of "heads of households" throughout selective portions of the city. Since they were in the business of linking people's names with geographical locations, city directory publishers soon realized that their job would be much easier if houses were numbered in some systematic manner.

Before the publication of the first city directory for Norfolk, Virginia, in 1801, the publisher Charles Simmons successfully petitioned the local authorities to allow him to number the houses and designate official street names in the city. The official government statement, issued prior to the publication of the directory, stated that "[Simmons] is about to compile and publish a Directory, in which he proposes to designate the places of abode of the inhabitants of this borough, and solicits the countenance and patronage of the court, setting forth that it would greatly aid and facilitate the undertaking by being duly authorised to number the buildings" (as quoted in Simmons 1801, iii). In this example, the government did authorize Simmons in his "undertaking," but this was not always the case in other cities.

In many American cities, the proposition to begin numbering houses was often initially posed by the editors of city directories, and the governing authorities were not always immediately receptive to their requests. Throughout the nineteenth century, city directory publishers across the United States constantly complained that the local authorities had failed to properly number houses. It was often viewed as "a great disadvantage and embarrassment to us, so that it has rendered it impossible to give a minute direction to some of the residences" (*Bridgeport and East Bridgeport Directory* 1858, n.p.). It was not uncommon for the publishers of city directories to decry "[t]he neglect of the proper authorities, in not numbering the buildings" (*Cleveland City Directory*

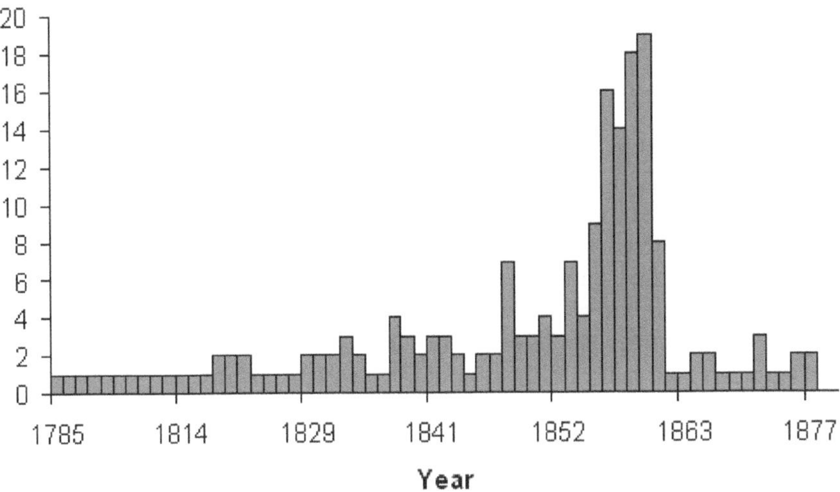

Figure 10.1 The number of US cities publishing a city directory for the first time per year

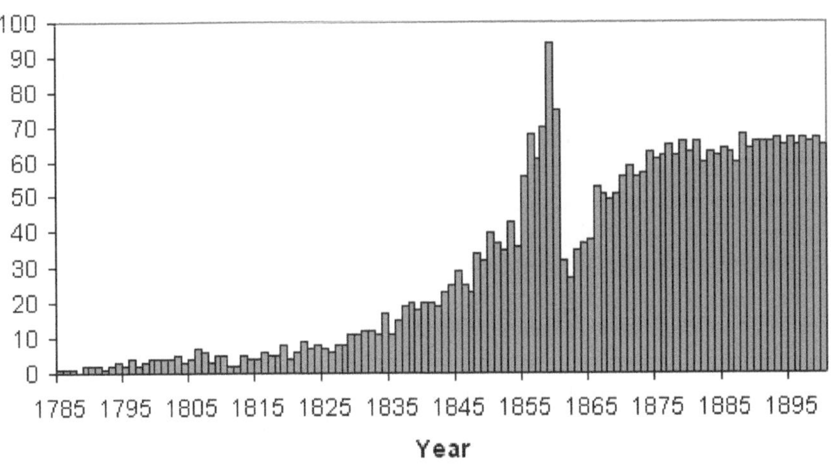

Figure 10.2 The total number of US cities publishing a city directory per year

1856, 3). This claim was made well into the nineteenth century in many US cities and towns.

Sometimes city directory publishers—in a move similar to Kreenfelt's in Paris—took it upon themselves to number the buildings without getting an official stamp of approval from the city government. This is precisely what happened in Mobile, Alabama, in the late 1830s. When Mobile's first city directory was published in 1837, none of the city's houses had been numbered. Prior to the publication of the second directory in 1838, its publisher "presented the city council with a petition of some seven or eight hundred names, principally property holders and business men, praying that honourable board to number the houses, which was rejected." The publisher then proceeded to number the houses "at his own expense," hoping that he would make up the cost with the sale of his directories. One year later, the same publisher considered numbering the houses again but became discouraged when he found that many of the numbers that he had placed upon houses the previous year had either been "torn down, or painted over." He was not willing to re-number the houses and noted that "some measures, either by the council or property holders should be taken to secure their permanency" (*Mobile Directory or Strangers' Guide* 1839, n.p.).

Even in smaller towns, such as Dover, New Hampshire, city directory editors complained that visitors and those "engaged in business" were constantly "subjected to perplexity" due to the lack of proper street signage and house numbering (*The Dover Directory* 1830, n.p.). The numbering of a city's houses was represented as an indispensible aid to commercial development, so it is little surprise that the business sections of a city were often the most comprehensively numbered (Thale 2007). Urban space was often compared to a gigantic recordkeeping book. Just as a book needs page numbers, the argument went, so too does a cityscape need serialized house numbers. Likewise, if a book needs an index of names and subjects, a city-text would also need what one city directory editor referred to as "an index to the great ledger of the community" (*The Beloit Directory, and Business Mirror* 1858, n.p.). This was precisely what city directories claimed to provide to the subscribing public (e.g., *Boston Directory* 1855; *Boyd's Philadelphia City Business Directory* 1858).

Not only was the city directory to serve as an index in a functional sense, it was also seen as a trustworthy "index to the growth and prosperity of a city" (*The Charlestown Directory* 1852, n.p.). In other words, their publishers viewed city directories as indices of "progress," both in demographic and economic terms. According to one directory editor, the organized contents of a city or business directory offered "a sure and reliable index to the character and resources of the places spoken of" (*Finning's Bridgeport Town and City Directory* 1855, n.p.). One turned to city directories for so-called "facts" about the city. As a different directory editor put it, "a publication of this order, allow[s] no scope for the play of the imagination, being a simple exponent of facts" (*Boyd's Lancaster Directory* 1857, iii). Another commented, "I have often been asked, "Where do you get those facts?" I cannot give you the details, but it is enough that you have them, so that they may be transmitted to posterity" (*Drown's Record and Historical View of Peoria* 1850, n.p.). Directories produced authoritative "facts" that provided a record of urban life for contemporaries and for posterity as they shaped interpretations of the geographies of the past.

Once collected and compiled, these "facts" could then "essentially aid the man of business, the capitalist and the stranger in their pursuits, and place before each, at a glance, much useful information" (*A Directory for the City of Buffalo* 1837, n.p.). The editor of the 1838 Philadelphia directory suggested that the publication was essential for "economising time, and expediting his general business transactions" (*O'Brien's Wholesale Business Intelligencer and Southern and Western Merchants' Pocket Directory*, n.p.). By minimizing delays in obtaining commodities or pursuing one's desired destination, the city or business directory—and the city's corresponding house numbers—resulted in the "economising" of time through the production of urban space as a city-text, fully equipped with "page numbers" and a bona fide alphabetical "index."

Directory publishers also argued that it was necessary for the whole city as well as its individual parts to be accessible "at a glance" to the public. In short, the directory sought to totalize and individualize at one and the same time. These dual strategies were deemed essential:

> in larger communities where the members are many—where the diversity of interest, pursuits and of property, is disconnected and great—where the sphere of the whole is extensive and grand, but the space of the several parts contracted and small, it becomes highly important to commerce at large, that the place and the business of each individual should be easily known. Intelligent experience will suggest the value of time, in the conduct of prosperous affairs. Its economy in business is the source of all wealth. Punctuality is the life—the soul of commercial transactions. None are above its demands however exalted or proud—none so humble and low but may serve its request. To assist the citizen and the stranger in the proper use of his labor and time in the pursuit of his interest, is the purpose of our Directory. (*Daughdrill and Walker's General Directory for the City and County of Mobile* 1856, 3)

The production of legible urban spaces was chiefly a way in which to improve "the value of time"—with punctuality serving as the "soul" of the circulation of capital. The publication of a city directory during the eighteenth and nineteenth centuries most often led a city to number its houses—if it had not already done so—thereby initiating the production of abstract space and the economization of time.

City Directories and the Strategies of Spatial Knowledge Production

Although some directory compilers utilized tax assessment records to produce their directories, the majority of publishers hired a team of men to "canvass" the city door-to-door, or did so themselves. In short, they conducted a privately-financed census of the city, often on an annual basis. It was claimed that city directories, and the related practice of house numbering, benefited a city's business community at large. Free copies of city directories were commonly placed in public spaces, such as post offices, hotels, railroad depots, libraries, banks and on steamships, thereby drawing attention to a city from those at a distance. Local boosterism was often framed within a discourse of national and imperial progress. Although city directories were most often produced for local markets, they also allowed cities to boast their achievements to the nation and project their future greatness in order to encourage capital investment.

City directories provided a highly selective depiction of the "public," which was often confined to white, male heads of households from the so-called "better" portions of a city (Burton 1956, 6). Working-class residents and people of colour were often systematically excluded from directories. As a directory publisher in New York City arrogantly professed, "the names of laborers, colored people, persons in low obscurity who rent tenements by the week or month, may be excluded without impairing the utility of the work" (*Longworth's American Almanac, New-York Register and City Directory* 1829, 638). Such racist and class-biased procedures of exclusion were cloaked in the guise of a parsimonious objectivity.

The gendered dimensions of directory publishing are also worth considering with respect to both the content of the directory as well as the knowledge gathering strategies of the directory canvassers. Women, domestic servants, and children—that is, the legal "dependents" of the "head of household"—were generally excluded from the directory, which was not a mere oversight on the part of city directory publishers. Rather, the directory can be seen as part and parcel of an attempt to naturalize and legitimize patriarchal family relations that were then codified by the American legal system as a form of what Carole Shammas (2002) has called "household government." Although not explicitly discussed by Shammas or other feminist historians, the practice of house numbering was, at least in part, a strategy of pinpointing the "household" as a site of government, with the male head of household then serving as the *representative* of the familial unit to the public at large, as depicted in the city directory.

Although servants, women, and children were generally excluded from the city directory, they were ironically the main sources of the information that filled its pages. It was rare for a directory canvasser to actually acquire information directly from the "head" of each household when going door-to-door. Most of the information was obtained from servants, as well as the wives and children, of the familial patriarchs. "The head of the family cannot always be at home," noted the Baltimore city directory of 1853, "and we must take the required information from illiterate servants or children" (*Matchett's Baltimore Director* 1853, n.p.) The city directory constructed an image of the "private sphere" for public display in which heads of households were located both socially and geographically as legitimate members of the community. However, those members of the household who were excluded from the finished product of the directory nevertheless proved indispensable to the actual *production* of spatial knowledges of individualization.

The Tactics of Resistance to Spatial Individualization

The publishers of city directories saw themselves as offering a public good that benefited the entire community. They were quite perplexed, then, to find that a significant number of inhabitants commonly refused to give the requested information or attempted to provide "false" responses to the inquisitive canvassers. "In some instances," explained the Memphis city directory of 1855, "the persons applied to were improperly reluctant in giving their names and places of business, seeming to think it an evidence of smartness to annoy the canvasser, others were too busy to attend to the matter" (*W.H. Rainey and Co.'s Memphis City Directory* 1855, 29). The directory canvasser was commonly taken to be an agent of the government, such as a tax assessor, a census taker, or a military drafter. "Sometimes people would refuse to give their names," noted one directory, "fearing it would cause them to be taxed, or stand a draft in the militia, or for the jury. Some would not give their full christian names, and many gave names without a residence" (*Mobile Directory or Stranger's Guide* 1839, n.p.). The linking of name and residence could serve various governmental purposes, and not all urban inhabitants wished to voluntarily offer such information.

Some would go to great lengths to obstruct the directory canvassers. "We have known wives to forget that they had husbands," exclaimed a directory editor, "sisters to forget that they had brothers … especially when they supposed that a tax bill or a notification to do military duty might be in the offing!" (*The Portland Reference Book, and City Directory* 1850, n.p.). This led one directory editor to respond by remarking that "[w]e advise every man to let his wife know his business, and learn her how to spell his name" (*Williams' Cincinnati Directory* 1849, 4). Women themselves were only deemed worthy of entry into a city directory if they had been widowed and were thus "heads of households."

Those who refused to cooperate were depicted as being "ignorant" and "unreasonable" (*Matchett's Baltimore Director* 1842, iii–iv). Yet, there was nevertheless considerable resistance to spatial individualization. As one directory editor explained:

> "But," says one, "will not every person gladly furnish such information for a work so useful?" To such enquirer we will only say, let them make the experiment. In hundreds of instances it is found very difficult to obtain the information wanted; and in many cases the applicant has the door rudely shut upon him before he can ascertain the names of such of the family as should be inserted. What is he to do in this case? He is not, like those employed to take the census, clothed with any power by which he can compel these churlish persons to furnish the desired information. He must therefore seek it from other sources, or fail in obtaining it. (*The Providence Directory* 1841, iv)

Quite revealing is the directory editor's comment, which contrasted his project to that of the government census taker by acknowledging that he had no official authority to "compel" a city's inhabitants to provide "the desired information." If some felt that the directory canvassers were "meddlers with that which was none of [their] business," the canvasser could not force them to respond (*The Charlestown Directory* 1831, 10).

Although directory compilers depicted the inhabitant's concerns respecting tax collection and jury duty as irrational fears and misconceptions, there was actually good reason to make such connections. Government employees, such as postal workers, certainly made use of directories to serve their own ends. Additionally, there were numerous instances when someone who was hired to conduct an official census for the government then used the "data" collected to publish their own city directory for private profit (*Clinton Almanac and Directory* 1856). Even when the directory canvassers were not directly linked to the state, it was not uncommon for them to refer to their canvassing as a sort of "census" conducted by those "who shall prosecute the details of this department of the work with as much care and precision as an agent of the government would take the census" (*New Orleans Directory* 1842, 5).

City directories were indeed powerful technologies of individuation, which many in the South recognized during the Civil War. While many Northern cities continued to publish city directories throughout the war, most cities in the South discontinued their publication. The year before the outbreak of war, fears developed among the community in Charleston, South Carolina, that the canvassers for the city directory were part of a "Northern enterprise," leading the editor of the directory to assure his readers "that no Northern men, either as printers, or otherwise, have had or have any connection with this publication" (*Directory of the City of Charleston* 1860, n.p.). In 1864, the Lexington city directory noted that its canvassers experienced considerable difficulty in obtaining names and addresses from residents:

> Many persons fearing that our canvassers were "assessing officers," striving to obtain their names for the purpose of fixing an onerous tax upon their property, refused to give their names until lengthened explanations were

> given them; others, believing that we were collecting matter preparatory to the "draft," indignantly denied us the information which we sought. (*Lexington City Directory* 1864, n.p.)

Even in Northern cities, such as Philadelphia, fears of the draft were widespread:

> The "terror" of the "Draft" is more extensively than ever on the people. Every person carrying a City Directory in his hand, is suspected of being an enrolling officer, either military or civil, and the very sight of this otherwise useful medium of information, creates suspicion of ulterior designs, and makes it impossible, in many cases, to get correct information. Spurious names are given, and the head of the family said to be in the army or navy, when the contrary is the fact. Doors are often closed and no names at all obtained; besides, other vexatious obstructions are thrown in the way. (*McElroy's Philadelphia City Directory* 1865, n.p.)

There were similar concerns expressed during times of "peace," yet the war clearly intensified people's fears of the "ulterior" motives of directory canvassers, and thereby increased the likelihood of mass resistance to the spatial individualization of the population. The resistance encountered by directory publishers in their attempt to link name and street address supports Christopher Thale's (2007, 126) recent claim that house numbering, and the gathering of spatial information, was not merely a matter of satifying an "unproblematic public interest." To the contrary, such practices of socio-spatial ordering were part of a much broader politics of spatial calculation.

Creative Destruction, Spatial Permanence and the Philadelphia Block Decimal System

During the first half of the nineteenth century, the practice of house numbering in US cities and towns was riddled with contradictions. The odd/even system, which was widely adopted in the 1790s, consisted of the enumeration of existing houses, so that when new structures were built or old ones torn down, the entire system was disrupted. In the mid-1850s, a new system was proposed and came to be known as the "Philadelphia System," or the block decimal system, of house numbering (Figure 10.3). The block system was first introduced in Philadelphia in 1856, and it provided one hundred numbers for each city block, regardless of the actual number of buildings existing at the time of enumeration (Morbito 1965; Thale 2007).

This system was ideally suited for gridiron cities, where streets were laid out perpendicular to one another. In Philadelphia, the numbers began at the Delaware River in one direction, and Market Street provided a baseline

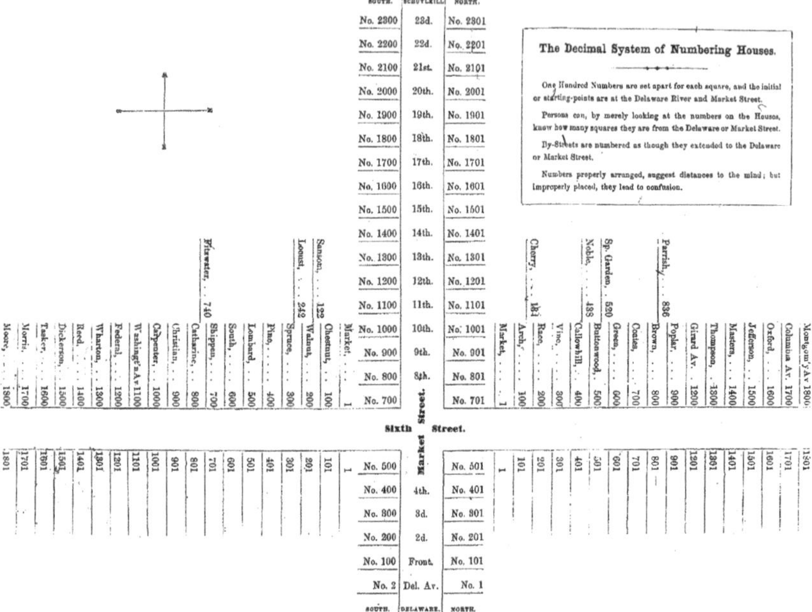

Figure 10.3 The Philadelphia Block System and the emergence of "theoretical" house numbers

for numbering houses along streets that ran perpendicular to the river. The Philadelphia city directory of 1861 noted that, "Persons can, by merely looking at the numbers on the Houses, know how many squares [i.e., blocks] they are from the Delaware or Market Street … Numbers properly arranged, suggest distances to the mind; but improperly placed, they lead to confusion" (*McElroy's Philadelphia City Directory* 1861, n.p.). In gridded cities that also used numbers as street names, the Philadelphia System enabled the systematic coordination of street and house numbering.

By the early twentieth century, most US cities either used the block system of numbering or the related equal interval method, with street frontage being numbered at designated intervals. It was common practice to assign a new number every 20 to 25 feet along a street, although the interval in business sections was often smaller than that in residential areas. This shift away from numbering discrete "objects" (i.e., existing buildings), and towards the establishment of an invisible house numbering grid of equal intervals, may at first appear little more than a mere technicality. However, I contend that it marks a major turning point in the spatial history of the concrete production of abstract urban space. It was one way of attempting to establish permanence and spatial fixity while also contending with the dialectic of creative destruction unleashed by modern capitalism. The continuous construction and demolition of buildings according to the dictates of the real estate market dramatically transformed the built environment of US cities during the nineteenth and twentieth centuries (Page 1999; Scobey 2002). Any attempt to construct a fixed spatial order based upon the numbering of discrete buildings that existed at one given point in time was futile and instantaneously became outdated and obsolete.

What the equal interval method did was construct a "theoretical space" of permanence that remained conceptually unaffected by the concrete changes in the built environment. Buildings could go up and come down, yet the theoretical space of equal intervals was consciously designed to anticipate and enable such transformations to occur in the first place. Both the Philadelphia System and the equal interval method of house numbering were governmental attempts to construct spatial "permanences" (Harvey 1996) in order to administratively rationalize the chaos of the real estate economy. The reorganization of a city's spatial regime of inscriptions—the shift from numbering existing "objects" to providing theoretical numbers by intervals—signalled a major reconfiguration of urban spatial governance in nineteenth-century America. Although such techniques of spatial ordering were first developed during the nineteenth century, they became widespread in twentieth-century cities and provided a basic framework for the emerging professionalized discourse on street and house numbering.

The Concrete Production of Abstract Space and the Grid of "Municipal Orderliness"

In the spring of 1909, the Clearing House of the American Society of Municipal Improvements (ASMI) conducted a survey on the house numbering strategies employed by cities across the United States. Over a hundred cities responded to the ASMI's request, and the results were published in an article entitled, "Systems of House Numbering" (1909), in the *Municipal Journal and Engineer*. The article opens by confidently declaring that "[t]he numbering of houses in a city is a municipal function," yet the author immediately qualifies this initial statement by maintaining that the issue of house numbering "too frequently does not receive the attention from municipal officials which it should" (1909, 563). Nevertheless, of the 136 cities that responded to the ASMI's survey, nearly all had already established some type of standardized house numbering "system" and had formally delegated the task of numbering houses to a municipal agency. In most cities, the numbering of houses was the responsibility of the city engineer, although in a few cases it was delegated to the city's building inspector, city clerk, assessor, superintendent of streets, board of public works, or even the mayor himself. In short, by the beginning of the twentieth century, the spatial practice of house numbering had become a taken-for-granted "function" of municipal government.

Gone were the days when the maverick city directory publisher went out and numbered a city's houses at his own expense without official approval from the city council. In fact, the only mention of city directories in the ASMI's article was a brief remark that the City Council of Beverly, Massachusetts, still hired "the publisher of the local directory" to affix numbers on the houses and allowed him to use "his own judgment in assigning them" (1909, 569). This example was the exception which proved the rule that city directory publishers had ceased to play the same pivotal role in house numbering that they had during the eighteenth and nineteenth centuries. By the turn of the twentieth century, professional engineers and city planners had taken over the role of devising systematic house numbering plans for America's cities.

The aim of this new professionalized discourse on street and house numbering was, as one anonymous author in the *Engineering News* put it, to create a sense of "municipal orderliness" (1915, 757). While the postal service was often responsible for devising route and box numbering schemes in rural communities, the engineering profession attempted to monopolize this spatial practice in large cities and smaller towns. This claim to professional territory was reiterated in the pages of engineering journals throughout much of the twentieth century to establish its legitimacy. "The marking of streets and the numbering of houses," noted one author, "are very proper functions of a city engineering department" (1915, 757). Early twentieth-century engineering textbooks, such as A. Prescott Folwell's *Municipal Engineering Practice* (1916), also began to formalize the methods of house numbering applied to American cities and towns.

Systematic street and house numbering had, by many accounts, become "a vital part in everyday life" (Schwada 1932, 186). This, in turn, gave rise to an institutionalized discourse on the virtues of constructing "scientific" street numbering systems based on the principle of functional efficiency. Journals such as *Civil Engineering*, the *Municipal Journal and Engineer*, *Public Management*, *The American City*, and *Public Works* served as a medium for the articulation and circulation of the discourse on uniform street and house numbering. Many of the contributors to these journals were practicing city engineers or city managers from across the United States who were in the midst of devising street and house numbering systems in their respective cities and towns. These journals offered a forum within which street numbering projects and plans could be compared and discussed across the nation.

Although numerous street and house numbering plans were described and proposed in these journals, most conformed to the norm of establishing two perpendicular baselines that would serve as the coordinate axes of a grid plan. The basic aim was to make "a trail that any one can follow, and find any address without map or street guide, provided he knows his alphabet and can count" ("A Solution to the Street Maze Problem" 1939, 18). The underlying assumption of this discourse was that the rectilinear geometry of the grid in itself was inadequate to produce a rationalized space if not combined with a "system" of alphanumerical ordering. In other words, it was not simply any type of grid but rather the *coordinate grid*—which united rectilinear form with sequential numbering—that became the standard for spatial organization within the field of civil engineering.

Although urban morphologists and urban planning historians tend to emphasize the geometric form of different settlement patterns, the engineers in charge of rationalizing urban space in twentieth-century America were concerned not only with the city as form, but also with the city as "text." As noted above, city directory publishers had viewed the city as a large record-keeping book (a "ledger"), in which house numbers and the directory functioned as a city's "page numbers" and "index," respectively. This analogy of the city as a "text" was drawn upon in twentieth-century engineering journals as well. "Like the paging of a book," noted Elmer Lee (1901, 78-79) in the *Municipal Journal and Engineer*, "street signs are a necessity ... No publisher would think of omitting the numerals at the head of the pages of a book. None the less imperative is the importance for well ordered and sightly street signs." Engineers insisted that the physical typography of the city should be systematic, legible, and "so simple that a child can easily master it" (Shibley 1938, 91).

There were many plans proposed to coordinate the numbering of streets with the numbering of houses. Three early street naming plans—those of Philadelphia (1682), Washington, DC (1791), and New York City (1811)—came to be seen as models for other American cities and towns (Mencken 1948; Stewart 1982). All three were grid-plan cities that adopted the practice of street numbering. Philadelphia had numbered streets in one direction and named streets in the other, whereas the New York grid plan of 1811 consisted of numbered east-west "streets" as well as numbered north-south "avenues." In Washington, however, north-south streets were numbered, east-west streets were lettered, and diagonal avenues named after the states of the Union were overlaid upon this alphanumerical grid. Moreover, the Washington coordinate grid was divided into four quadrants (NE, NW, SE, SW), with its origin at the Capitol.

Although most cities had begun numbering houses as early as the eighteenth and nineteenth centuries, movements arose in a number of cities during the first half of the twentieth century—such as in Chicago, Salt Lake City, and Milwaukee—to systematize what were taken to be the "haphazard" and "hodgepodge" methods of street naming and house numbering currently in use (Schwada 1932, 186; also, see Thale 2007). Most city engineers looked favourably upon the "quadrant" system as a means of systematically ordering urban space. Two of the most widely cited twentieth-century models for street and house numbering—the Schwada plan and the Lyman plan—have their origins in this period, and both were based upon some type of quadrant coordinate system (Corwin 1978).

During the second half of the twentieth century, a number of "how-to" technical manuals were published that formalized the various street and house numbering practices used across the country in order to provide city planners with a general set of rules that could be applied to particular situations. Such technocratic accounts also attempted to establish street and house numbering as a legitimate body of governmental knowledge deemed essential to the production of spatial "legibility." While most engineers framed their arguments for uniform street and house numbering in terms of its benefits to the "average citizen," it is often evident that such systems were designed principally to simplify record-keeping practices for city governments, public utilities, and private businesses. One author noted that unsystematic house numbering plans resulted in a "waste of time and loss of patience ... [which] slow down the governmental and business operations of a city" (Martin 1951, 53). What was clearly needed, from this perspective, was "some systematic and business-like arrangement in ... city numbering," which would, as another author noted, "be worth thousands of dollars a year to merchants whose employees are called upon to make deliveries both night and day" (Moffett 1917, 428; Marple 1922, 40). Here again is a call to economize time by rationalizing space in a "business-like" fashion (Figure 10.4).

The business-like management of a city street system was to be based upon the establishment of technical criteria for the construction of a "legible" landscape. A number of key variables were utilized to "scientifically" determine a suitable system of street signs: legibility, visibility, durability, appearance, and cost ("Selection and Placing of Street-Name Signs" 1916; "Street Signs—Their Character and Location" 1909; Nicholson 1931). These variables were defined in such a way that they could be measured in quantitative terms with respect to any street sign or house number under scrutiny. Here were the mundane mechanics of engineering a "legible" landscape in action. Legibility could be measured by the contrast between the colour and size of letters or numbers and the background colour of the sign, while the "visibility test" was defined as "the distance which the observer was from the sign at the time the sign could be read" (Rockwood 1937, 170). Competing signs were ranked according to these criteria and then the rankings for each variable were totaled up to form an overall assessment of the sign (Figure 10.5). The key issue was making street signs and house numbers "legible at a considerable distance at all times" ("Street Signs that are Readable Night and Day" 1929, 197). The call for legibility was central to the modernist management of the physical typography of the cityscape.

When influential urban theorist Kevin Lynch advocated that visual "legibility" should be a fundamental principle of urban design in post-World War II America, therefore, he was merely extending the well-established discourse on the legibility of street typography to a consideration of other

A HOUSE NUMBER WHERE THE DELIVERY MAN CAN SEE IT

Figure 10.4 The production of spatial legibility and the "business-like arrangement" of house numbering

Figure 10.5 The "Visibility Test" and the making of a geo-coded world

aspects of the "city image," such as edges, districts, nodes, and landmarks. In fact, Lynch (1960, 3) even draws upon the same modernist metaphors of the city-as-text, when he notes that: "Just as this printed page, if it is legible, can be visually grasped as a related pattern of recognizable symbols, so a legible city would be one whose districts or landmarks or pathways are easily identifiable and are easily grouped into an over-all pattern." Lynchian cognitive mapping traces its intellectual roots back, at least in part, to the modernist engineering discourse on the virtues of legibility in street signage and systematized house numbering. This goes a long way toward explaining why cognitive mapping theorists aligned themselves during the 1990s with positivist spatial analysts in their quest to geo-code the world (Gold 1992; Kitchin and Freundschuh 2000).

Conclusion

Geographers often note that the determination of "where" something is spatially located is one of the basic questions of geography. It is all the more perplexing, then, that so little has been written on the historical geography of house numbering—a spatial practice that has arguably been one of the principal ways in which the ordering of geographic space has served as a means of "locating" people and property over the course of the last two centuries. In this chapter, I have explored to connections between the spatial practice of house numbering and the publication of city directories as well as the institutionalization of house numbering in municipal government within the United States.

As Lefebvre (1991) and others have argued, the production of abstract space was a key strategy to contain the dialectical processes of capitalist urbanization within the fixed order of logic and number. The development of house numbering systems was central to the project of producing abstract spaces, which in turn was directly linked to the construction of spatial rationalities of legibility. These practices sought to remake urban space as a legible "text" in an attempt to rationalize social relations through the process of spatial individualization. The image of the city-as-text, therefore, is not limited to postmodern accounts of the urban. The fundamental question for historical geographers is not: "Should we conceptualize the city as a 'text' or not?" but rather, "In what ways and for what purposes has the city been conceptualized as a 'text' historically and how have such conceptualizations sought to render urban space intelligible?" At the risk of oversimplification, it could be argued that the modernist view of the city-as-text sought to delineate a fixed and coherent textual "order," whereas postmodernists have pointed to the instability of any such attempt to impose a transparent legibility by emphasizing the multiplicity of narratives that make up the textuality of the cityscape.

The shift from one spatial regime of inscriptions to another did not "resolve," once and for all, the contradictions of urban space. Lefebvre (1991, 142) rightly suggests that if urban spaces are indeed to be seen as "texts," then they must surely be "over-inscribed" where "everything therein resembles a rough draft, jumbled and self-contradictory." From a dialectical standpoint, abstract spaces are not coherent finished products but always in the process of becoming. "They are contingent," as Harvey (1996, 261) puts it, "on the processes that create, sustain and dissolve them." Codifying a new regime of spatial inscriptions is one thing, yet ensuring its "permanence" cannot be achieved by top-down imposition alone.

One of the most fascinating aspects of the history of house numbering, which deserves further analysis in future studies, is the manner in which such practices have actually been *internalized* by urban inhabitants themselves. Individual property owners are often responsible for actually inscribing numbers upon their own buildings. This process of self-objectification in spatial terms can help us better understand how "the establishment of calculation as a norm for guiding social organization [was] facilitated by individuals who themselves performed calculations as they consider[ed] their "place" in the world around them" (Crampton and Elden 2006, 683; also, see Cohen 1999; Joyce 2003). Those who are "governed" by the superficial transparency of abstract space also utilize this spatial façade for innumerable purposes—whether conformist, subvertive, or otherwise.

Future researchers may actually look back on the end of the twentieth century as marking another pivotal moment of spatial rationalization. In the United States and elsewhere, recent house numbering reforms have been directly linked to the implementation of emergency response management systems that rely upon GIS technology (Lucy 1995; Curry, Phillips, and Regan 2004). At the global scale, the World Bank has highlighted street addressing as an indispensable strategy of urban governance in the "developing world" and has funded numerous urban house numbering projects since the mid-1980s, particularly in Sub-Saharan Africa (Farvacque-Vitkovic et al. 2005). To the extent that the numbering of houses has played a fundamental role in reshaping the landscapes of everyday life, a historical geography of house numbering can shed light on the spatial politics of both the past and present.

References

A Directory for the City of Buffalo (1837), (Buffalo, NY: Sarah Crary).

"A Solution for the Street Maze Problem" (1939), *Public Works*, March, 18.

Alotta, R. (1990), *Mermaids, Monasteries, Cherokees and Custer: The Stories Behind Philadelphia Street Names* (Santa Monica: Bonus Book, Inc.).

Azaryahu, M. (1996), "The Power of Commemorative Street Names," *Environment and Planning D* 14: 3, 311–30.

Biddle, Clement (1791), *The Philadelphia Directory*.

Boston Directory (1855), (Boston, MA: Geo. Adams).

Boyd, William (1857), *Boyd's Lancaster Directory* (Lancaster, PA: Sprenger and Westhaeffer).

Boyd's Philadelphia City Business Directory (1858), (Philadelphia: Wm. H. Boyd).

Burton, R. (1956), *City Directories in the United States, 1785–1820: A Bibliography with Historical Notes* (Ann Arbor: University of Michigan).

City Directories of the United States, Pre-1860 through 1901 (Woodbridge: Research Publications).

Cleveland City Directory (1856), (Cleveland: Spear, Denison, and Co.).

Clinton Almanac and Directory (1856), (Clinton, MA: E. Ballard).

Cohen, P. (1999), *A Calculating People: The Spread of Numeracy in Early America* (New York: Routledge).

Coles, T. and Shaw, G. (1997), "A Sign of the Times: Scandinavian Town Directories as Sources for Urban Historical Geography," *Geografiska Annaler* 79B: 2, 65–81.

Corwin, M. (1978), *Street-Naming and Property-Numbering Systems: Report No. 332, Planning Advisory Service* (Chicago: American Society of Planning Officials).

Crampton, J. and Elden, S. (2006), "Space, Politics, Calculation: An Introduction," *Social & Cultural Geography* 7: 5, 681–5.

Curry, M. et al. (2004), "Emergency Response Systems and the Creeping Legibility of People and Places," *The Information Society* 20: 5, 357–69.

Daughdrill and Walker's General Directory for the City and County of Mobile (1856), (Mobile, AL).

Dean, M. (1999), *Governmentality: Power and Rule in Modern Society* (London: Sage Publications).

Directory of the City of Charleston (1860), (Charleston: W. Eugene Ferslew).

Drown's Record and Historical View of Peoria … Also, an Almanac for 1851 … To which is added a Business Directory of the City, with Business Cards, (1850), (Peoria, IL: S. De Witt Drown).

Duncan, J. (1990), *The City as Text: The Politics of Landscape Interpretation in the Kandyan Kingdom* (Cambridge: Cambridge University Press).

Farvacque-Vitkovic, C. et al. (2005), *Street Addressing and the Management of Cities* (Washington, DC: World Bank).

Finning's Bridgeport Town and City Directory and Annual Advertiser (1855), (Bridgeport, CT: Pomeroy and Morse).

Folwell, A. (1916), *Municipal Engineering Practice* (New York: John Wiley and Sons, Inc.).

Foucault, M. (1980), "Truth and Power," in Gordon, C. (ed.), *Power/Knowledge: Selected Interviews and Other Writings, 1972–1977* (New York: Pantheon Books), 109-133.

——(1991), "Governmentality," in Gordon, C. and Miller, P. (eds), *The Foucault Effect: Studies in Governmentality* (Chicago: University of Chicago Press), 87-104.

—— (2007), *Security, Territory, Population: Lectures at the Collège de France, 1977–78*, ed. M. Senellart, trans. G. Burchell (New York: Palgrave Macmillan).

Garrioch, D. (1994), "House Names, Shop Signs and Social Organization in Western European Cities, 1500-1900," *Urban History* 21: 1, 20–48.

—— (2002), *The Making of Revolutionary Paris* (Berkeley: University of California Press).

Gold, J. (1992), "Image and Environment: the Decline of Cognitive-behaviouralism in Human Geography and Grounds for Regeneration," *Geoforum* 23: 2, 239–47.

Harvey, D. (1996), *Justice, Nature, and the Geography of Difference* (Oxford: Blackwell Publishers).

Henkin, D. (1998), *City Reading: Written Words and Public Spaces in Antebellum New York* (New York: Columbia University Press).

Hoffmann, H. (1937), "Changed House Numbers and Lost Street Names in New York of the Early Nineteenth Century and Later," *New-York Historical Society Quarterly Bulletin* XXI: 3, 67–92.

Hunt, J. (1995), "House Numbering in Revolutionary Paris," *Bulletin of the Institute of Mathematics and its Applications* 31: 9–10, 145.

Isaacs, J. (1940), "The History of Street Numbering in New York," *Real Estate News*, January, 19, 28.

Joyce, P. (2003), *The Rule of Freedom: Liberalism and the Modern City* (London: Verso).

Kain, R. and Baigent, E. (1992), *The Cadastral Map in the Service of the State: A History of Property Mapping* (Chicago: University of Chicago Press).

Kitchin, R. and Freundschuh, S. (eds) (2000), *Cognitive Mapping: Past, Present and Future* (London: Routledge).

Latour, B. (1999), *Pandora's Hope: Essays on the Reality of Science Studies* (Cambridge, MA: Harvard University Press).

Lee, E. (1901), "Municipal Street Signs," *Municipal Journal and Engineer*, March, p. 78-80.

Lefebvre, H. (1991), *The Production of Space*, trans. D. Nicholson-Smith (Cambridge: Blackwell).

Lexington City Directory (1864), Lexington.

Longworth's American Almanac, New-York Register and City Directory (1829), (New York: Thomas Longworth).

Lucy, W. (1995), *Addressing Systems: A Training Guide for 9-1-1* (Coshocton, OH: National Emergency Number Association).

Lynch, K. (1960), *The Image of the City* (Cambridge, MA: MIT Press).

MacPherson's Directory, for the City and Suburbs of Philadelphia (1785), (Philadelphia: John MacPherson).

Marple, A. (1922), "House Numbers on the Curb," *The American City* 27: 1, 40.

Martin, D. (1951), "Uniform House Numbering and Street Naming System," *Public Works*, June, 53–4, 72–4.

Matchett's Baltimore Director (1842), (Baltimore: Baltimore Director Office).

—— (1853), Baltimore.

McElroy's Philadelphia City Directory (1861), (Philadelphia: E.C. and J. Biddle & Co.).

—— (1865), (Philadelphia: A. McElroy).

Mencken, H.L. (1948), "American Street Names," *American Speech* 23: 2, 81–8.

Miles, J. (1972), *House Names Around the World* (Detroit, MI: Gale Research Co.).

Mobile Directory or Strangers" Guide (1839), Mobile, AL.

Moffett, C. (1917), "Our Metropolitan Labyrinth," *The Art World*, March, 428–9.

Monmonier, M. (2002), *Spying with Maps: Surveillance Technologies and the Future of Privacy* (Chicago: University of Chicago Press).

Morbito, J. (1965), "Physical and Social Implications of Street Naming and House Numbering Systems in Today's Urban Environment," Master's Thesis, School of Urban Planning and Landscape Architecture, Michigan State University, unpublished.

New Orleans Directory (1842), (New Orleans: Pitts and Clarke).

Nicholson, J. (1931), *House Number Signs* (New York: Municipal Administration Service).

O'Brien's Wholesale Business Intelligencer and Southern and Western Merchants" Pocket Directory (1838), (Philadelphia: John G. O'Brien).

Page, M. (1999), *The Creative Destruction of Manhattan, 1900–1940* (Chicago: University of Chicago Press).

Pickles, J. (2004), *A History of Spaces: Cartographic Reason, Mapping, and the Geo-coded World* (New York: Routledge).

Pronteau, J. (1966), *Les Numérotages des Maisons de Paris du XV*[e] *Siècle a Nos Jours* (Paris: Ville de Paris Commission des Travaux Historiques, Sous-Commission de Recherches D'Histoire Municipale Contemporaine).

Rockwood, B. (1937), "Selecting a Street Name Sign," *Public Management*, June, 169–71.

Rose, N. (1999), *Powers of Freedom: Reframing Political Thought* (Cambridge: Cambridge University Press).

Rose-Redwood, R. (2006), "Governmentality, Geography, and the Geo-coded World," *Progress in Human Geography* 30: 4, 469–86.

Schwada, J. (1932), "Street Names and House Numbers: Review of Systems Proposed and in Use, with a Recommended Uniform Plan," *Civil Engineering* 2: 3, 186–9.

Scobey, D. (2002), *Empire City: The Making and Meaning of the New York City Landscape* (Philadelphia: Temple University Press).

Scott, J. (1998), *Seeing Like a State: How Certain Schemes to Improve the Human Condition Have Failed* (New Haven, CT: Yale University Press).

"Selection and Placing of Street-Name Signs" (1916), *Municipal Journal* XLI: 4, 93–7.

Shammas, C. (2002), *A History of Household Government in America* (Charlottesville: University of Virginia Press).

Shaw, G. and Coles, T. (1995), "Directories as Elements of Town Life: The Case of National Socialist Germany," *The Geographical Journal* 161: 3, 296–306.

—— (1997), *A Guide to European Town Directories, Vol. 1: Germany, Austria, Switzerland and Scandinavia* (Aldershot: Ashgate).

Shaw, G. and Hudson, P. (2002), "Edge of Empire: Transnationalism and Identity in Wellington, New Zealand, *ca.* 1860—*ca.* 1920," *Landscape Research* 27: 1, 51–66.

Shibley, M. (1938), "Sane and Simple Street Naming and Marking," *The American City*, March, 91.

Simmon's Norfolk Directory (1801) (Norfolk: Charles Simmons).

Smail, D. (2000), *Imaginary Cartographies: Possession and Identity in Late Medieval Marseille* (Ithaca, NY: Cornell University Press).

Stewart, G. (1982), *Names on the Land: A Historical Account of Placenaming in the United States* (San Francisco: Lexikos).

"Street-Name Signs" (1915), *Engineering News*, 22 April, 757.

"Street Signs That Are Readable Night and Day" (1929), *The American City*, April, 197.

"Street Signs—Their Character and Location" (1909), *Municipal Journal and Engineer* XXVI: 17, 710–15.

"Systems of House Numbering," *Municipal Journal and Engineer* 26: 14, 563–9.

Tantner, A. (2007), *Ordnung der Häuser, Beschreibung der Seelen: Hausnummerierung und Seelenkonskription in der Habsburgermonarchie* (Innsbruck: StudienVerlag).

Thale, C. (1986), "Milwaukee's Street Numbering System," *Milwaukee History* 9: 3, 66–76.

—— (2007), "Changing Addresses: Social Conflict, Civic Culture, and the Politics of House Numbering Reform in Milwaukee, 1913–1931," *Journal of Historical Geography* 33: 1, 125–43.

The Beloit Directory, and Business Mirror (1858), (Beloit, WI: Wright and Newcomb).

The Charlestown Directory (1831), (Charlestown, MA: Waitt and Dow).

—— (1852), (Charlestown, MA: Nathan Merrill).

The Portland Reference Book, and City Directory (1850) (Portland, ME: Thurston abd Co.).

The Providence Directory (1841), (Providence, RI: H.H. Brown).
Trow's New York City Directory (1878), (New York: Trow City Directory Co.).
Tscherny, G. (1992), *Odd & Even: A Study of House Numbers* (New York: Champion).
W.H. Rainey and Co.'s Memphis City Directory, and General Business Advertiser (1855), (Memphis, TN: E.R. Marlett and W.H. Rainey).
Williams, A. (1913), *The Development and Growth of City Directories* (Cincinnati: Williams Directory Co.).
Williams' Cincinnati Directory and Business Advertiser (1849), (Cincinnati: C.S. Williams).

Chapter 11

Planning and Revamping Urban Toponymy: Ideological Alterations in the Linguistic Landscaping of Vuosaari Suburb, Eastern Helsinki

Jani Vuolteenaho and Terhi Ainiala

Introduction

Urban place names have a capacity for bringing hegemonic ideologies into mundane existence. Most dramatically, the power-embeddedness of toponymic city-texts is unveiled after changes in regimes of governance. As Gill (2005, 480) has put it, "the reworking of language (through the injection of new words, the changing of the meaning of existing terms, and the elimination of some words) in order to invest it with a new ethos [is] important to the creation of a new regime's symbolic culture." Of course, under more steady urban conditions the ideological transformation of toponymies typically takes place more incrementally. As essentially palimpsest-like constructions, urban namescapes develop as "sums of additions and erasures" along with the protracted processes of societal development (Azaryahu, this volume). Until now, the critical conceptualizations of urban place naming have focused on nationalist, socialist and post-socialist street name revisions (e.g. Azaryahu 1997; Faraco and Murphy 1997; Pinchevski and Torgovrik 2002; Light 2004; Gill 2005), or the role that spatial addressing systems have played in the rationalization of urban space (e.g. Rose-Redwood, this volume). In European urban settings, in particular, much less attention has been paid to the ways the present-day marketers, developers, and indeed city-authorities increasingly brand places by bestowing them with nomenclatures that are attractive and marketable.

In this chapter, we tackle the rarely scrutinized subject of "linguistic landscaping" (Rapoport 1977, 110) of urban space through a case history of naming practices in Vuosaari, a seaside suburb in easternmost Helsinki. Inhabited today by 35,000 residents, our study area's fitful urban growth started roughly coincidentally with its annexation to the Finnish capital in 1966. Adding a hitherto understudied symbolic dimension to research on Vuosaari as an exemplary case of the neoliberalization of urban development in the "social-democratic" Finnish context (e.g. Haila 2005; Vuolteenaho et al. 2007; on the general effects of neoliberalization in Finland, see Patomäki 2007), we track processes, actors and name-motifs behind the development of the district's authoritatively planned and housing corporation nomenclatures since early 1960s. Interestingly, the initial building and naming projects in Vuosaari occurred at the zenith of authoritative modernist planning in Finland. In the recent development of the district, by contrast, place-promotion strategies—*à la* neoliberal urbanism—have reigned supreme. In this chapter, our presumption is that the shifts in Vuosaari's naming practices mirror this planning ideological transformation.

To introduce the chapter's central concepts and to embed Vuosaari's case history in the larger ideological context of urban (name-)planning, we will first focus on pivotal tendencies in the evolution of authoritative baptizing practices especially European cities. In the section preceding our empirical analyses, we also conceptualize challenges that the recent neoliberal turn in urban development is posing to the now canonized modernist practices of linguistic landscaping.

Toponymic Modernization in European Cities

> What was otherwise reserved for only a very few words, a privileged class of words, the city has made possible for all words, or at least a great many: to be elevated to the noble status of name. This revolution in language was carried out by what is most general: the street.—Through its street names, the city is a linguistic cosmos. (Benjamin 1999, 522)

In place name literature, the outset of systematic management of urban toponymies is commonly dated back to the second half of the eighteenth century. Prior to this historical watershed, as "no communal authority cared" (Langenfelt 1954, 331), relatively unstable toponymic textures based mainly on prominent buildings and topographic features dotted European townscapes (Harling-Kranck 2006). The now nearly universal addressing system based on fixed street signs and house numbers was unknown (Farvacque-Vitkovic et al. 2005), and neither was any systematic separation between the names of physical constructions and commercial agencies made: "in the cities of early modern Europe the houses and shops almost all had names and signs … leaning over the street or set into the facade" (Garrioch 1994, 20). Since the age of the Enlightenment, however, authority-led nominatory reforms, and especially the street-name based registering of addresses began to spread across Europe and other continents. At stake was a "pivotal historical moment in the rationalization of urban space" (Rose-Redwood, this volume). In the following decades, the administrative benefits of a principally flawless registration system for all roads, buildings and other constructions in a city became apparent with an array of functions: from modern postal services to property records management, the control of illegal housing construction, the recruitment of military troops, tax-collection, fire protection, policing, and so on (Farvacque-Vitkovic et al. 2005, 9–10). In short, the "official" toponymies became technocratic instruments in steering and keeping track of urban systems, functions and people.

While the systematic identification of spatial objects in the urban fabric was the technocratic *raison d"être* behind these reforms, administrative efficiency went hand in hand with more explicit ideological goals (Parkhurst Ferguson 1988, 391). Customarily, Paris is regarded as the foremost path-setter in this respect. As Parkhurst Ferguson (ibid., 390–91) depicts the convoluted impacts of the French revolution, setting a model for further toponymic changes across European regimes:

> The Revolution took politics onto the streets and set a pattern for the century to come by "secularizing" the "sacred geography" that had been delimited by the monarchy. [...] Grégoire's report, adopted by the Assemblée in January 1794, reviewed the history of street naming from Peking to Philadelphia and made several proposals. First, names should be short, comprehensible, euphonic, and appropriate, and second, they should be morally correct... These plans for global nominatory reform testify to the pedagogical thrust evident in so many enterprises of the Revolution and also to the demands of a modernizing state.

In the tumultuous nineteenth century, the influx of people to the industrializing cities and the rise of European nation-states provided the context for a crucial ideological-pedagogic aspect of toponymic modernization: the utilization of urban linguistic landscaping as an additional symbolic means towards the nationalization of people's collective consciousness. As Azaryahu (1996, 314; this volume) has put it, honorific (re-)naming, no more "limited to celebrating dynastic glory but ... increasingly associated with nation-building measures," became a prominent socialization strategy through which ideological messages were introduced into the settings of everyday human experience. Prestigious avenues, central squares and monumental buildings, in particular, were exploited as the stages of nationalist toponymic enlightenment (see Therborn 2002). Via their names, a nation's people-uniting symbols and constitutive narratives—various emblems of its societal progress and victorious battles, its great artists, politicians and war heroes, as well as places from its provinces and allied regimes—could be made not only centrally visible, but also a part of the mundane communication about locational facts among urban dwellers.

Notably, the so-called model of group-naming (based on a thematically coherent set of invented names affixed to a cluster of contiguous landscape elements) was also initiated in Paris in the early nineteenth century

(Langenfelt 1954, 337; Harling-Kranck 2006, 217). It was subsequently exploited throughout Europe to serve ideological-pedagogic ends. During a major name revision in late nineteenth century Stockholm, for instance, a new "official geography was conferred upon the entire city" on the basis of designedly nationalistic name-motifs of "patriotic and historical names," "Nordic mythology," "famous places near the city," "the southern provinces," "the northern provinces," "famous Swedish authors" and "prominent men within technology and engineering" (Pred 1990, 126–9).

However, any account of a name-planning legacy in Europe would be incomplete without recognizing a persistent presence of locally derived names (Langenfelt 1954, 334–5, 338). Actually, even the issuing of honorific names has been very often "biased in favour of local history" (Azaryahy 1996, 326). Neither has the coining of group-names been restricted to explicitly patriotic themes. Rather, the obvious appeal of the "category principle" has been based on its flexibility in the face of on-going urban growth and a consequent need for ever more name-ideas (Harling-Kranck 2006). Through it, a principally limitless number of both local (from topography to cultural traditions) and non-local (e.g., distinctively "urban" motifs derived from the names of foreign places, classical mythologies and fine arts) themes have been "elevated to the noble status of name" (Benjamin 1999, 522). In addition, there has always existed an inherently local toponymic category: linguistically autochthonous place names or "autochthonyms" (Faraco and Murphy 1997). Whether adopted into use in original or corrupted forms, these kinds of officialized "local borrowings" have been favoured in bestowing places with identities rooted in local history (in different national contexts, see: Leenen 1946; Stemshaug 1985; Room 1992; Fuchshuber-Weiβ 1995; Faraco and Murphy 1997; Paikkala et al. 1999; Johansson 2007).

At this point, a couple of remarks regarding the institutionalization of "street naming" as a prerogative of local authorities in the European context deserve attention. Unlike in the United States, for instance, where private-sector actors have conventionally held key roles in name-planning (Algeo 1978; Tarpley 1995, 1494–6), a major European tendency has been to firmly incorporate this charge into the regulative procedures of public-sector town planning. Even since the nineteenth century in many bigger cities, municipal planning departments (equipped typically with engineering and architectural expertise) and advisory toponymic boards (with expertise also focused on local history and native languages) have often cooperatively designed the make-up of official toponymies (e.g. Langenfelt 1954, 338). Understood as an in-built aspect of (otherwise visually and technocratically oriented) local planning, it has become a commonplace in name-planning circles in different European countries to perceive various functions of urban toponymy as mutually complementary (see e.g. Paikkala et al. 1999; Boerrigter, forthcoming). If coherently organized and appropriately chosen, so the established presumption goes, official place name systems not only facilitate the efficient functioning of urban systems as administrative units, but can also act as enlightening keys to history and local environment, and aid people to orient themselves in urban space (Langenfelt 1954, 338; cf. on the modernist ideal of the "legible city" from a perspective of a famous American urbanist: Lynch 1960, especially page 108; see also Rose-Redwood, this volume).

To conclude our historically grounded overview with aspects of modernist nominatory regulation that bear direct relevance to the naming policy of the City of Helsinki, commemorative names, group-names and autochthonyms have all been utilized in the forging of its official nomenclature in the course of the last two centuries (Helsingin kadunnimet 1970/1981). Importantly, the temporally and spatially varying criteria used for place-baptizing have resulted in a distinctive kind of spatial hierarchy: whilst prestigious and openly patriotic toponyms figure prominently in the older parts of Helsinki, rather prosaic group-names have predominated naming practices in its more newly urbanized outskirts. In this respect, it is symptomatic that invented group-names—differentiating neighbourhoods and their parts through name-motifs that range from folk music to agricultural livelihoods and more modern professions, aspects of Finnish wildlife, physical geography and culture, and so on—currently cover almost 90 percent of the planned nomenclature in its suburbs (Carpelan and Schulman 1996, 51; Närhi 1999). Notably, an explicit name-planning guideline applied in the immediate post-war decades (at the time when Vuosaari was annexed into Helsinki) was to eschew too elitist or explicitly ideological connotations in naming the neighbourhoods outside the central parts of the city (see parallel findings in other national settings: Faraco and Murphy 1997; Pinchevski and Torgovrik 2002). Witness how this sort of spatially selective avoidance of "smart" or "non-neutral" toponyms defined Helsinki's Toponymy Committee's naming policy at the turn of the 1970s:

> [D]ifferent street names may be valued differently, varying according to time, place and the social status of the residents. The majority of the names used in Helsinki belong to ones that, due to their stout pertinence, are suitable in any era or environment whatsoever. These kind of "colourless" names do not in any way irritate any side; inter alia, they are names that cling to surrounding nature or the history of a place … Instead … "smart" street names such as Apollokatu ["Apollo Street"] … Hallituskatu ["Government Street"], Mannerheimintie ["Mannerheim Road"], Bulevardi ["Boulevard"] and Senaatintori ["Senate Square"] suit well the city centre and neighbourhoods with splendid stone buildings … (Pesonen 1970/1981, 80–81) [translation from Finnish by JV]

In the case of Helsinki, the local name-planners' efforts towards sensitivity to local history, the cognitive clarity of toponymic patterns as well as predilection for "modest" names in suburban areas have thus all complemented the technocratic and nationalistic maxims of toponymic modernization. In the upcoming empirical sections, our aim is to shed light on the impacts of these modernist principles on Vuosaari's linguistic landscaping. In addition, we will be also interested in the ways the authoritative naming practices inherited from the modernist era have been recently challenged by market-responsive naming strategies in our study area.

From Modernist Name-planning to Neoliberal Geo-branding?

In cities throughout the older industrialized world, the Fordist-Keynesian forms of regulative planning have been increasingly impugned by the triumph of neoliberal policies since the 1970s (Brenner and Theodore 2002; Harvey 2005). Also in Europe, the socially egalitarian goals of local governmental control have been often overshadowed by economic boosterism. In the face of tightened inter- and intra-urban competition, the break with modernist ideal "that planning and development should focus on large-scale, metropolitan-wide, technologically rational and efficient urban plans, backed by absolutely no-frills architecture" (Harvey 1989, 66) has gone hand in hand with the establishment of strategic flagship projects in specific inner city and suburban areas, as well as "an eclectic planning style where attention to design, detail, morphology and aesthetics is paramount" (Swyngedouw et al. 2002, 567). Intriguing questions in the context of this chapter arise: as *ad hoc* public-private partnerships devoted to commercial attractiveness and "trend planning" (Brindley et al. 1996) are now key actors in making local planning decisions, are the holistically planned modern urban toponymies now on the verge of being "creatively destroyed" as part of a move to much more market-responsive and fragmentary namescapes? In other words, and to paraphrase Brenner and Theodore (2002), has naming become an "actually existing" cultural manifestation of neoliberal urbanism? Or do there exist important "path-dependencies" between the strictly modernist and the newer urban naming practices (cf. ibid.)?

Obviously, one important tendency in the current development of urban toponymies concerns the relative blurring of the paradigmatic dichotomy drawn between the spheres of official and commercial naming in the antecedent two and half centuries (see Garrioch 1994; Harling-Kranck 2006). Current marketing literature sheds light on this on-going toponymic commodification. Thus, Freire (2005, 348) asserts that "today there is a general agreement among academics and practitioners that places can be branded in much the same way as consumer goods and services." In the typically fatalistic neoliberal vein, he seems to see "no alternative" to the deliberate branding of places through naming and other means of marketing. Quite simply, toponyms differentiate places as an inescapable part of what he dubs "geo-branding." According to this conception embedded in free-market ideology, place names are (and should be) forged with an eye to people as consumers:

> the geo-brand concept … is probably the strongest tool society will ever have to protect and maintain its identity and fight the trend of modern standardization … For a consumer a place will always mean something, implying a place name which will function as a brand, even if not managed under a branding conceptual framework. (Freire 2005, 348)

A noteworthy paradox in the rhetoric that branded toponyms facilitate anti-standardization lies, of course, in the oft-remarked fact that stereotypical, globally approved imageries are regularly favoured in the current regime of place-marketing (e.g. Ritzer 1996; Hannigan 1998). Not the least because the influences from English (and to a more minor degree from other "global" languages) in the form of lexical, orthographic and pronunciative borrowing are currently rampant across all cultures and continents (e.g. Gorter 2006), it is difficult to espouse the neoliberalist reasoning. Nonetheless, there certainly are distinctive processes at work in the toponymic distinguishing of urban namescapes in the pursuit of competitive advantages. American studies on the language used to market cities, neighbourhoods and properties, make it clear that naming frequently plays a key role in the "theming" of urban space. Shopping malls, for instance, tend to adopt names that evoke a sense of local, exotic, culturally refined, spectacular, or otherwise appealing identity among their desired users (Kowinski 1985; Crawford 1992, 24). Likewise, the developers of residential estates typically resort to some overarching name-motif to attract a specific niche of home-buyers (Tarpley 1995; Gottdiener 1997, 73–4; Norris 1999). Due to a lack of holistic regulation, in turn, these catchy and variegated naming strategies tend to produce fragmented linguistic landscapes. As commentators on the postmodern aspects of American urban development have stressed, in the commodified interior spaces of shopping malls as well as in many metropolises at large, toponymic-semiotic flamboyancy runs rife without any apparent spatial hierarchy or ordered re-presentation of local history (e.g. Soja 1989; Goss 1999; Raento and Douglass 2001).

Arguably, the above mentioned tendencies towards linguistic globalization, market-niche based theming, and fragmentation of urban toponymies are contributing to the neoliberalization (if not "Americanization") of urban namescapes also in European regimes. Inter alia, authoritative name-planners in Sweden have lamented how construction companies increasingly "foreground their own ideas regarding 'seductive' neighbourhood names" and "estate agents stretch the limits of neighbourhoods by suggesting for them prestige-enhancing appellations" (Wahlberg 2003; quoted in Harling-Kranck 2006, 243). Also in Finland, name-planners have bemoaned how contemporarily fashionable names have often "nothing whatsoever to do with local geographical circumstances" (Viljamaa-Laakso 1999, 82; Harling-Kranck 2006, 243–4). According to Viljamaa-Laakso's (1999) observation made in reference to "the return of the Middle Ages" (Eco 1986) in the context of metropolitan Helsinki, there are even signs that the modern addressing system is currently in the process of being substituted by a landmark system of "market temples, shopping malls and arenas with their novel castle walls, as represented by *Maxis*, *Vexis* and *EuroSpars*, and *Columbus* in the east, and *Lippulaiva* [Flagship] in the west" (Viljamaa-Laakso 1999, 83). She continues her trenchant appraisal of the fate of modern name-planning in the face of contemporary urban development:

> In terms of urban development generally and the "history of ideas" behind local nomenclature in particular, we are living in the era of market economy, consultants and estate-based planning. In business, the importance of names has been increasingly recognized ... Companies often also seek to change [planned] toponyms: construction firms and even municipalities view their production areas through the lenses of marketing, which often means that old nomenclature of places is seen as non-valid ... Novel names are often of foreign origin, pronouncing the advent of new times. (Viljamaa-Laakso 1999, 82; translation from Finnish by JV)

Yet, it has to be borne in mind that the outcomes of neoliberalization generally vary based on locally embedded interactions between "inherited regulatory landscapes" and "market-oriented restructuring projects" (Brenner and Theodore 2002, 351). In this respect, it is important to note that the local authorities still have legal autonomy (or indeed monopoly) in submitting name proposals for areas under planning in most European regimes at least. From an administrative point of view, official name-planning has thus not yet suffered as dramatic setbacks as suggested by Viljamaa-Laakso (1999) and other critics: there still "remains a widespread acceptance of the need to control" also toponymic development (cf. Taylor 1998, 152). Despite apparent continuities, however, the enhanced influence of economic boosterism has inevitably altered the ideological context in which town planning and regulative name-planning as a part of it take place. Due to the lack of previous research on the subject, the only way to grasp the actual effects of neoliberal market-responsiveness to toponymic landscapes in Finnish and other European cities goes through concrete case studies.

Research Questions

The ideological legacy of name-planning in European cities and its recent neoliberal challenges inspire a number of questions with respect to toponymic transformation in the Vuosaari suburb. As Figure 11.1 illustrates, the urban growth of the district since early 1960s has bestowed its landscape not only with architecturally distinctive neighbourhoods, but also with chronologically layered toponymic city-texts. In the following case history, we utilize 389 planned names—retrieved from various sources (Helsingin kadunnimet

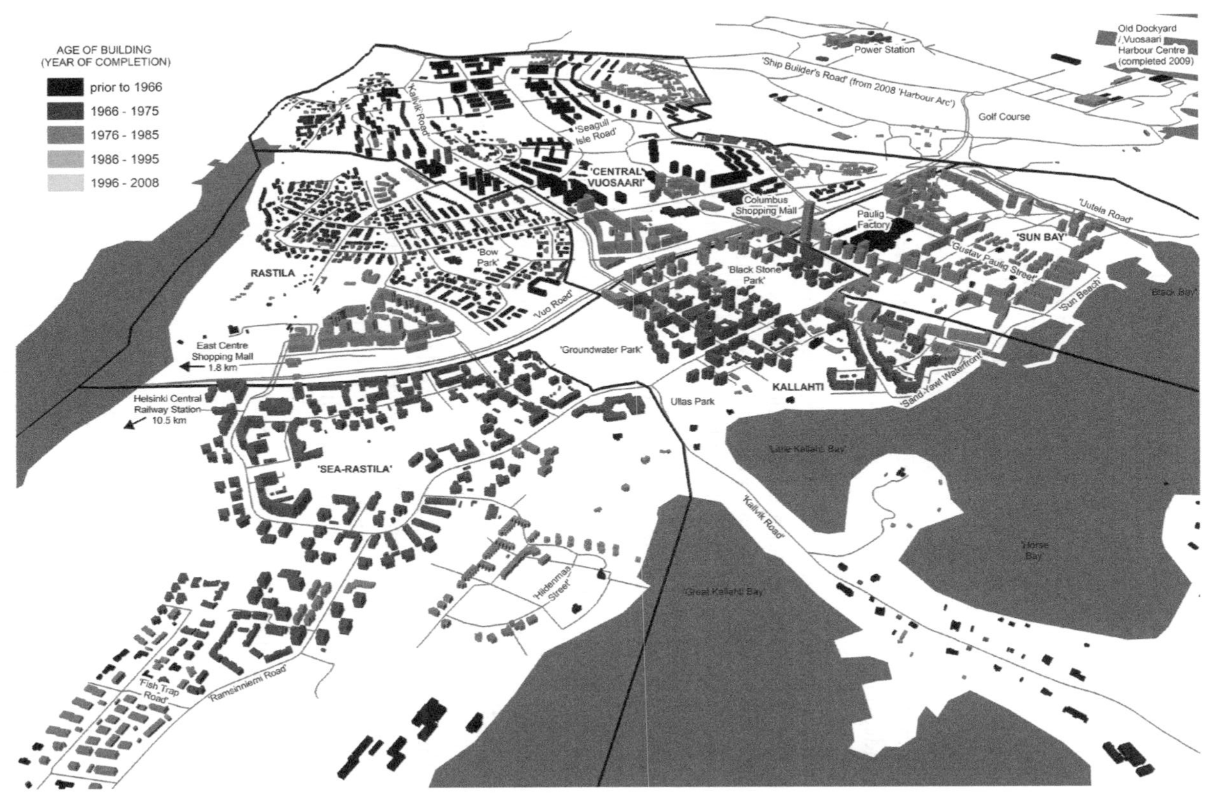

Figure 11.1 The building stock in Vuosaari's neighbourhoods in 2008

1970/1981; Helsingin kadunnimet 2 1979; Helsingin kadunnimet 3 1999; Keski-Vuosaaren rakennusinventointi 1999)—in tracing planning processes, principles and name-motifs involved in development of the suburb's linguistic landscaping. Regarding this corpus of officially ratified or suggested names—given in both Finnish and Swedish, as provided by the law—we focus on the following research questions:

1. In which ways have the principles of official name-planning and modernistic planning in general guided the make-up of local toponymy?
2. Besides municipal planning-authorities, what other actors have contributed to the evolution of Vuosaari's official nomenclature?
3. Have there occurred shifts towards place-promotion or other characteristics of neoliberal urbanism in the suburb's linguistic landcaping?

In addition, we juxtapose the development of Vuosaari's planned toponymy with that of 259 local housing corporation names ratified during the same period (the web-database of National Board of Patents and Registration and the Tax Administration at http://www.ytj.fi). In Finland, a housing corporation (Hsg. Ltd) is a standard legal arrangement of private home-owning regarding especially flats in apartment blocks and terraced houses. While the naming of housing corporations is controlled by different (trade registration) authorities than the so-called street toponymy, the rationale of including them in the analysis is that they represent a group of place-bound names that is not only more directly influenced by estate and housing markets, but whose language is also less strictly controlled.[1] Our interpretations focus on three issues:

1. What kinds of intended functions (pedagogical-edificational, spatially organizing, commercially attractive, etc.) have characterized Vuosaari's planned and housing corporation toponymies?
2. What types of symbolic registers have been utilized in the two name categories?
3. What languages have been represented in them?

1 In official terms, each and every planned place name in Helsinki is a duplicate name given in both Finnish and Swedish. In contrast, the majority of local housing corporations have in recent decades preferred Finnish-based names as the law provides them a right to this (quite understandably given that 87 per cent of Helsinkians spoke Finnish as a mother tongue in 2004). Equally importantly, the rules regarding the use of foreign terms (not based on the two aforementioned national languages of Finland) are also more liberal with regard to housing corporation names.

Vuosaari's Linguistic Landscaping until Late 1980s: Combining Systematicity and Traditionalism

> As from 1960, the bestowing of names on the new town plans, as well as the revising and replenishing of nomenclatures of the already-built neighbourhoods, have proceeded in the conventional way ... In this way things proceeded, inter alia, when the new area of Vuosaari was incorporated into Helsinki's 54th city district. (Pesonen 1970/1981, 84; translation from Finnish by JV)

In contrast to its more or less spontaneously born, predominantly vernacular Swedish pre-annexation toponymy, Vuosaari's absorption into the urban planning system of Helsinki around the mid-1960s was a prelude to a wholesale systematization of its nomenclature. In the following decades, not only the City of Helsinki, but also a couple of other economic and non-governmental actors played major roles in this process. Actually, it was Saseka Ltd—a major local construction material manufacturer—that consulted Helsinki's city-planners to draw up the district's first town plan, as the enterprise expected to extract profit from its lands in central Vuosaari. In 1963, the County Administrative Board ratified this plan (e.g. Schulman 2005, 10–11). It and its pre-annexation supplements comprised more than sixty duplicate (both Finnish and Swedish) street names (Helsingin kadunnimet 1970/1981, 236–40). Even a few years prior to its annexation in 1966, Vuosaari was parting from the anything but meticulous (name-)planning system of Helsinki Rural Municipality, its former parent commune.

Somewhat paradoxically, the urban growth of Vuosaari did not at first lead to a symbolic upheaval in the prosaic tenor of its traditional toponymy (see Kepsu 2005, 133–5). In line with the hereditary and anti-elitist suburban naming policy of the City—and as if also underscoring its built form's spaciousness and nature-relatedness (both ideals embraced by the Finnish architectural modernism of the time; see e.g. Lampi 2005)—the district's newly ratified nomenclature echoed an essentially pre-modern ethos. In particular, name-motifs to do with local coastal topography and wildlife, as well as the area's traditional livelihoods of seafaring and fishing (and to a more minor degree agriculture) dominated the suburb's bilingual street nomenclature from the 1960s even to late 1980s (Table 11.1).

One major category in introducing town plan nomenclature comprised of autochthonous derivations from often centuries-old local names. A couple of toponyms were formed on the basis of the district's name itself (Nordsjö in Swedish; originally Norsö "Strait Isle"), as with the main roads *Vuosaarentie* ("Vuosaari Road") and *Vuotie* ("Vuo Road").[2] Similarly, the old appellation

2 Throughout our analyses, we present English translations besides the actual Finnish (and more occasionally Swedish) toponyms. All translations are by the authors.

Table 11.1 A sample of Vuosaari's typical names, ratified prior to 1990

Official names and English translation	Neighbourhood(s), year of ratification	Locational information	Symbolic associations
Kallvikintie Kallviksvägen "Cold Bay Road"	Several neighbourhoods (thoroughfare), 1963	Based on traditional name	Coastal topography Coldness Local history
Mastotie Martvägen "Mast Road"	Rastila 1963; change in 1966 to "Main Mast Road"	Cluster-based	Seafaring Local history
Pitkänsiimantie Långrevsvägen "Long Line Road"	Sea-Rastila 1963	Cluster-based	Fishing Local history
Kivisaarentie Stenholmsvägen "Stone Isle Road"	Central Vuosaari 1963	Cluster-based, based on traditional name	Maritime topography Local history
Vuosaaren silta Nordsjö bro "Vuosaari Bridge"	West from Rastila 1967	Based on traditional name	Local history
Airoparinkuja Årparsgränden "Pair of Oars Lane"	Rastila 1973	Cluster-based	Rowing boats Local history
As. Oy Lokkisaarentie 1–3 "Hsg. Ltd Seagull Isle Road 1–3"	Central Vuosaari 1964	Address-based	Maritime nature Local history
As. Oy Säästöpurje "Hsg. Ltd Budget Sailcloth"	Rastila 1965	Based on adjacent street name	Home-saving Seafaring Local history

for the village Rastböle (Rastila in Finnish) lent its name to *Rastilantie*, as did the cape Leppäniemi ("Alder Cape") to *Leppäniementie*, the Uutela farm to *Uutelantie* and the Bay Mustalahti ("Black Bay") to *Mustalahdentie* (Figure 11.1). Even more common practice was to employ three sea-related, essentially nostalgic name-motifs newly tailored for the city's easternmost suburb by planning authorities. In Rastila, roadways and parks were group-named after "rowing and sailing vessels" (*Airoparintie* "Pair of Oars Road," *Peräsintie* "Rudder Road," *Melontakuja* "Canoeing Alley") or "seafaring and spar-buoys" (*Merikorttitie* "Chart Road," *Isonmastontie* "Main Mast Road"). In Meri-Rastila ("Sea-Rastila"), names were derived from "fishing tackle" (*Katiskatie* "Fish Trap Road"). In a third dominant name-category, applied in Keski-Vuosaari ("Central Vuosaari"), the above principles of local linguistic borrowing and thematic clustering were combined, as the christening of its several streets was based on the names of islands across the surrounding archipelago (e.g. *Lokkisaarentie* "Seagull Isle Road," *Tammaluodontie* "Mare Islet Road"). The guiding influence of the three naming models also persisted in the supplementation of the local stock of light traffic route and recreational area names (*Märssypolku* "Topsail Walk," *Keulapuisto* "Bow Park"). Notably, a majority of the local housing corporations also followed suit by basing their names on their location along the road network (*As. Oy Pienivillasaarentie 1* "Hsg. Ltd Small Wool Isle Road 1," *As. Oy Airoparintie 13* "Hsg. Ltd Pair of Oars Road 13"). In this way, the city-planners' nostalgically motivated maritime name-choices were effectively duplicated in the commercial nomenclature of the district.

During the first three decades of Vuosaari's urbanization, its linguistic landscaping involved only a few aberrations from the rigidly traditionalist canons depicted above. In the early 1960s, its planners derived some exceptional name-coinages or -suggestions from local geology (*Punapaadentie* "Red Stone Slab Road"), the relative location of streets (*Pohjoinen rivitalokuja* "Northern Terrace"), and the area's newer soil-based industry (commissioned as the pre-annexation town plan was by the Saseka Ltd). Obviously for systematicity's sake, however, almost all of these vicissitudes were either erased (*Kivimyllyntie* "Stone Mill Road") or revised (e.g. *Länsikuja* "West Lane" > *Kallvikinkuja* "Kallvik [Cold Bay] Lane") in subsequent years. From the 1970s onwards, a handful of light traffic routes were given terse names that addressed their present-day urban functions: *Kuntopolku* ("Fitness Path") instructed new suburbanites to a local sports field, and *Puhospolku* ("Shopping Centre Walk") to a cluster of premises to do their daily acquisitions. In a similar vein, a major artery running to a new *Valmet* shipyard (1974–1987) on Vuosaari's eastern periphery, *Laivanrakentajantie* ("Ship Builder's Road"), signified presentist meanings. In the light of the concurrent geopolitical context, the putting of the state-owned shipyard (serving chiefly Finland's bilateral trade with the Soviet Union) on a toponymic pedestal can be even interpreted as a politically motivated "leftist" manoeuvre by the city-fathers of the time.

For an interesting contrast, the arrival of big private-sector employers to Vuosaari—most notably *Paulig* coffee and seasoning refinery (1968–) and the national headquarters of *Shell* (1969–1992)—did not at this stage inspire any expressions in the planned nomenclature of the district.

Among Vuosaari's housing corporations, curiosities in naming were equally time-specific. In the 1960s, many housing companies equipped themselves with compound names that resonated simultaneously with coastal nature or livelihoods, and more idiosyncratically, the pursuit of "home-saving." The housing units in question—take *As. Oy Säästöruori* "Hsg. Ltd Budget Helm," *As. Oy Säästösaari* "Hsg. Ltd Budget Isle," *As. Oy Säästölokki* "Hsg. Ltd Budget Seagull"—were built under the auspices of the Home-Savers' Association, an important non-governmental actor behind Vuosaari's early urbanization (see e.g. Hankonen 1994, 370–78; Schulman 2005, 11). Odd as it may sound from the vantage point of today's neoliberalized housing markets, the trend was to immortalize the association's puritanical ideology (as represented by the so-called "shoulder-banks" in which intending home-owners saved money by participating in cooperative building operations) into the very namescape of the rising suburb. In the 1970s and 1980s, a few housing corporations also began to embrace hitherto unforeseen spatial expressions and metaphors that were sometimes at odds with the anti-elitist mainstream of local nomenclature. Suffice it to mention two exceptions of this sort: *As. Oy Aurinkopelto* ("Hsg. Ltd Sunfield") played with the meanings of sunshine without making any explicitly local references, and a generic term "court" in the name *As. Oy Poijuhovi* ("Hsg. Ltd Buoy Court") was obviously chosen to conjure up prestigious connotations with regard to the quality of housing.

Nonetheless, the above minor aberrations do not undermine the fact that toponymic hints at the area's maritime nature and pre-urban history reigned supreme across Vuosaari's namescape until late 1980s. Why was Helsinki's (name-)planning machinery almost exclusively stuck with such locally introverted, societally obsolete, and indeed rather narrow historical register in its naming policy applied in Vuosaari? Why were not, inter alia, less obviously masculine traditional local livelihoods, or the area's past in the nineteenth and early twentieth century as a recreation zone of upper-class Helsinkians (as encapsulated in the romantic summer abode names such as *Villa Solvik* "Villa Sun Bay" and *Villa Bardoft* "Villa Pineneedle Scent"; see e.g. Lassenius 1997), allowed any room in the official local nomenclature? We contend that the answer lies in the interplay of technocratic-cognitive and cultural-pedagogic functions that the suburb's toponymy was intended to serve within the wider context of the regulative planning policy of Helsinki. In the first place, the way the systematization of Vuosaari's toponymy took place was tied up with the "modernist project of constructing legible urban spaces" (Rose-Redwood, this volume). Not only was the local built environment given an organized appearance by the thematic subdivisions of its nomenclature, but the new satellite suburb as a whole was also differentiated from the other suburbs of Helsinki through its concise toponymic identity. At the same time, the selective nominatory re-presentation of "the" local past was intended to provide its new residents a materially grounded symbolic register to both learn from and identify with. Meanwhile, the absence of elitist connotations gave its toponymy a fittingly neutral and egalitarian flavour (cf. Pesonen 1970/1981). At this stage, the utilization of names as the tools of place-marketing was clearly an unthinkable option for the local planning authorities.

Vuosaari's Linguistic Landscaping from Early 1990s Onwards: A Breakthrough of Toponymic Place-promotion

> In Vuosaari, the City has created outstanding prerequisites for building new residences. Yet even a good product will not sell without marketing. So far, the involved enterprises have marketed the new residences only very modestly. (Vuosaari ... 1998; translation from Finnish by JV)

> Very close to wonderland. It is difficult to speak about Aurinkolahti modestly. It's enough to see the neighbourhood's own little riviera, wide sea with its islands and sunrises and your voice will tremble. [...] Top-quality urban facilities by the sea. The sea, nature. Splendid architecture, engrossing apartments. Beach & Boulevard. What else does one need to enjoy life? (Aavan meren tällä puolen ... *Aurinkolahti*, Helsingin uusin kaupunginosa [On this side of the open sea ... Sun Bay, Helsinki's newest neighbourhood]. An advertising brochure by the City of Helsinki, Pro Paulig Ltd, Sato-Developers Ltd, Skanska Ltd and YIT-developers, no date; translation from Finnish by JV)

In the mid-1990s, a market-led turnabout occurred in the planning of Vuosaari. Crystallizing this change, Vuosaari's newest residential neighbourhood—bestowed in earlier planning documents with a somewhat sombre name *Mustalahti* ("Black Bay")—was re-baptized as *Aurinkolahti* ("Sun Bay") after the *Villa Solvik* built in early 1900s (Helsingin yleiskaava 1992, 98; Raportti Vuosaaren... 1995; Lassenius 1997, 42). Negotiated between the city-authorities and four building companies, the aim was to sell Aurinkolahti's estates exclusively to private-sector developers, and built the neighbourhood "full of hard-money [not publicly funded] apartments" (Haila 2005, 35; Ikonen 2005). Image construction efforts in the "exceptionally fascinating and personal seaside town" were targeted at well-to-do home-seekers who "are no longer content with being able to buy or rent a flat wheresoever, but who strive by their best possible means to find a living place that pleases them" (Raportti Vuosaaren ... 1995). If anything, Aurinkolahti's marketing echoed the market-responsive, now internationally fashionable rhetoric of neoliberal place-promotion.

Crucially, the planning of Aurinkolahti and the adjacent commercial centre (with its extrovertedly named *Columbus* shopping mall)—in both of which Pro Paulig Ltd, an affiliate of the "local coffee roaster" Paulig Ltd, was pivotally involved—preceded dramatic shifts in the implementation of both planned and housing corporation nomenclatures all across Vuosaari. What were the reasons behind this sudden deluge of openly commodified names in the peripheral suburb of Helsinki? In particular, why did the legal holder of the planning monopoly in the district, the City of Helsinki itself, change its policy and start to trumpet—through toponymic and other means—the trendiness of Vuosaari even more enthusiastically than its private-sector partners (e.g. Vuosaari… 1998, see the quote above)?

As has been regularly the case with the legitimatization of neoliberal reforms in different parts of the world (Harvey 2005), the toponymic sea-change on the eastern edge of Helsinki was spurred by a drastic economic depression that Finland and its capital were facing in early 1990s. The ill-starred building projects of Meri-Rastila and Kallahti (see Figure 11.1) explain a great deal of the subsequent march of local events. In the unquestioned lead of the planning machinery of the City, the landscaping of these locales had already introduced some novelties to the built form and nomenclature of Vuosaari. In both neighbourhoods, an enhanced historical appreciation of the local legacy as a villa area of the upper classes had surfaced in the quaint guise of Swedish- and German-originated autochthonyms (*Harbonkuja* "Harbo Lane," *Ullaksenpuisto* "Ullas Park") derived from the seafront's old recreational abodes (Table 11.2). In addition, three new name-groups based on "local geology" (*Mustankivenpuisto* "Black Stone Park"), "groundwater resources" (*Kaivonkatsojantie* "Diviner Road") and "former concrete manufacturing" (*Kalkkihiekantie* "Lime Sand Road") were introduced. Of particular interest is the last name-group as it reflected a recent adoption of the re-developing of the city's seafronts into a cornerstone of its planning policy (Pennanen 2003). In line with this policy, the previous location of the Saseka factory (1938–1978) in Kallahti provided a fitting reservoir for a fashionable postindustrial tenor in its waterside toponymy (*Hiekkajaalanranta* "Sand Yawl Shore," *As. Oy Hiekkalaituri* "Hsg. Ltd Sand Pier"). In the midst of the depression, however, these and other novelties did not count much. In early 1990s, all private-sector developers withdrew from their commitments in Meri-Rastila and Kallahti, which worsened their reputation abruptly. Not the least because the City of Helsinki held ambitious plans for Vuosaari's new harbour, it had drifted into an awkward situation in developing its easternmost flagship area. It could not but start to search for more market-based strategies to prevent the heralded "New Vuosaari" from becoming a total planning failure.

By a stroke of fortune, the City found a devoted partner for its investments and branding efforts especially from Paulig Ltd, a local company that shared a "patron's" vested interest in halting the district's development failure (Kangas 1993; Haila 2005). Consequently, around the mid-1990s floodgates were

Table 11.2 A sample of Vuosaari's typical names, ratified since 1990

Official names and English translation	Neighbourhood(s), year of ratification	Locational information	Symbolic associations
Hiekkajaalanranta Sandskutsstranden "Sand Yawl Waterside"	Kallahti 1992	Cluster-based	Unloading of seasand Local industrial history
Kahvikatu Kaffegatan "Coffee Street"	Sun Bay 1998	Cluster-based Nearness of Paulig factory	Culinary culture Local industry
Helletori Värmetorget "Hot Spell Square"	Sun Bay 1999	–	Sunshine Southern exoticism
Mailapojanpolku Caddiestigen "Caddie's Trail"	East from Sun Bay	Nearness of golf course	Spare time opportunities Upscale sport
Astrakaaninkuja Astrakangränden "Astrakhan Lane"	Central Vuosaari 2002	Cluster-based	Traditional horticulture Garden idyll Local history
As. Oy Helsingin Rantahiekka "Hsg. Ltd Helsinki's Beach Sand"	Kallahti 1992	Seaside location	Sandy beach Local industrial history
As. Oy Helsingin Auringonkulta "Hsg. Ltd Helsinki's Sun Gold"	Sun Bay 2002	Loosely name-based	Sunshine Prestigious housing
As. Oy Helsingin Triton "Hsg. Ltd Helsinki's Triton"	Sea-Rastila 2003	–	The sea Greek mythology Royal prestige
As. Oy Helsingin Corretto "Hsg. Ltd Helsinki's Corretto"	Sun Bay 2005	Nearness of Paulig factory	Culinary culture Local history Internationalism

opened to the formation of powerful public-private partnerships for housing developments and often anything but modest marketing of them. In the specific case of Aurinkolahti's coastal zone, the City proved its commitment to revamp the neighbourhood's maritime image by building a 700 metres long swimming beach, a premier "Venetian-type" channel lining the area in the east, piers for yachts and dinghies, and impressive watery vistas opening from the neighbourhood's luxury apartments. The trendy ambience of the area was underscored through "southerly" toponymy, built around the themes of "sun," "sea" and "sand" (interestingly enough, "sex" was here the only motif missing from the four paradigmatic S's of global tourism!). Besides the name "Sun Bay"— in itself unlikely name for a northern European suburb—a number of subsequent street (*Hellekuja* "Hot Spell Lane"), promenade (*Aurinkoranta* "Sun Beach") and park (*Kultahiekanpuistikko* "Gold Sand Square," *Kauniinilmanpuisto* "Park of Beautiful Weather") names invested the area's seaside location with metaphoric allusions to much warmer southerly climes (Table 11.2). As an instance of semiotic multiplication of a toponym, not a few of Aurinkolahti's housing corporations also inspired their names from the same "sun-urban" imagery (*As. Oy Helsingin Auringonpaiste* "Hsg Ltd Helsinki's Sunshine" and *As. Oy Helios* "Hsg. Ltd Helios," *As. Oy Corona* "Hsg. Ltd Corona"). Blatantly, the rationale of the above kind of "re-conceptualization"—with no counterparts in the district's earlier nomenclature—of the local coastal geography was to be responsive to the potential home-buyers' expected preferences and market Aurinkolahti as a fashionable lifestyle space by the sea.

In other spatial settings as well, a profusion of various types of allusive name-motifs became a new hallmark of linguistic landscaping in Vuosaari. To start with exotically named housing corporations, a miscellany of maritime and climatic motifs have been utilized across its different neighbourhoods (*As. Oy Pasaatituuli* "Hsg. Ltd Trade Wind," *As. Oy Helsingin Cirrus* "Hsg. Ltd Helsinki's Cirrus"). For another novel focus in both planned and housing corporation naming practices, in Aurinkolahti flourish now retro-spirited references to rose breeds of the neighbourhood's villa era (*Juhannusruusunkuja* "Midnight Summer Rose Lane," *As. Oy Nukkeruusunkuja 3* "Hsg. Ltd Helsinki's Doll Rose Lane 3"), and the adjacent Omenamäki's ("Apple Hill") toponymic idyll has been constructed around a historical garden and its apple breeds (*Keltakanelinkuja* "Yellow Cinnamon Lane," *As. Oy Omenankukka* "Hsg. Ltd Appleflower"). Name-choices have been also based on fashionable sports (*Urheilukalastajansilta* "Sport Fisher's Bridge," *Golfaajanraitti* "Golf-Player's Main Street," *As. Oy Helsingin Swingi* "Hsg. Ltd Helsinki's Swing") and high-cultural spare time opportunities (as with "theatrical" street names *Aitiopaikka* "Box Seat" and *Ensirivi* "Dress Circle" in north-western Vuosaari). Increasingly many names have "followed fiction" in the manner of postmodern urbanism (see Ellin 1998). A prominent instance in this respect is Porslahti neighbourhood's hilariously patriotic street toponymy, derived from fantastic figures (*Kultakutrinkuja* "Goldilocks Lane") and places (*Pilvilinnankuja* "Castle in the Clouds Lane") taken from the Finnish national author Topelius's (1818–1898) children's stories. By the same token, certain erasures in the planned toponymy pay witness to the ideologically motivated prettification of the local namescape. Nearby the Vuosaari Harbour Centre and its Business Park, the re-naming of the aforementioned *Laivanrakentajantie* ("Ship Builder's Road") to *Satamakaari* ("Harbour Arc") implies a rhetorical polishing of the former name's suspicious stigma with the Finnish-Soviet trade by a more fashionable terminology of today's cargo industry.

Importantly, hardly any notion summarizes the changed tenor in Vuosaari's toponymy better than its symbolic urbanization—a multifaceted process that has taken place through distinctively urban types of name-choices, increased multilingualism, and above all the wholesale diversification of the local namescape. It is not just that "roads" have changed to more urban-sounding "streets" in the newer nomenclature of the district. Equally symptomatically, hitherto very rare commemorative names[3] (an urban name category *par excellence* in historical terms) salute now the area's recreational and industrial past: historical villa-owners have lent their names to parks and lanes across the southern seafront (*Ivan Falinin kuja* "Ivan Falin's Lane"), and roadways in Aurinkolahti have been baptized after Gustav (1850–1907) and Bertha (1857–1923) Paulig, a German-origin founding couple of the district's long-term employer and current developer. A further instance of toponymic urbanization is a cluster of streets in the vicinity of Paulig refinery that carries now names evocative of its seasoning and coffee products: *Inkiväärikuja* "Ginger Lane," *Mokkakuja* "Mocha Lane," *Pomeranssipuisto* "Seville Orange Park," *Sokeritori* "Sugar Square" and parallel names form an unmistakeably urban motif in which local and global associations of industrial heritage and consumption-oriented culinary delights run together.

In tandem with the proliferation of distinctively urban name-motifs and -categories, linguistic diversification of Vuosaari's namescape has been a particularly striking tendency in recent years. In particular, this notion applies to the linguistic make-up of housing corporation names as these are subjected to less strict controlling by authorities than the planned nomenclature. Around the Paulig factory, a number of *en vogue* Romanic loan names have been adopted from the cultivated terminology of Italian special coffees (*As. Oy Helsingin Cappuccino, As. Oy Helsingin Macchiato, As. Oy Helsingin Corretto*) and Mediterranean tourism destinations (*As. Oy Helsingin Casa Mare, As. Oy Helsingin Gondolieeri* "Hsg. Ltd Helsinki's Gondolier"). Also English-language terms have added local linguistic diversity through allusions

3 Actually, the only veritable person-referring honorific name given in Vuosaari before the 1990s was ratified in 1978, when a park in central Vuosaari was dedicated to Martti Ilveskorpi (1908–1973), the primus motor of the then defunct Home-Savers' Association.

to gentrified industrial urbanity (*As. Oy Helsingin Dockside*, *As. Oy Helsingin East End*), as have done Estonian loan names (*Pärnunkatu* "Pärnu Street," *Hiidenmaankatu* "Hiidenmaa Street") and the "classical" evocations of sea-related goddesses and gods of Greek and Roman mythology (*As. Oy Helsingin Afrodite* "Hsg. Ltd Helsinki's Aphrodite," *As. Oy Helsingin Poseidon* "Hsg. Ltd Helsinki's Poseidon") in a couple of grandiloquent housing corporation names. Unquestionably, the neoliberalization of naming practices in Vuosaari has pronounced its "cityness" in multiple ways, increasing the variety of evocative terms "elevated to the noble status of name" (Benjamin 1999, 522) in order to market the suburb to trend-knowledgeable home-buyers.

Conclusions

Against the backdrop of the ideological hallmarks of modern name-planning in European cities, and challenges that present-day neoliberal urbanism is currently posing to them, our chapter has endeavoured to make sense of the processes, motifs and actors behind linguistic landscaping in Vuosaari, a suburb annexed into Helsinki in 1966. We began our case history from the early 1960s, when Vuosaari was incorporated into the town planning system of the capital of Finland. In tune with the hierarchical planning maxim of bestowing suburbs with fittingly modest nomenclatures, the urbanization of the district was long tied up with a toponymic identity constructed around its sea-related, pre-urban past. Both most of its inherited and invented names served a double function of spatial systematicity and the remembrance of the past uses of local landscape through toponymic hints. Conceived as a satellite suburb among others in the holistic planning system of Helsinki, the local authorities' insistence on the introverted, historically narrow symbolism of its nomenclature continued up to late 1980s. Indirectly, this preoccupation also guided the development of local housing corporation nomenclature.

All this changed in the 1990s not least due the economic depression in the beginning years of the decade. Urged by a will to imitate regeneration projects on overseas urban waterfronts, the City's ambitious schemes around the Vuosaari's Harbour Centre, and the unfortunate development projects in Meri-Rastila and Kallahti, an outright neoliberal turn in the tenor of the planning of one of Helsinki's flagship areas took place. Urbane plurality and allusive name-motifs gained in popularity in Vuosaari's linguistic landscaping. In Aurinkolahti's marketing, for instance, toponymic references to the "southern" exoticism, the neighbourhood's upper-class villa heritage, post-industrial patina, trendy culinary culture, and horticultural heritage have all been utilized side-by-side. In an unmistakably postmodern fashion, the neolibralization of (name-)planning ideals in Vuosaari has yielded both fictionalization and fragmentation (see Harvey 1989; 2005). Indeed, even though local (name-)planning in Helsinki is still a city-authorities' prerogative in legal terms, it appears as if the narrowly traditionalist monotony of the suburb's once-standardized namescape has been lately subjected to "creative destruction" under the sway of market-responsive aesthetization.

How was the modern name-planning maxim of preferring modest names in suburban areas cast aside so abruptly and with such apparent ease in Vuosaari's case? Undoubtedly, the private-sector developers' sway over local toponymic matters has strengthened radically in recent years, specifically through their engagement in public-private partnerships. While real estate operators like Saseka Ltd and the Home-Savers' Association had earlier at best subordinate influence on local toponymy, it is likely that current developers (most notably Pro Paulig Ltd) have on occasions masterminded appealing "geo-brands" to their target estates. At the same time, the city-authorities' (neo)liberalized attitude towards market-responsiveness and the developers' stipulations have been of paramount importance. As the Helsinki City planning machinery started to develop Vuosaari in a more market-led fashion, its local naming policy was quite simply adjusted to the exigencies of the situation. Last but not least, an important undercurrent of the recent changes in linguistic landscaping of Vuosaari has concerned the presumed "readers" of its toponymic city-texts. In this respect, a contrast between pre-1990s and newer naming practices seems particularly striking: an important rationale behind the authorities' earlier preoccupation with "traditional" names was related to their civic merits in enlightening people about a (canonized) local history, whereas the bulk of recent names have been plainly forged with an eye to potential home-buyers.

While more studies on contemporary urban naming practices are needed, it seems that our case history of Vuosaari has unveiled a burgeoning neoliberal tendency in suburban linguistic namescaping, ultimately intended to enthral people as consumers: forget modesty, be creatively destructive and open to market-trends!

References

Algeo, J. (1978), "From Classic to Classy: Changing Fashions in Street Names," *Names* 26: 1, 30–95.

Azaryahu, M. (1996), "The Power of Commemorative Street Names," *Environment and Planning D: Society and Space* 14: 3, 311–30.

—— (1997), "German Reunification and the Politics of Street Names: The Case of East Berlin," *Political Geography* 16: 6, 479–493.

—— (2009), "Naming the Past. The Significance of Commemorative Street Names," in Berg and Vuolteenaho (eds).

Bäcklund, P. and Schulman, H. (eds) (2005), *Suunnittelun Kumppanuudet. Tapaus Vuosaari* (Helsinki: Helsingin Kaupungin Tietokeskus).

Benjamin, W. (1999), *The Arcades Project* (Cambridge: The Belknap Press of Harvard University Press).

Berg, L.D. and Vuolteenaho, J. (eds) (2009), *Critical Toponymies: Contested Politics of Place Naming* (Aldershot: Ashgate).

Bladh, G. and Kuvaja, C. (eds) (2006), *Kahden Puolen Pohjanlahtea I. Ihmisiä, Yhteisöjä ja Aatteita Ruotsissa ja Suomessa 1500-luvulta 1900-luvulle* (Helsinki: SKS).

Boerrigter, R. (forthcoming), "Municipal Policy Regarding Street Names in the Netherlands," *Proceedings of the XXII International Congress of Onomastic Sciences* (Pisa: Università di Pisa).

Brenner, N. and Theodore, N. (2002), "Cities and the Geographies of 'Actually Existing Neoliberalism'," *Antipode* 34: 3, 349–79.

Brindley, T., Rydin, Y. and Stoker, K. (1996), *Remaking Planning* (London: Routledge).

Carpelan, A. and Schulman, H. (1996), *Helsinki Kadunnimissä* (Helsinki: Helsingin Kaupunki).

Crawford, M. (1992), "The World in a Shopping Mall," in Sorkin (ed.).

Eco, U. (1986), *Travels in Hyperreality* (San Diego: Harcourt Brace Jovanovich).

Eichler, E. et al. (eds) (1995), *Namenforschung/Name Studies/Les Noms Propres*, Volumes I–III (Berlin: Walter de Gruyter).

Ellin, N. (1996), *Postmodern Urbanism* (Oxford: Blackwell).

Evans, G. (2003), "Hard-branding the Cultural City – from Prado to Prada," *International Journal of Urban and Regional Research* 27: 2, 417–40.

Faraco, J.C.G. and Murphy, M.D. (1997), "Street Names and Political Regimes in an Andalusian Town," *Ethnology* 36: 2, 123–48.

Farvacque-Vitkovic, C. et al. (2005), *Street Addressing and the Management of Cities*. (Washington D.C.: The World Bank).

Fuchshuber-Weiβ, E. (1995), "Straβennamen: Deutsch," in Eichler et al. (eds).

Freire, J.R. (2005), "Geo-Branding, Are We Talking Nonsense? A Theoretical Reflection on Brands Applied to Places," *Place Branding* 1: 4, 347–62.

Garrioch, D. (1994), "House Names, Shop Signs and Social Organization in Western European Cities, 1500–1900," *Urban History* 21: 1, 20–48.

Gill, G. (2005), "Changing Symbols: The Renovation of Moscow Place Names," *The Russian Review* 64: 3, 480–503.

Gorter, D. (2006), "Introduction: The Study of the Linguistic Landscape as a New Approach to Multilingualism," *International Journal of Multilingualism* 3: 1, 1–6.

Goss, J. (1999), "Once-upon-a-Time in the Commodity World: An Unofficial Guide to Mall Of America," *Annals of the Association of American Geographers* 89: 1, 45–75.

Gottdiener, M. (1997), *The Theming of America* (Boulder, CO: Westview Press).

Haila, A. (2005), "Kahvinpaahtaja Kaupunkisuunnittelijana," in Bäcklund and Schulman (eds).

Hankonen, J. (1994), *Lähiöt ja Tehokkuuden Yhteiskunta* (Helsinki: Gaudeamus).

Hannigan, J. (1998), *Fantasy City* (London: Routledge).

Harling-Kranck, G. (2006), "Kaupunkinimistö. Kaupunkiasutus ja Kadunnimet Ruotsissa ja Suomessa," in Bladh and Kuvaja (eds).

Harvey, D. (1989), *The Condition of Postmodernity* (Oxford: Blackwell).

—— (2005), *A Brief History of Neoliberalism* (Oxford: Oxford University Press).

Helsingin Kadunnimet (1970/1981), 2nd rev. edn (Helsinki: Helsingin Kaupunki).

Helsingin Kadunnimet 2 (1979), (Helsinki: Helsingin Kaupunki).

Helsingin Kadunnimet 3 (1999), (Helsinki: Helsingin Kaupunki).

Helsingin Yleiskaava 1992: Selostus (1992), *Kaupunginkanslian Julkaisusarja* A 31 (Helsinki: Helsingin Kaupunki).

Ikonen, J. (2005), "Hyvin Lähellä Satumaata:" Kaupungin Imago ja Toimet Sen Rakentamiseksi. Tapaus Vuosaaren Aurinkolahti, unpublished Master's thesis (Helsinki: Department of Geography, University of Helsinki).

Johansson, C. (2007), *I Gatuplanet. Namnbrukarperspektiv på Gatunamn i Stockholm* (Uppsala: Uppsala Universitet).

Kangas, A. (1993), "Case Vuosaari," *Yhteiskuntasuunnittelu* 31: 4, 43–50.

Kepsu, S. (2005), *Uuteen Maahan. Helsingin ja Vantaan Vanha Asutus ja Nimistö* (Helsinki: SKS).

Keski-Vuosaaren Rakennusinventointi 1999 (1999), (Helsinki: Helsingin Kaupunkisuunnitteluvirasto).

Kowinski, W.S. (1985), *The Malling of America* (New York: William Morrow).

Langenfelt, G. (1954), "Streets-names, Old and New," in Sahlgren et al. (eds).

Lassenius, T. (1997), *Elämää Vuosaaren Kallahdessa* (Helsinki: Helsingin Kaupunginmuseo).

Leenen, J. (1946), *Theorie en Praktijk van de Straatnaamgeving* (Leuven: Instituut voor Naamkunde).

Light, D. (2004), "Street Names in Bucharest 1990–1997: Exploring the Modern Historical Geographies of Post-Socialist Change," *Journal of Historical Geography* 30: 1, 154–72.

Lynch, K. (1960), *The Image of the City* (Cambridge, MA: MIT Press).

Norris, D.A. (1999), "Unreal Estate: Words, Names and Allusions in Suburban Home Advertising," *Names* 47: 4, 365–80.

Närhi, E.M. (1999), "Ratamestareita, Pasuunoita, Orpaanportaita: Aihepiirien Käyttö Nimistönsuunnittelussa," in *Helsingin Kadunnimet 3*.

Paikkala, S. et al. (eds) (1999), *Yhteinen Nimiympäristömme. Nimistönsuunnittelun Opas* (Helsinki: Kotimaisten Kielten Tutkimuskeskus).

Parkhurst Ferguson, P. (1988), "Reading City Streets," *The French Review* 61: 3, 386–97.

Patomäki, H. (2007), *Uusliberalismi Suomessa* (Helsinki: WSOY).

Pennanen, P. (2003), "Policies and Impacts of Urban Regeneration – Waterfront Development in Helsinki, Finland 1980–2000," unpublished PhD thesis (London: Department of Geography, King's Collage London).

Pesonen, L.A. (1970/1981), "Kadunnimistön Tarkistus Vuoden 1946 Alueliitoksen Johdosta ja Nimistön Laatiminen Uusiin Kaupunginosiin," in *Helsingin Kadunnimet*.

Pinchevski, A. and Torgovnik, E. (2002), "Signifying Passages. The Signs of Change in Israeli Street Names," *Media, Culture and Society* 24: 3, 365–88.

Pred, A. (1990), *Lost Words and Lost Worlds* (Cambridge: Cambridge University Press).

Raento, P. and Douglass, W.A. (2001), "The Naming of Gaming," *Names* 49: 1, 1–35.

Rapoport, A. (1977), *Human Aspects of Urban Form* (Oxford: Pergamon).

Raportti Vuosaaren kehittämisestä (1995), (Helsinki: Helsingin Kaupunki).

Ritzer, G. (1996), *The McDonaldization of Society* (Thousand Oaks, CA: Pine Forge Press).

Room, A. (1992), *The Street Names of England* (Stamford: Paul Watkins).

Rose-Redwood, R.S. (2009), "Indexing the Great Ledger of the Community: Urban House Numbering, City Directories, and the Production of Spatial Legibility," in Berg and Vuolteenaho (eds).

Sahlgren, J. et al. (eds) (1954), *Quatrième Congrès International de Sciences Onomastiques. 2. Actes et Mèmoires* (Uppsala and København).

Schulman, H. (1990), *Alueelliset Todellisuudet ja Visiot. Helsingin Kehitys ja Kehittäminen 1900-luvulla* (Espoo: Teknillinen Korkeakoulu).

—— (2005), "Vuosaari—Helsingin Tytärkaupunki?," in Bäcklund and Schulman (eds).

Soja, E.W. (1989), *Postmodern Geographies* (London: Verso).

Sorkin, M. (ed.) (1992), *Variations on a Theme Park* (New York: Hill and Wang).

Stemshaug, O. (1985), *Namn I Noreg*. (Oslo : Det Norske Samlaget).

Swyngedouw, E. et al. (2002), "Neoliberal Urbanization in Europe: Large Scale Urban Development Projects and the New Urban Policy," *Antipode* 34: 3, 542–77.

Tarpley, F. (1995), "Street Names as Signposts of World Cultures," in Eichler et al. (eds).

Taylor, N. (1998), *Urban Planning Theory since 1945* (London: Sage).

Therborn, G. (2002), "Monumental Europe: The National Years. On the Iconography of European Capital Cities," *Housing, Theory and Society* 19: 1, 26–47.

Viljamaa-Laakso, M. (1999), "Markkinavoimat, Virtuaalitodellisuus ja Paikannimet," in Paikkala et al. (eds).

Vuolteenaho, J. et al. (2007), "Asuntosäästäjien Tapiolasta Helsingin East Endiksi," *Yhdyskuntasuunnittelu* 45: 4, 6–29.

Vuosaari. Suunnittelu ja Rakentaminen 1998–2000 (1998), (Helsinki: Helsingin Kaupunki).

Ward, S.V. (1998), *Selling Places. The Marketing and Promotion of Towns and Cities 1850–2000* (London: E. and F.N. Spon).

Chapter 12
Toponymic Silence and Sámi Place Names during the Growth of the Norwegian Nation State

Kaisa Rautio Helander

In the last few decades the indigenous peoples around the world have begun to discuss the questions connected to their naming traditions and as a whole to their toponymy. The main issue is the right to their own names as a part of wider rights to language, identity and history. Inextricably, the discussion concerns also the question of official toponymy. As soon as one considers indigenous naming traditions from the perspective of the definitions of official nomenclature, a number of questions arise: Which authorities have the power of defining the place names adopted into official use? How have state-sanctioned naming practices and legislation changed historically with respect to indigenous naming traditions? Which names in which language or languages are used in the authorized toponymy? Through what types of strategies of naming and renaming are the indigenous toponymies marginalized? Regarding the consequences of official renaming strategies for the indigenous cultures, Linda Tuhiwai Smith (2004, 51) writes:

> Renaming the land was probably as powerful ideologically as changing the land. Indigenous children in schools, for example, were taught the new names for places that they and their parents had lived in for generations. These were the names which appeared on maps and which were used in official communications.

In empirical terms, in this chapter I will deal with the questions of how Sámi place names have been handled by Norwegian authorities in the county of Finnmark, the northernmost and youngest county of Norway (Figure 12.1). I will pay special attention to the decades that preceded independence, which Norway gained from Sweden in 1905. This period is part of the era of Norwegian nationalism prior to independence, and it was an important period for solidifying a uniquely Norwegian national identity that included the country's Sámi regions. In my case study, I will analyze the state authorities' renaming strategies in northern Norway and how these were utilized to

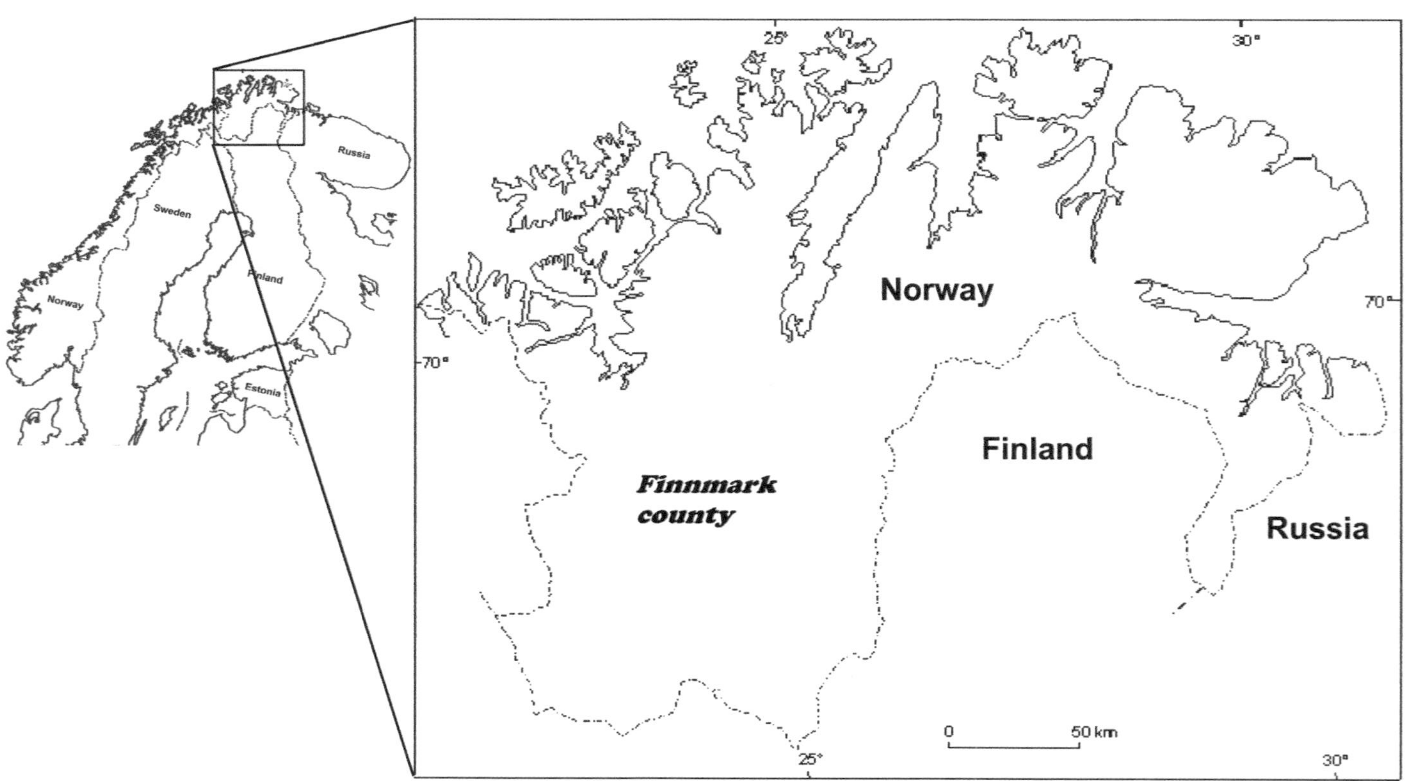

Figure 12.1 The county of Finnmark

build up a representation of a homogenous Norwegian nation state, without acknowledging the linguistic and historical settlement legacy of Sámi areas.

During the nineteenth century, the building of the nation state of Norway was characterized by strong nationalist policies, with a special focus on linguistic nationalism that underscored the distinctiveness and historical roots of Norwegian language in various ways (Østerud 1992, 1996; Thorkildsen 1994). At the same time, the narrative of Norwegian settlement history was also bestowed with highly nationalistic meanings—a process in which the Norwegian and *Norwegianized* place names played a very significant role. Accordingly, an assimilation policy against northern minorities began in the middle of the nineteenth century and it continued even until the 1980s. Called Norwegianization, this assimilation policy focused especially on language. Indeed, language became a measure and a symbol of the failure or success of the policy of Norwegianization (Minde 2005, 7). While the authorities exerted a multitude of measures to suppress the Sámi language in general, also place names were changed with different linguistic strategies from the original Sámi into the Norwegian language (Helander 2008). Through state-led institutional practices such as mapping and land registration, toponymic strategies were used in the assimilation policy as a means of constructing a Norwegianized settlement history, and as well for creating and strengthening the territorial idea of the nation state.

In Finnmark as well as throughout the Sámi region that covers the northernmost parts of Scandinavia, Finland and Cola Peninsula in Russia, the establishment of modern nation states has affected Sámi people in many ways. The states have, for example, had a profound influence on how Sámi place names are treated and accepted in official use in their respective national territories. In essence, the Sámi languages and their dialectic variations do not coincide with the borders of the nation states. Instead, Sámi place names are part of language and dialectic regions that are historically a great deal older than the state borders that now divide the Sámi-speaking area (see also Helander 2006). Only very recently have Sámi people started to attain some of the linguistic rights that have long been self-evident for speakers of the majority languages of the Nordic countries (ibid., 52). Within this context, it is important to note that Sámi languages are still spoken; they are not dead or dying languages as is often thought by non-Sámi speakers, because of the way that official policies have sought to silence them. Crucially, all this means that Sámi place names cannot be considered as substrates merely loaned and used in other neighbouring languages. On the contrary, Sámi place names are still in oral use, but they appear only sporadically in official use due to the historical reasons I will elucidate through the case study of Finnmark outlined in this chapter.

Methodologically, I will apply the ideas of the historian of cartography, the late J.B. Harley (2001), who has utilized deconstruction to illustrate historical and cultural power relations at work in supposedly power-neutral mapping practices. In his view, the "task is to search for social forces that have structured cartography and to locate the presence of power—and its effects—in all map knowledge" (ibid., 152). As defined by Harley, deconstructing the map urges us to discover the silences and contradictions that challenge the apparent objectivity and honesty of the cartographic image (ibid., 152–3). Although some scholars have been sceptical about whether it is possible to figure out intentions of a cartographer or to show the principles of choice of the toponymy (e.g. Andrews 1994; see also de Hóir 1972), Harley's notions prove insightful for documenting and analyzing empirically how the Norwegianization of Sámi toponymy was actually carried out in northern Norway (see further Helander 2008).

In fact, the politics of naming behind topographic mapping and land registration processes in Norway is readable from a number of archival records. In this respect, not only topographic maps, land purchasing documents and land registers of the time, but also governmental regulations and other background material related to their production provide valuable pieces of evidence on the process of creating Norwegianized toponyms. Inter alia, there exist records from the period under study that include place name lists (showing how place names in Sámi languages were deliberately and systematically changed) or correspondence between various authorities (such as cartographers and name consultants discussing how to deal with—whether to silence or subjugate—Sámi names). Based on these types of material, I will concentrate my analysis on specific toponymic strategies that were utilized to Norwegianize Sámi place names, especially in Finnmark.

Toponymic Silencing—Ignoring the Oral Traditions of Sámi Names

Harley's work provides a clear analysis of the role of *silences* in the production of mapping. As I will apply the theoretical concept of silencing as a starting point in my analyses of the politics of naming in Finnmark, it is important to remark that both silences and utterances are constituent parts of map language, each necessary for the understanding of the other. In a fundamental sense, silences on maps (i.e., what is left unmarked on them) are active performances in terms of their social and political impact and their potential effect on the consciousness of both powerful and marginalized groups (Harley 2001). One category of political and social silences utilized in cartographic representations that Harley stresses is the *toponymic silence*, a power strategy used by a majority to control and dictate which names in which language can or cannot be used in official contexts. While toponymic silence has been usually regarded with respect to cartography-related naming, it is possible to apply this notion also to other types of official place names use such as on road signs and in land purchasing documents (Helander 2008).

According to Harley (ibid., 100), manipulations with silences are at one level the result of deliberate censorship or policies of acculturation and at another level unconscious rejection of "Other" people by those belonging to politically more powerful groups. In Norway, toponymic strategies of silencing Sámi place names were not at all an "unconscious rejection of others" (ibid. 100) but very consciously regulated governmental policy against the use of Sámi toponymy. As will be seen, toponymic silences were active performances based on deliberate exclusion of the "Other" that was seen as incommensurate with the hegemonic historical settlement narrative of the Norwegian nation state. Actually, there were (and still are) various ways to silence a Sámi name: (1) to replace it with a Norwegian name created by the authorities; (2) to choose a Norwegian name already in parallel use as a part of oral tradition; or, (3) to leave a place in question without any official name even though there existed a Sámi name for it in local oral usage.

From the point of view of Sámi name traditions, all these methods amounted to silencing. However, there were also obvious differences between the acts of silencing in which a Norwegian name for a toponymically designated location was taken into official use (strategies 1 and 2), and those in which a vernacularly named location was left without any name in official toponymy (strategy 3). In the latter case, the toponymic silence also meant geographical silence, which has had long-standing consequences for people's notions of the importance of specific places and landscapes. Indeed, locations without names in official use are easily regarded as inferior locations without meaning. As Berg and Kearns (this volume) remark: "Place names play a central role in mapping, for they provide the means of filling ostensibly empty space (on maps) with meaningful points of interest." Or as Tilley (1994, 18–19) puts it: "Without a name culturally significant sites would not exist, but only as a raw void, a natural environment." These notions are relevant to underscore the nature of Norwegian name policy in Sámi areas: it produced deliberately nationalistic representations, created by silencing the original toponymy and effectively de-culturalizing the Sámi landscape. As Figure 12.2 illustrates, in the first topographic maps of Finnmark made at the end of nineteenth century, for example, there are crucial resource areas (such as large mountain areas and river valleys) without official names despite the fact that Sámi people had longstanding names for such places.

From the Norwegian point of view the priority was in settled areas, to create a representation that matched the narrative about Norwegian settlement history. In contrast, the tracts outside the settled areas were also very important resource areas for local Sámi people. However, without names the unsettled areas could be regarded as meaningless or useless in the mainstream Norwegian imagination. In turn, this representation also strengthened an image of which areas in Norway were regarded as cultural areas. An old western notion of "culture" (and later also of "cultural landscape" and "cultural place name") was very strongly associated with typically Norwegian forms of permanent,

Figure 12.2 A portion of a topographic map sheet Z4 on Finnmark

Source: Norwegian Mapping Authority 1900.

agrarian settlement and culture. Through cartographic-toponymic means, culture was posited as a binary opposition to nature (see e.g. Barsh 1999). However, for the Sámi people landscape was and continues to be intimately connected to resource sites such as mountain areas, valleys and fjords, as well as to periodical use of these areas. Not just permanent settlement and agricultural landscapes matter for the Sámi in "cultural" terms (Helander 2008).

Toponymic Subjugation—Imposing a Hierarchy of Place Names

Toponymic silence has been just one power strategy used by various Norwegian authorities. Of course, cases in which a Sámi place name is taken into consideration along with a Norwegian place name, cannot be called toponymic silence. During the Norwegianization period, Sámi place names could also be used along with their Norwegian counterparts on maps and in official documents. Tellingly, the Sámi name was in these cases usually bracketed or written after or under the Norwegian parallel name. I call this kind of deliberate hierarchy of majority and minority names *toponymic subjugation*. Both in mapping and land purchasing activities, this strategy was intentionally established through governmental regulations, especially at the end of the nineteenth century. In Norwegian topographic maps, for example, toponymic subjugation was visually strengthened with brackets around a Sámi name and by using smaller or thinner lettertype for Sámi place names. The portion of a topographic map in Figure 12.3 illustrates this subjugation strategy. In it, the Sámi name *Bigganjargga* (*Bigga* of *Biggá* "Sámi personal name of a woman," *njargga* of *njárga* "promontory") is an original name and *Birgitnes* (*Birgit* "Norwegian personal name of a woman," *nes* "promontory") the Norwegian name translation. The Norwegian place name *Birgitnes* is not in oral use even today and hence, it is without any tradition. However, the Norwegian name is prioritized on maps and the Sámi name subjugated.

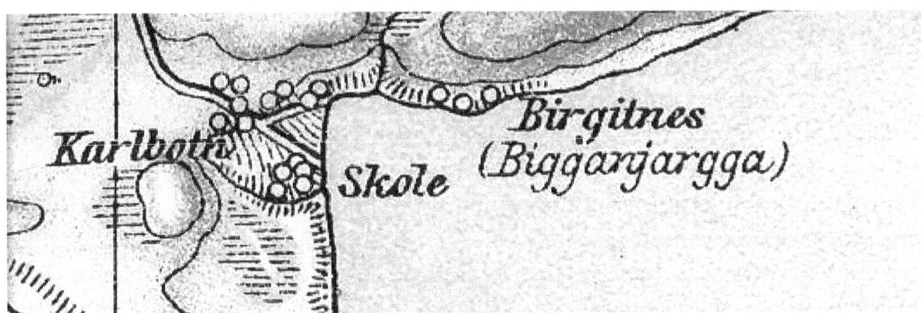

Figure 12.3 A portion of a topographic map sheet Z4

Source: Norwegian Mapping Authority 1900.

The village name *Karlbotn* also found in Figure 12.3 is a Norwegian name first adopted for official use at the end of the nineteenth century (see Helander 2008). The original Sámi name for this village is *Stuorravuonna* (*stuorra* "big," *vuotna* "fjord") but it is silenced on the map sheet. Also, the Norwegian word *skole* "school" provides an iconic example of the policy of Norwegianization, not the least because the state educational system has had the most significant role in Norwegianizing generations of Sámi people. Both strategies, toponymic silence and subjugation, were thus effective visual means of giving priority to the majority toponymy.

Toponymic Strategies in Governmental Regulations

The official use of place names in Finnmark county began to be strictly regulated by Norwegian authorities from the 1870s. Through this process the strategies of silencing and subjugating Sámi toponymy became embedded in intentional government policy in Norway. Legislation concerning choice and use of place names was regulated especially in the production of topographical maps but also in connection with land purchasing and land leasing activities. In this section, I will draw upon examples of late nineteenth and early twentieth regulations with respect to both land purchasing and mapping practices in Norway to illustrate processes of silencing and subjugation.

In 1863, a new act that concerned land purchasing especially in Finnmark county, was implemented. In 1876, the *Land Purchasing Act* was given a new Royal decree where minority place names were mentioned for the first time. This statute regulated how Sámi and Kven[1] place names were supposed to be considered in land purchasing situations. Symptomatically, section 3f of the Royal decree stated: "The farm shall be given a Norwegian name with the possible current Sámi or Kven name added in brackets" (*Reglement* 1876). In 1895, a new Royal decree came into force in the Land Purchasing Act and the same regulation was repeated. This regulation resulted in *toponymic subjugation* for minority farm and meadow names. Even so, it was not uncommon for existing Sámi names not to enter into official use at all—the recommended toponymic subjugation easily became toponymic silence. The formulation of the Royal decree "the farm shall be given a Norwegian name" suggests that the farms in the Sámi area typically had no Norwegian names. Henceforth, however, Norwegian names were deliberately tailored for them with the aid of many different linguistic methods. Overall, the use of Norwegian settlement and land property names has strongly contributed to create a false image of Norwegian settlement in traditional Sámi settlement areas (see further Helander 2008).

Interestingly, many linguistic strategies were used in creating Norwegian names on the basis of a Sámi name. First, I will briefly mention *partly integrated loan names* where specifics are in Sámi but generics translated into the Norwegian language. As an illustration of this strategy I will use names mentioned in the letter of 1866 from the Department of Financial Affairs (DSS 2021), and compare these with the names in land register of 1884 (Matr. 1884) (see Table 12.1).

The examples in Table 12.1 illustrate a typical group of loan names in Norwegian language in which the specific is in Sámi and generic in Norwegian language. In practice, this loan strategy has been an effortless way for the

1 Place names in Kven language refer to names of the regional Finnish-speaking minority in specific areas in northernmost Norway.

Table 12.1 Changing the generic parts of the original Sámi compound names into the Norwegian language was one of the linguistic strategies to solve a lack of Norwegian names

1866 list of land properties		**1884 land register**
Original Sámi names	→	*Generics changed in Norwegian*
Hajagedde	→	Hajaengen
Ravdegedde	→	Ravdejord
Aldogedde	→	Aldojorden
Nirgojavri	→	Nirgovandet
Marmenjarggæčče	→	Marmenæsodden

Norwegian authorities to resolve a lack of Norwegian names especially through the system for registering land purchase activities.

A second common group of loan names was especially used in the division of land properties. To make up for the lack of a majority name, a Norwegian variation word such as *Øvre, Nedre, Østre, Vestre* (which mean Upper, Lower, East, West, respectively) was merely added to the Sámi name. For example, a property called *Galbajohka* (with the spelling of the time *Galbajok*; *galba* "shield," *johka* "river") in Unjárga in East Finnmark was parcelled out into two properties in 1882. The new names for the divided properties—*Østre* and *Nedre Galbajok* with Norwegian variation words combined with the original Sámi name—were added to the land register of 1884 (see Figure 12.4). In English these names would thus be with the current Sámi spelling: *East Galbajohka* and *Lower Galbajohka*.

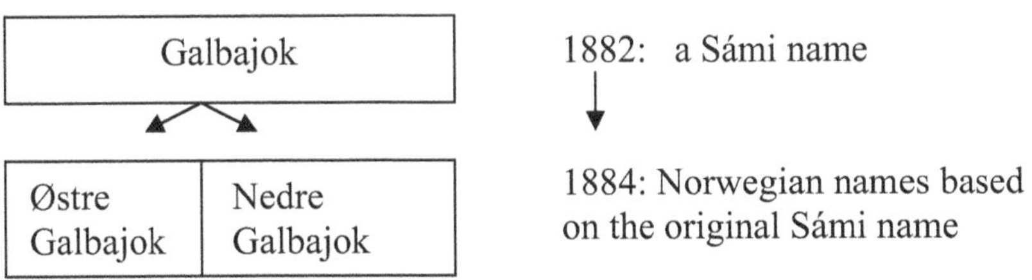

Figure 12.4 In division of land properties the lack of the pre-existing Norwegian names was often solved by adding a Norwegian variation word to the original Sámi name

By their linguistic characteristics, these types of loan results can be described as *toponymic mixtures*. Mixing the languages has been a very easy way to compensate for the lack of a Norwegian name especially in land purchasing but also in mapping, particularly if it was difficult to translate the meaning of a specific into Norwegian. In essence, mixing the languages meant that Sámi names were taken into the Norwegian nomenclature. Yet were these Sámi names actually silenced, given that it is (often) possible to recognize the Sámi name as a part of the Norwegian loan name? From the standpoint of linguistics, these types of loan names can be seen as a form of toponymic silence in a sense that mixed forms are not used in the Sámi language. They do not simply fit the linguistic rules of naming in the Sámi language. Accordingly, the original name used in Sámi was silenced in official contexts in such cases.

A *Land Purchasing Act* in effect only in Finnmark was passed in 1902 with the new amendments. The requirements for the Norwegianization of the county's toponymy became stricter under this law. Thus, section 5 of the Act stated very influentially that "the land property shall be given a separate, Norwegian name" (Ot.prp. 1901). Crucially, the place name statute in the Act meant a total silencing of all Sámi settlement names, farm names and village names based on farm names. This regulation was also taken into consideration during subsequent exercises in both mapping and assigning names on road signs, with toponymic silence as a result. In effect, a total Norwegianization of Sámi settlement names in official use was achieved by the *Land Purchasing Act* of 1902.

Settlement names in the Sámi language have been afforded legal status only very recently, in the *Place Name Act* of 1990. Importantly, however, due to the very successfulness of toponymic silencing in official contexts in the preceding period, there are significant difficulties in gaining recognition for Sámi settlement names. This is especially the case with village names, in particular in coastal areas where the Norwegian assimilation policy has had very strong consequences. Today, local authorities in numerous municipalities, as well as road administration authorities in counties, continue official silencing of Sámi settlement names (see Helander 2006, 2007).

At the same time as the regulations for land purchase were introduced, the first language regulations for topographical mapping were also developed. In 1886, The National Land Survey of Norway wrote to the Department of Defence that for mapping in the county of Finnmark the following rules applied:

> Instead of foreign place names, a Norwegian name should be used where existing; if there is no Norwegian name or it has not been established, a Norwegian translation will be registered as an original name of the location and the foreign name will be written in brackets. (NGO 1886)

Remarkably, indigenous Sámi names were considered "foreign" under these *ad hoc* regulations. In Norway, toponymic strategies were very deliberately used to exclude the Sámi "Other" and at the same time to strengthen the Norwegian majority's sense of "Self." In subsequent Norwegian mapping regulations from 1895, Sámi place names were expressly defined as the Other which should be subjugated:

> When a Sámi name exists, attention should be paid collecting a Norwegian name or a translation of a Sámi name on the spot. The Norwegian name will be written as a main name and the Sámi name will be written in brackets and without stress. (Instr. 1895, 23)

These regulations involved a deliberate Norwegianization in the mapping process, resulting in toponymic silencing of most settlement names and names for prominent natural features (such as names of fjords, large islands, and mountains) that already had a Norwegian parallel name in use. Toponymic subjugation was the main result of the names of those locations that had no Norwegian name in oral use (see further Helander 2008). After independence in 1905, several regulations concerning Sámi place names were further developed for the purposes of state map production. All these regulations stressed the priority of Norwegian toponymy, including rules on how to change Sámi names linguistically into Norwegian. At the early phases of the nation building process, most of the Sámi areas were still monolingual Sámi speaking regions and thus most of these locations did not have Norwegian parallel names. In the course of decades, however, the official toponymic policies discussed above (predicated as they were on Norwegianization of the landscape) constructed Sámi regions in Finnmark into Norwegian space.

Consequences of Toponymic Silencing and Subjugation

Even today, it is a common perception among native Norwegian speakers that Sámi place names that are not in official use are not in oral use either. Toponymic silence has thus been quite effective—among Norwegians at least—in erasing Sámi signifiers from the landscape. In that way the state achieved its objective of Norwegianizing Sámi nomenclature in official use. This erasure has also been strengthened through a wider language policy that saw the Sámi language banned from all official use as well from settings such as schools (see e.g. Minde 2005). While this powerful legacy has not meant language loss in all Sámi areas, it has unfortunately lead to loss of the language especially in many coastal areas.

By all accounts, the strategies of silencing and subjugation have divided Sámi settlement and natural feature names into two different groups. The settlement names were totally silenced but names of natural features could appear on maps in subjugated form (Helander 2008). Over time, this division in official contexts has contributed to the perception among the majority that settled areas are exclusively Norwegian or Norwegian-speaking, whereas Sámi language and people belong to uninhabited regions (such as mountain areas).

This is, of course, part of a longstanding culture/nature discourse that places white Westerners in settled areas (above all cities), and indigenous peoples in a pre-modern state of nature (Berg and Kearns, this volume; Herman, this volume).

The Place Name Act has been in force in Norway since 1990. In principle, the Act guarantees minorities the right to official recognition of their place names. At the same time, however, this Act still reflects the hegemonic belief that every place in Norway has a Norwegian name. The Act guarantees Norwegian names priority protection. This is true even for Norwegian names that were either consciously constructed or translated (without any connection to oral traditions) during the Norwegianization process. *The Place Name Act* does not require that Norwegian names must be in use among the local people to retain their official status, even though this is required for Sámi names. Hence, the Act maintains hegemonic power relations. Even though one objective of the recently adopted official place name policy in Norway is toponymic equality, there is still much work to do before indigenous toponymy gains equality with majority Norwegian names.

Decolonization: Restoring Sámi Toponymy

The history of Norway's treatment of indigenous toponymy in official use also raises the question of renaming and its connection to the politics of naming. This question is very important to indigenous peoples, because their lands and nomenclature have for centuries been the subject of colonial naming with colonial forces defining the right to name.

The objective of the very conscious silencing and subjugation of Sámi toponymy has been to create a representation fitting with an imagined territory of the Norwegian nation state and to create an image of a unified Norwegian nation with a Norwegian language. Norwegianizing original Sámi place names was one means to create an imagined territory and community of the uniform nation state. The consequence of this kind of state place name policy was a very deliberate *renaming*: the Sámi toponymy was intentionally changed into Norwegian both through land purchase activities and through official mapping.

When Sámi people today demand equal rights for their language and place names as a part of the decolonization process, it is often argued that they are *renaming* locations. Local authorities in Norway can still argue that Sámi people are making "new names" when requesting that their traditional place names should be accepted in official use. However, if a place has a Sámi name in oral tradition, it is not a question of renaming it, but this process is better described as *restoration*. The difference between renaming and restoration is that in the former new names are taken into use, but in restoration a place name primarily used orally will also become part of official (written) use. This kind of restoration of Sámi toponymy in official use has a crucial meaning for Sámi people in the toponymic decolonization process.

Acknowledgements

I am grateful to Anja Kaunisoja, of Oulu, Finland for the map illustrated in Figure 12.1. I am also grateful to Johan Isak Siri, Guovdageaidnu, Norway for map editing.

Maps and Archival Documents:

DSS (2021), *Jordsalget, Finnmarkens Amt*. Abbroberte Skjøter I til og med 1868 (40/1866), Direktoratet for statens skoger (2021; 4A03814) (Oslo: Riksarkivet).
Instr. (1895), *Instruks for detaljemåling*, Norges geografiske oppmåling, Kristiania (Hønefoss: Statens Kartverk, Kartsamlingen).
Matr. (1884), *Matricul over jordeiendommene i Tanens Sorenskriveri*. Christiania 1884.—Tana fogde-sorenskriverembete (1855–1959), 254 (Tromsø: Statsarkivet i Tromsø).
NGO (1886), Brev fra Norges geografiske oppmåling 23.12.1886 til Den Kongelige Norske Regjerings Forsvarsdepartements arméafdeling (Hønefoss: Statens Kartverk, Hovedarkiv).
Ot.Prp. (1901) = Odelstingsproposisjon nr. 20, 1901–1902, Angaaende udfærdigelse af en lov om afhændelse af statens jord og grund i Finmarkens amts landdistrikt.
Reglement (1876), Reglement angaaende Fremgangsmaaden ved Afhændelse eller Bortforpagtning af Statens Jord og andre den tilhørende Herligheder i Finmarkens Amts Landdistrikt i Henhold til Lov af 22de Juni 1863. Givet ved Kongelig Resolution af 6te Mai 1876.
Z4 Nesseby (1900), Gradteig Z4 Nesseby. Topografisk kart over Norge. Norges geografiske oppmåling (Hønefoss: Statens Kartverk).

References

Andrews, J.H. (1994), "Meaning, knowledge and power in the map philosophy of J.B. Harley," *Trinity Papers in Geography*, No. 6 (Dublin: Department of Geography, Trinity College), 1–27.
Barsh, R.L. (1999), "How Do You Patent a Landscape? The Perils of Dichotomizing Cultural and Intellectual Property," *International Journal of Cultural Property* 8: 1, 14–47.

Berg, L.D. and Kearns, R.A. (2009), "Naming as Norming: 'Race,' Gender and the Identity Politics of Naming Places in Aotearoa/New Zealand" (this volume).

Harley, J.B. (2001) [1988], "Silences and Secrecy. The Hidden Agenda of Cartography in Early Modern Europe," *The New Nature of Maps*, 83–107 (originally: Imago Mundi 1988: 40, 57–76).

Helander, K.R. (2006), "The Legalization of Saami Place Names in Norway," *GeoNames 2005. Proceedings of the International Conference on Minority/Indigenous Names and Multilingual Areas* Ljouwert/Leeuwarden. 52–8.

—— (2007) [2004], "Treatment of Saami Settlement Names in Finnmark in Official Norwegian Place Name Policy," in M. Jones and A. Schanche (eds), *Landscape, Law and Customary Rights*, Dieđut 2004: 3, 2nd edn (Guovdageaidnu: Sámi Instituhtta/Nordic Sámi Institute), 102–21.

—— (2008), *Namat dan nammii*, Dieđut 2008: 1 (Guovdageaidnu: Sámi allaskuvla).

Herman, D. (2009), "The Aloha State: Place Names and the Anti-conquest of Hawai'i" (this volume).

De Hóir, É. (1972), "The Anglicisation of Irish Placenames," *Onoma* 17, 192–204.

Minde, H. (2005), *Assimilation of the Sami. Implementation and Consequences*, Gáldu čála Journal of Indigenous Peoples Rights 2005: 3 (Guovdageaidnu: Gáldu Resource Centre for the Rights of Indigenous Peoples).

Østerud, Ø. (1992) [1991], *Nasjonens selvbestemmelsesrett. Søkelys på en politisk doktrine* (Oslo: Universitetsforlaget).

—— (1996), "Norwegian Nationalism in a European Context," *Nationalism in Small European Nations.* KULTs skriftserie 47 (Oslo: The Research Council of Norway), 29–39.

Smith, L.T. (2004), *Decolonizing Methodologies. Research and Indigenous Peoples* (London and New York: Zed Books).

Thorkildsen, D. (1994), "Skadinavismen—en historisk oversikt," *Nasjonal identitet—et kunstprodukt?* KULTs skriftserie 30 (Oslo: The Research Council of Norway), 191–209.

Tilley, C. (1994), *A Phenomenology of Landscape* (Oxford: Berg).

Chapter 13
Virtual Place Naming, Internet Domains, and the Politics of Misdirection: The Case of www.martinlutherking.org

Derek H. Alderman

Early advocates of the Internet predicted that cyberspace would transcend the constraints and conflicts of the physical, material world (e.g., Spring 1999). Critics of this perspective have since argued that on-line activities "remain embedded within the context of the off-line spaces and the social relations of everyday life" (Valentine and Holloway 2002, 304). The same tensions shaping non-virtual activities, such as racism and sexism, are also at work in organizing activities in cyberspace (Kitchin 1998). For example, Barney Warf (2001, 6) has challenged utopian interpretations of the Internet by suggesting that "social categories of wealth, power, and place are inevitably reinscribed in cyberspace." With the reinscription of these categories and distinctions, life on-line is not a socially unified realm but just as potentially fragmented and divided as life in the physical world (Graham 2000). Adams and Ghose (2003, 419) characterized new information and communication technologies as "contested terrains in which various groups compete for a strategically advantageous position." As other scholars have established, the Internet has a transformative quality in which its speed and reach allows people to expand the scale of their activities and activism (Adams 1996; Froehling 1997; O'Lear 1997; Alderman 2008). In this respect, cyberspace is not merely an extension or mirror of politics in the physical, offline world, but also a medium for redefining (and even intensifying) lines of social struggle. Perhaps no group knows this better than African Americans, who have witnessed in cyberspace a "reconcentration [rather than a redistribution] of power along class and racial lines" (Beckles 2001, 311).

Domain names provide a window into studying the socially contested and divided nature of cyberspace. More than the innocent technical process of registering a World Wide Web address or URL, acquiring a domain is about claiming territory online. Since domains are assigned on a first come, first served basis, the naming process has led to numerous disputes as people compete to establish a recognizable identity in cyberspace (National Research Council of the National Academies 2005). Internet domains have been studied from a variety of perspectives, such as their role in e-commerce development

(Zook 2000); management and governance of cyberspace (McDowell, Steinberg, and Tomasello 2008); violations of trademark law (Durie 1997; Sharrock 2001); and online branding and marketing (Hashim and Murphy 2007). Despite their importance, domain names have received little attention from scholars interested in the cultural politics of place naming. This neglect has been fueled, in part, by popular claims that cyberspace makes geography unimportant (e.g., Cairncross 1997). In reality, the web is not a placeless social space. Several scholars have begun mapping the contours of cyberspace, such as the external distribution of computers and infrastructure, divides and inequalities in Internet use and access, the internal arrangement and spatial structure of hyperlinks, the development of virtual communities, and the inscription of place-specific cultures and politics into online spaces (Alderman and Good 1997; Walmsley 2000; Brunn and Dodge 2001; Dodge and Kitchin 2001; Holloway and Valentine 2001; Longan 2002; Crampton 2003; Wang, Lai, and Sui 2003; Alderman 2006b; Warf and Vincent 2007).

In analyzing the spatiality of the Internet, it is important to recognize that place can be conceived of without necessarily limiting it to physical location. Even in the absence of a conventional material location, media technologies constitute "gathering places" because of the important social and symbolic functions they provide (Adams 1992). Language and communication are central to the construction of place (Tuan 1991). According to Adams (1998, 89): "a communication system is to communicators as a place is to inhabitants." He asserted that cyberspace is composed of "virtual places" that—like physical, offline places—are defined by communication flows or topologies that can either open up or close down avenues for social interaction. While domain names in and of themselves are not virtual places, they are essential in directing and structuring the patterns of communication and interaction that create place online and literally inscribe meaning into cyberspace. I suggest that we view domain names as the virtual equivalent of place names and attempt to understand how they operate culturally and politically.

Rony and Rony (1998, 15) described domains as "the new meta symbol of the information landscape" because they serve as identifiers as well as locators of individuals, companies, and organizations in cyberspace. A domain name "gets an Internet user to a specific web site and hints at what content will be found there." The navigation and orientation provided by Internet names has social implications as well as informational consequences. People use domains to seek out and locate web sites that match their needs, interests, and ideologies—contributing to the development of an online sense of place and community. In exploring the topologies of the Internet, one should note that the power and influence of domain names comes not only from their ability to direct Internet users to certain virtual places and perspectives. They also have the ability to misdirect people, either accidentally or by design. In some instances, this misdirection takes users to web sites that significantly challenge (and even offend) their worldviews. I am interested in misdirection as a political strategy, how domain names are used to facilitate (or hinder) flows of information and communication, and how these flows and directionalities shape the construction of virtual places and the fulfilment of certain social goals.

The purpose of this chapter is to explore domain naming as a new critical toponymy and thus prompt us to rethink what constitutes a place name within the Internet age and outside traditional, physically-based definitions of geography. The chapter consists of three parts. First, I discuss the technical and cultural importance of domains, suggesting that they represent important symbolic points of identity for people and, for some, valuable real estate. Second, I delve into why naming the web is a politicized act and provide examples of when domain names become sites of economic and political contest. Third, I use the case of martinlutherking.org to illustrate how Internet domains can be appropriated and manipulated by social actors and groups to give visibility to their perspectives and de-legitimize the views of others. Stormfront, a white supremacist organization, controls the King domain and has used the power of misdirection, the unique information structure of the Internet, as well as their own rhetorical strategies to assist them in defaming the slain civil rights leader and gaining a highly visible platform for their cause. Stormfront's attack on King is indicative of the blatant racism that Internet users can find online, despite utopian hopes that cyberspace would transcend the tensions of the physical, offline world. Counter-intuitively, the King domain name is implicated in the construction of "places of hate" and the intensification (rather than the de-escalation) of racial conflict.

What's in a Domain?

Domains translate the numeric addressing system used by computers into structured strings of characters and words, thus providing a more intuitive way of naming and finding web sites. For instance, typing www.food.com in one's web browser is an easier way of accessing the Internet site associated with the television channel Food Network than remembering its server's IP (Internet Protocol) number of 204.78.38.16. The domain name system is hierarchically structured and every domain contains two or more levels of names separated by periods. At the upper level of this hierarchy are top level domains (TLDs). For instance, the web address www.food.com consists of the .com [commercial] generic domain name. There are 18 other generic top-level domains intended for different kinds of registrants and operating at varying levels of restricted and unrestricted access. They are .aero, .biz, .cat, .coop, .edu, .gov, .info, .int, .jobs, .mil, .mobi, .museum, .name, .net, .org, .pro, and .travel. The newest TLD is .asia, introduced in 2007 to serve the growing economies of Asia and to create a common, Pan-Asia regional identity. Individual countries have been assigned their own top-level domain codes such as .bz for Belize, .uk for

United Kingdom, and .ca for Canada. Beneath top level domains (in other words, to the left of the dot or period) are second level domains. This would be the "food" in food.com. A sub-domain (or third-level domain) can exist beneath or to the left of a second level domain.

As mentioned earlier, domain names play a critical role in directing communication in cyberspace, a fact not fully realized until a mistake occurs. Consider, for example, how the seemingly innocent practice of misspelling a web site's address can have embarrassing consequences. In 2001, a North Carolina newspaper published an article entitled "Peeps Shows," about the popular marshmallow candies shaped like bunnies and peeps (baby birds). A list of web sites related to the candy figures accompanied the story (*The Daily Reflector* 2001a). Unfortunately, the domain name of one of the web sites was misprinted. The newspaper left off one "s" from the correct address, www.peepsshow.com. The misprinted address led to a web page that was an off-colour, adult pornographic site—prompting the publication of a correction two days later (*The Daily Reflector* 2001b).

Domains are not just web addresses. As with traditional place names, they can be viewed as symbolic texts embedded in larger systems of meaning and ideology that are read, interpreted, and acted upon socially by people. Adopting a semiotics approach, Phil Steinberg and Stephen McDowell (2003) have encouraged us to see the Internet as an arena of signification and domain names as signifiers involved in the production and consumption of politically-important images and identities. As they found, the manipulation of country-level domain names provide insight into the ongoing redefinition and reproduction of the idea of modern statehood. Wass (2003) also views the domain name system as a means of communicating cultural values. According to her, nation-states use their unique country code domain names to achieve certain priorities. While some countries restrict use of their codes to citizens in an attempt to forge a national online identity, others have been more liberal in allowing access to parties outside the host nation. Responding to a shortage of commercially valuable web addresses, the small Pacific island nation of Tuvalu leased rights to register domains under its Internet country code to a corporation. Tuvalu's domain extension (.tv) is the same as the abbreviation for television, making it an easily recognizable identifier for media-related enterprises. Ironically enough, before striking this lucrative deal, Tuvalu had limited telephone service and no television service (Wilson 2001).

The importance of domain names as a point of identification is not limited to countries, however. This became particularly evident when Beaver College, near Philadelphia, Pennsylvania, officially changed its name to Arcadia University. The school's name change had been prompted by years of derogatory references to rodents and the TV show "Leave it to Beaver" as well as vulgar allusions to the female anatomy. Exacerbating these image problems was the fact that Internet filters in many schools and homes blocked access to the college's web site, beaver.edu. Filter programs even censored email messages from Beaver College. According to school administrators, the name issue forced the college "to send out three times as many mailings as similar schools to attract equal number of students" (Abel 2000, B1). While many schools benefit from carrying their traditional name recognition over into cyberspace, Beaver College discovered that cyberspace made their already difficult identity even more problematic when competing for public attention and legitimacy.

As illustrated above, people have a vested interest in acquiring a domain name that will promote a positive image as well as assist Internet users in easily finding them. Wilson (2001) compared the appropriation of domain names as a search for highly visible real estate on the Internet. Specifically, he wrote: "commercial success in cyberspace takes on some similarities to high street retailing in that good location is an essential element to attracting good customers and capturing passing traffic. The Internet equivalent of a main street address is a domain name that is easily recognized and remembered by the public" (Wilson 2001, 60). Wilson's comparison of a domain name to a street address inspires us to reflect on the potential similarities between naming in online versus offline situations; however, an important distinction should be made. While it is possible to have hundreds of Main Streets in the many cities of the US, it is impossible to have multiple Internet sites with the same name. One can use the same second level domain with a different top level domain (e.g., there can be www.food.com, www.food.org, www.food.net), but there are only a limited number of these top level domains and not all are the same in terms of access, purpose or popularity. For example, the .com top-level domain remains the most coveted and visible name space online.

The Politics of Domain Names

Because it is necessary for each domain to be a globally unique identifier, there is often intense competition over the registering of domain names. One never "owns" an Internet domain. Rather, one obtains exclusive rights to use a name through a registration and renewal process with an authorized registrar or reseller. Registration is competitive because there are a limited number of words or word combinations to use in one's domain. As early as 1999, a survey found that 93 per cent of words in a standard 25,500-word English dictionary had already been claimed in naming domains in the .com realm (Weise 2000). Moreover, not all Internet addresses are equally memorable and domains are often bought, sold, and fought over. Sales of domain names in 2005 generated $29 million, almost double the figure for 2004 (Swartz 2006). Corporations and other organizations realize the potential advertising value of domains as slogans and brand names. For example, the recruiting efforts of the United States Army are written directly into its domain, goarmy.com.

The desire to garner attention for a domain name is so strong that some companies have used financial incentives to persuade small towns to change their monikers for the sake of advertising. Examples in the United States include the renaming of Halfway, Oregon to half.com and changing Santa, Idaho to Secretsanta.com. In New Zealand, officials in Pokeno renamed their town JenniferAnn.com for a year to promote the web site for Jenny Hannah's new online lingerie business. According to Bell and Lyall (2002), the renaming brought millions of visitors to Hannah's domain and helped an economically depressed town regain some needed publicity after being bypassed by a newly built highway. These cases illustrate that domains are part of a larger political economy of place naming and identity that is not strictly online or offline, but which is co-evolutionary and recombinant in nature (Graham 2000).

In the competition to register Internet domains, some people appropriate domain names that are highly sought after by other parties, either for the purpose of eventually reselling and profiting from them or using them as a platform for bringing visibility to their own agenda. For example, the past several years have seen a growing number of legal disputes over the protection of corporation names, celebrity names, and trademarks in cyberspace, leading ICANN (International Corporation for Assigned Names and Numbers) to adopt a uniform domain name dispute resolution policy to arbitrate ownership issues (Mueller 2001). Some critics, such as the Domain Name Right Coalition, have expressed concern that ICANN's decisions, which have tended to favor trademark holders, have unnecessarily facilitated the corporatization of the web at the sacrifice of the rights and free speech of individuals, non-commercial organizations, and small businesses.

One of the most obvious examples of what one reporter (Leonard 2002) called "corporate brand name bullying" is the dispute over Nissan.com. In 1999, the Nissan car company sued Israel-born entrepreneur Uzi Nissan in an attempt to gain control of the Nissan domain name. Mr Nissan first registered the domain in 1994 for the purposes of promoting his Nissan Computer Company in Raleigh, North Carolina (Matthews 2000). The auto manufacturer claims that it is an infringement and dilution of its trademark while Mr Nissan asserts the historical legitimacy of his family's name and the fact that he was using the name Nissan commercially long before the car manufacturer. Domain name rights activists have characterized this as a case of "reverse domain hijacking," when "a trademark owner asserts a spurious claim of trademark infringement and/or dilution to seize a domain name similar or identical to the registered mark away from a legitimate domain holder" (Rony and Rony 1998, 616). However the matter is finally settled in court, the ultimate loser in the struggle may very well be Uzi Nissan, who has spent millions of dollars fighting the larger, stronger corporation.

Controlling domain names is not just economically driven but also open to manipulation by political and social campaigns. This became painfully evident to Dick Cheney during his televized vice presidential debate with

John Edwards in 2004. In attempting to refute accusations of wrongdoing at his former company Halliburton, Cheney referred viewers to factcheck.com. He meant to point the public to factcheck.org, a nonpartisan voter advocacy group, but got the domain name wrong. The company that owned factcheck.com took advantage of this error by automatically routing web users to a page titled, "why we must not re-elect President Bush: a personal message from George Soros." Soros, a billionaire, is a major Democratic donor. It is probably just as well that Cheney misspoke about factcheck.org since the web site did not provide the exonerating evidence that he claimed it did (Suellentrop 2004). Name misdirection, whether intentional or accidental, is a common theme in cyberspace. Rather than simply creating confusion, domain name misdirection can be used strategically to direct communication toward certain web sites and away from others.

The term "typo squatting" has been used to refer to the practice of registering domain names that closely resemble other people's domains but contain a common typographical error or variant. The domain name is intended to take advantage of the misspelling of popular Internet addresses in browser windows, leading users to alternative web sites that mimic the intended site or, in some cases, offer a message different than what people expect. The practice is often associated with attempts to disrupt commercial activity of legitimate domain name registrants, but it can also be used for political reasons. One of the most famous instances involved Christopher Lamparello, a gay activist who was allowed by the US Supreme Court to keep the domain and associated web site, fallwell.com. Lamparello's domain, which relies on a misspelling of the official Internet address of the deceased conservative, anti-gay pastor Jerry Falwell (falwell.com), is used to direct users to a web site that challenges the religious leader's views on homosexuality. Lamparello's lawyer, Paul Levy, reflected on how domains can be used in questioning the reputation and views of a public figure: "A domain name is not just the source of a web site, but the substance of a Web site … You can say the name of the person you're criticizing, and you can put their name in the domain name of your Web site" (quoted in Lisberg 2006, 7).

As suggested by Paul Levy and the efforts of Lamparello, the very domain names used to direct people to web sites can be used as tools of advocacy and protest. Obvious examples include the scripting of web addresses to explicitly challenge the dominance of corporations (e.g., walmart-really-sucks.com) and elected officials (e.g., stopdemocrats.com). Domain names can also be crafted to advocate for certain points of view (e.g., stopmercurypollution.org, milksucks.com, abortionismurder.org). Some individuals and organizations have realized that the politics of Internet domains is not simply a game of offense but also one of defence, such as when they "register domain names that may be used against them and to preempt opponents from using easily recognizable terms that may reference them in a poor light" (Wilson 2001, 63). For example, civil rights organizations such as NAACP and the Anti-

Defamation League have claimed racially and ethnically derogatory domain names before they can be used by hate groups. Virtual place naming, like the naming of physical places, is about claiming the power to control patterns of communication and representation. Yet, domain name appropriation is not limited to one side of the political spectrum. As Warf and Grimes (1997) have argued, the power of Internet can be wielded for reactionary purposes as well as progressive ones.

Defaming the King

One of the most controversial examples of the political appropriation of domain names is martinlutherking.org. The domain is not controlled by a civil rights organization, museum or even the King family. Stormfront, a white nationalist organization, first registered the domain name in 1999 and has used it since as the address for a web site that defames Martin Luther King Jr and espouses white supremacist views. They characterize the civil rights leader as a Communist traitor, a beastly womanizer, and a plagiarizing intellectual. Through the martinlutherking.org domain, Stormfront rewrites King's identity in negative, non-heroic terms. The site's creators carry on what sociologist Gary Fine (2000) has termed "reputational politics," the struggle to redefine an individual's memorial legacy. King's image remains an embattled symbol in America despite his ascension to the rank of historical icon and the focus of national and state holidays (Alderman 2006a). King's reputation, like the reputations of all historical figures, is not universally set but open to constant reinterpretation by people, particularly his detractors. Stormfront's use of the domain and associated web site is not confined to attacking King but includes casting illegitimacy on the reputation of other prominent African American leaders, promising to reveal the "truth" about figures such as Rosa Parks and Jesse Jackson. The web site's creators protest what they call the "civil rights industry," which allegedly hides the subversive and un-American activities of black leaders.

The King domain/web site is a result of the ongoing rise in online hate speech and comes in reaction to the growing social influence of minorities and the perceived loss of white power and privilege (Gerstenfield et al. 2003). No doubt, the extreme views of Stormfront are not shared by the majority of white Americans. Yet, as asserted by Flint (2004), dismissing members of the extreme right as simply social outsiders fails to recognize their involvement in everyday relations and the possibility of their messages, however unattractive, influencing the political mainstream. The King domain represents a clear attempt on the part of Stormfront to expand their influence so as to challenge mainstream social and political attitudes. For example, in calling for the repeal of the King Holiday, the web site's creators provide Internet users with flyers that can be downloaded and printed out for mass distribution. Many of these efforts are directed at children, who are instructed: "Take these flyers to school or around town and leave them in common places where people hang-out. You can even put them on cars if they are parked on a PUBLIC STREET." Stormfront's intent is to do more than simply create a virtual place of hate and intolerance. The organization also seeks to intensify the level of social conflict found in the offline world, prompting us to consider how virtual and non-virtual places are mutually constituted and constructed.

The King domain directs Internet users to a wide variety of information and communication resources that demonize the civil rights leader's reputation. As these resources illustrate, Stormfront's manipulation of King's name is not limited to the area domain naming. A freely downloadable video is available through the web site. It is a 23 minute, detailed presentation of evidence that supposedly reveals the immoral and phony character of Martin Luther King. Specifically, the video programme casts aspersions on the fact that King Sr and Jr adopted the name "Martin Luther" (after the famous religion leader) in place of "Michael." King is called "Michael" throughout the Stormfront web site, thus denying the civil rights leader the right to his own name. This is a particularly poignant issue when we consider the historical role of white masters in naming African slaves and how, even after emancipation, African Americans lacked the power to re-identify or re-imagine themselves in ways that broke away from white expectations and normative standards. When black Americans such as Muhammad Ali (formerly Cassius Clay) and Malcolm X (formerly Malcolm Little) challenged these standards and renamed themselves, they were confronted—as King is here—with suspicion and resistance from recalcitrant whites.

The defaming of King and his name is carried out through many discussion forums hosted through the Stormfront-run domain and web site. While there are occasionally comments posted that counter white supremacist ideas, these efforts are quickly and resoundingly attacked and discredited. In dismantling King's reputation, forum users take aim at the many streets and public places that bear the civil rights leader's name. Examples of these discussion threads include "MLK Streets around USA are Slums" and "I wouldn't want to end up at the Martin Luther King Hospital." Again, online and offline geographies do not exist in isolation of one another. These negative Internet comments mirror a wider stigmatization and marginalization of King-related naming that occurs offline. African American activists often complain that they struggle to control where the civil right leader's name is placed in terms of streets and other public places. While these activists seek to rename prominent roads that cut across racial communities, white opponents often want King's name confined to black neighbourhoods, thus limiting the ability of African Americans to remember him on their own terms. Arguably, the King domain should be studied within the context of this larger racial politics of place naming. The registering of martinlutherking.org by Stormfront represents the ultimate act of memorial appropriation by resistant whites, in effect keeping

African Americans from being able to claim and celebrate their heritage. As asserted by Marc Bernstein, a lawyer specializing in Internet issues, "It [Stormfront's use of the domain] definitely deprives the King estate of their ability to use the name for their own purposes" (quoted in Mack 2000, n.p.).

Although opponents have sought to limit the visibility of King's name within the physical, material world, the situation is quite the opposite on the Internet. Stormfront actively capitalizes on the recognizable nature of martinlutherking.org to help spread their views. It appears to have been a successful strategy; the media reported that the King domain attracted 2 million visitors in only a year after its establishment (Mack 2000). Underlying the popularity of the King domain name and web site is a politics of misdirection. When the white nationalist organization registered the domain, Internet filters were increasingly being used by schools, libraries, and homes to block out the web sites of hate groups. Appropriating King's name, because of its positive associations with civil rights, allowed Stormfront leaders to evade these barriers as well as provided them a way of tarnishing his good name, literally. Unsuspecting Internet users are often drawn into the domain name expecting to find martinlutherking.org associated with a web site that celebrates his legacy rather than one trying to destroy it. Alexa software, which analyzes traffic flows into and out of domains, reveals that martinlutherking.org is most frequently visited in January, the month in which the King Holiday is observed. Moreover, analysis finds that people who visit martinlutherking.org also visit more historically legitimate King web sites, leading one to speculate that the Stormfront-controlled domain is being accessed mistakenly in a search for general information on King. This misdirection of Internet users to martinlutherking.org potentially interferes with the flow of traffic to conventional web sites about King. It is an interesting corollary to the way in which radio operators, during times of war, used noise or misleading information to deliberately block out the reception of transmitted signals by the enemy. Stormfront is able to "jam" or disrupt the broadcast of a heroic account of King, not only through its negative online portrayal of the civil rights leader but also through its ability to shape patterns of communication among other MLK-related web sites.

The power of martinlutherking.org, or any domain for that matter, is not limited to the resonance of the name it employs but includes how it interacts with the unique information structure of the Internet as well as the search engines used in cyberspace. As Paul Carter (2006, 97) argued, the Internet is characterized by a "informatics" in which traditional thematic organizations of information, such as found in libraries, have given way to a net of signs and subjects that "generate[s] almost infinite connections" with no regard for the interrelationships between web sites and resources. In such an environment, martinlutherking.org is not indexed or categorized any differently than any other domain dealing with the civil rights leader. The King domain is allowed to exist in the very same technological and ideological space as other, more positive Internet references to the civil rights leader, a fact that contributes to the domain name's ability to facilitate misdirection. Search engines, such as google.com, play a central role in perpetuating this misdirection. These engines are not neutral technological conduits but can clearly favour certain sites over others in the ways in which they list or rank search results (Introna and Nissenbaum 2000). At the time of the writing of this chapter, the King domain appeared third in a result list when searching for "Martin Luther King" in Google. The two Internet sites ahead of martinlutherking.org on the Google search list were BuildtheDream.org, a sponsored web site devoted to the building of a King memorial in Washington DC, and a web site related to the Martin Luther King Jr Papers Project at Stanford University. The Stormfront-controlled domain was actually higher up on the results list than the web site for The King Center for Nonviolent Social Change in Atlanta, Georgia. In explaining the high ranking for martinlutherking.org, a participant in the Stormfront discussion forum pointed to the confusion caused by the domain rather than the message attached to it :

> Look at the name of the site ... Anyone looking for information on Martin Luther King would surely go to this Website not expecting it to be pro-white. The more visits the site gets, the more the site goes up in rank. Common Sense. I am sure it is not because everyone is coming to the conclusion that Martin Luther King was a communist and the world as a whole is beginning to change its views on King.

The King domain consistently appears near the top of Google's results list because of the search engine's PageRank algorithm. As explained by Nicholas Carlson (2007, n.p.), senior associate editor with internetnews.com: "Google's PageRank algorithm determines a web site's relevancy to a keyword by crawling the Internet to find out how many times that particular site is linked to other sites and how often that keyword is used to describe the link." Stormfront's dominant place on Google has not gone unchallenged, however. In 2006, a group of bloggers started a campaign of using "Google Bombs" to manipulate the PageRank algorithm and drop martinlutherking.org from the search engine's list of top results. In dropping these Google Bombs, bloggers posted links to more factual, less offensive King sites and labeled them "Martin Luther King" in hopes of changing the density of keywords associated with particular sites. While there were early signs of progress, Carlson (2007) reported that the Stormfront-run domain continues to register high on the Google list. In offering an explanation, Tom Hoffman, the blogger who led the Google Bomb initiative, suggests that the search engine's algorithm is attracted to "the character for character match between *Martin Luther King* and martinlutherking.org" (Carlson 2007, n.p.). It is also possible that the very people who are opposed to the King domain have helped to keep it popular within Google. In protesting martinlutherking.org, Internet users invariably provide a link to the web site and the Google PageRank algorithm is sensitive to the number of sites that link to a particular domain. In light of the Internet's system of information organization and retrieval, opponents have been encouraged to target the issue of links as they try to cut into the power of Stormfront to misdirect and offend. Carlson (2007) has asked users to resist the temptation of linking to martinlutherking.org, no matter how much they want to spread word of the danger of the domain to concerned citizens. Users are also asked to contact people with sites that already link to the King domain for the purpose of convincing them to remove these links. As illustrated here, domain names—like traditional place names—can become sites of social contest and debate. However, in cyberspace, the politics of influencing these lines of debate require that we address, more centrally, the flows of communications that create virtual places and how these flows can be controlled in reshaping the relative power of places.

The ability of martinlutherking.org to misdirect Internet users is also due, in part, to the representational tactics employed by Stormfront. The web site's description, which appears underneath the domain name on the Google search results list, states: "The truth about Martin Luther King: Includes historical trivia, articles and pictures. A valuable resource for teachers and students alike." When the King domain name and web site first appeared in 1999, it was crafted to look like an innocuous educational resource. Its adversarial position on King's life and legacy was only evident after closer inspection. The page's attack on King is more overt now, although it still retains a factual and encyclopedic look that easily draws in Internet users. The misleading nature of the King domain has been exposed by critics who have expressed concern about its potentially negative influence on the education of children (Mintz 2002; Gerstenfield et al. 2003; Williamson and Pierson 2003). These critiques have removed much of the deceptive potential of the King domain, although the attention has also fueled Stormfront's desire for publicity, which directly shapes it potential to recruit. Yet, the King domain continues to confuse some Internet users, even to the point of hindering attempts to protest the domain and its racist message. For example, a cyber-petition entitled "Stop martinlutherking.org" has received a limited number of electronic signatures of support and some of these signatures, ironically enough, have come from white supremacists who interpreted the campaign as a petition against King rather than as a petition against a site the defames King. One petition signer wrote: "I agree, we need to get this Website off the Internet. In fact, we should remove all Web sites about that maniac KING." Another supporter wrote: "I'm all for this petition. WHITE POWER!" Ironically, it appears that the King domain is so effective in misdirecting people that it may be escaping the interests of the very community it is intended to represent.

While it is important to understand how martinlutherking.org shapes online communication and social interaction, it is also necessary to consider the larger impact of martinlutherking.org on race relations in the physical, material world. Recent events in Warren County, Illinois are illustrative. In February of 2006, the Warren County Public Library came under public criticism for posting a link to martinlutherking.org on a web resource page for Black History Month. Officials claimed the action was a mistake, a result of careless investigation on the part of a library worker, and quickly removed the connection to the Stormfront site after a complaint uncovered the error. While the complainant, a black woman attorney, accepted the library's explanation, she did make a statement that gave insight into how the misdirection caused by the domain name worked to intensify feelings of racism and undermine her opinion of the town. Specifically, she wrote to library officials:

> Even though I frequently hear about Warren's racist environment and prejudicial attitude toward African-Americans, I continue to reside here because my experience has not yet been a negative one. When people have laughed and asked me how I can reside here, I have always held my head up high. Today, I hang my head in shame … not for myself, but for the city of Warren and you—its representatives. (Quoted in Franz 2006, n.p.)

In this respect, the King domain is not just involved in creating a highly charged, virtual place of hate. Rather, because of the politics of misdirection, it can also inflame race-based feelings in the physical, offline world, in effect creating a place-based experience that exacerbates existing fears among African Americans and challenges the representation of Warren County as tolerant and sensitive to minority sensibilities.

Concluding Remarks

Cyberspace is part of—rather than a transcendent of—the physical, material world and its long-standing social divides and conflicts. The Internet domain name system is often deeply involved in these conflicts since domains play a critical role in identity construction, the maintenance (and challenging) of economic and political boundaries, and the claiming of power online. Domain names are not simply the identifiers of web sites. By virtue of their importance in guiding flows of communication and social interaction across cyberspace, domains serve as conduits through which virtual places are constructed. Domains and their associated web sites also influence the social construction and contestation of place in the physical, material world.

Stormfront's martinlutherking.org illustrates the potential of analyzing domain names as powerful symbolic texts involved in larger political struggles over race, memory, and reputation in America. One of the most important lessons we learn from the King domain is the political value of misdirection in cyberspace, how the communicational power of domains can be harnessed and manipulated by social and political groups as they seek to be seen and heard. In the case of Stormfront, an otherwise fringe movement is allowed to occupy a prominent, albeit deceiving, position within the informational and ideological spaces of the Internet. The King domain provides white supremacists with a tool to create a place of hate and a platform for discrediting the teachings and activism of the civil rights leader. While the King domain reflects some of the tensions that have existed (and will probably always exist) in our offline social geographies, the real danger of the domain name comes from its potential to intensify social anxieties and further widen America's racial divide—a paradoxical and disturbing memorial to a person who was devoted to eliminating that divide and whose name, at least outside the context of martinlutherking.org, symbolizes social justice and racial equality.

References

Abel, D. (2000), "Some Colleges Struggle with the Name Game," *The Boston Globe* 2 August, B1.

Adams, P. (1992), "Television as Gathering Place," *Annals of the Association of American Geographers* 82(1), 117–35.

—— (1996), "Protest and the Scale Politics of Telecommunications," *Political Geography* 15(5), 419–41.

—— (1998), "Network Topologies and Virtual Place," *Annals of the Association of American Geographers* 88(1), 88–106.

—— and Ghose, R. (2003), "India.com: The Construction of a Space Between," *Progress in Human Geography* 27(4), 414–37.

Alderman, D.H. (2006a), "Naming Streets after Martin Luther King, Jr: No Easy Road," in R. Schein (ed.), *Landscape and Race in the United States* (New York: Routledge Press), 213–36.

—— (2006b), "Rednecks, Bluenecks, and Hickphonics: Southern Humor on the Electronic Frontier," in E. Piacentino (ed.), *The Enduring Legacy of Old Southwest Humor* (Baton Rouge, LA: Louisiana State University Press), 261–78.

—— (2008), "The Politics of Preserving the King's Courts: Why We Should Take Elvis Fans Seriously," *The Southern Quarterly* 46(1), 46–77.

—— and Good, D.B. (1997), "Exploring the Virtual South: The Idea of a Distinctive Region on the Web," *Southeastern Geographer* 37(1), 20–45.

Beckles, C.A. (2001), "Black Liberation and the Internet: A Strategic Analysis," *Journal of Black Studies* 31(3), 311–24.

Bell, C. and Lyall, J. (2002), "A Town called JenniferAnn.com," *Space and Culture* 5(3), 278–86.

Brunn, S.D. and Dodge, M. (2001), "Mapping the 'Worlds' of the World Wide Web," *American Behavioral Scientist* 44(10), 1717–39.

Cairncross, F. (1997), *The Death of Distance: How the Communications Revolution Will Change Our Lives* (London: Orion Publishing Group Limited).

Carlson, N. (2007), "Google PageRanks MLK Mistake," *internetnews.com* 12 January, accessed 25 January 2008 at http://www.internetnews.com/bus-news/article.php/3653831.

Carter, P. (2006), "Parrot Interpreter: Representation, Extinction, and the Electronic Information Environment," *Cultural Studies Review* 12(1), 89–103.

Crampton, J. (2003), *The Political Mapping of Cyberspace* (Chicago: University of Chicago Press).

The Daily Reflector (2001a), "Peeps Show," 15 April, E1.

—— (2001b), "Peeps Show Correction," 17 April, B2.

Dodge, M. and Kitchin, R. (2001), *Mapping Cyberspace* (New York: Routledge).

Durie, R. (1997), "Internet Law—The Internet, Trademarks, and Domain Names" *Computer Law and Security Report* 13(1), 29–33.

Fine, G.A. (2000), *Difficult Reputations: Collective Memories of the Evil, Inept, and Controversial* (Chicago: University of Chicago Press).

Flint, C. (2004), "Introduction: Spaces of Hate: Geographies of Hate and Intolerance in the United States of America," in C. Flint, (ed.), *Spaces of Hate: Geographies of Hate and Intolerance in the United States of America* (New York: Routledge), 1–20.

Franz, N. (2006), "Library Link to 'Racist' Web Site Called Mistake," *Macomb Daily* 28 February, accessed 25 August 2006 at http://www.macombdaily.com/stories/022806/loc_wrnlib001.shtml.

Froehling, O. (1997), "The Cyberspace 'War of Ink and Internet' in Chiapas, Mexico," *Geographical Review* 87(2), 291–307.

Gerstenfield, P.B., Grant, D.R. and Chiang, C.P. (2003), "Hate Online: A Content Analysis of Extremist Internet Sites," *Analyses of Social Issues and Public Policy* 3(1), 29–44.

Graham, S. (2000), "The End of Geography or the Explosion of Place? Conceptualising Space, Time and Information Technology," in M. Wilson and K. Corey (eds), *Information Tectonics: Space, Place and Technology in and Electronic Age* (London: Wiley), 7–28.

Hashim, N.H. and Murphy, J. (2007), "Branding on the Web: Evolving Domain Name Usage among Malaysian Hotels," *Tourism Management* 28(2), 621–24.

Holloway, S. and Valentine, G. (2001), "Placing Cyberspace: Processes of Americanization in British Children's Use of the Internet," *Area* 33(2), 153–60.

Introna, L.D. and Nissenbaum,H. (2000), "Shaping the Web: Why the Politics of Search Engines Matters," *The Information Society* 16(3), 169–85.

Kitchin, R.M. (1998), "Towards Geographies of Cyberspace," *Progress in Human Geography* 22(3), 385–406.

Leonard, A. (2002), "Nissan v. Nissan," *Salon.com*. 3 June, accessed August 2006 at http://dir.salon.com/story/tech/col/leon/2002/06/03/nissan/index.html?pn=1.

Lisberg, A. (2006), "Gay Man Wins Internet Fight vs. Falwell," *Daily News* (New York) 18 April: News, 7.

Longan, M.W. (2002), "Building a Global Sense of Place: The Community Networking Movement in the United States," *Urban Geography* 23(3), 213–36.

Mack, J. (2000), "Hate Site Distorts King's Dream," *ZDNet UK* 26 January, accessed August 2006 at http://news.zdnet.co.uk/internet/0,39020369,2076574,00.htm.

Matthews, J. (2000), "Will The Real Nissandotcom Please Stand Up?," *Newsbytes News Network* 3 August, accessed August 2006 at http://www.findarticles.com/p/articles/mi_m0NEW/is_2000_August_3/ai_63918846.

McDowell, S.D. and Steinberg, P.E. (2001), "Non-State Governance and the Internet: Civil Society and the ICANN," *Info: The Journal of Policy, Regulation and Strategy for Telecommunications Information and Media* 3(4), 279–98.

—, Steinberg, P.E. and Tomasello, T. (2008), *Managing the Infosphere: Governance, Technology, and Cultural Practice in Motion* (Philadelphia: Temple University).

Mintz, A.P. (2002), "Lies, Damned Lies, and the Internet," in A.P. Mintz (ed.), *Web of Deception: Misinformation on the Internet* (Medford, NJ: New Information, Inc.), xvii–xxv.

Mueller, M. (2001), "Rough Justice: A Statistical Assessment of ICANN's Uniform Dispute Resolution Policy," *The Information Society* 17(3), 151–63.

National Research Council of the National Academies (2005), *Signposts in Cyberspace: The Domain Names System and Internet Navigation* (Washington, DC: National Academies Press).

O'Lear, S. (1997), "Electronic Communication and Environmental Policy in Russia and Estonia," *Geographical Review* 87(2), 275–90.

Rony, E. and Rony, P.R. (1998), *The Domain Name Handbook: High Stakes and Strategies in Cyberspace* (Emeryville, CA: CMP Books).

Sharrock, L.M. (2001), "The Future of Domain Name Dispute Resolution: Crafting Practical International Legal Solutions from within the UDRP Framework," *Duke Law Journal* 51(2), 817–49.

Spring, T. (1999), "Cisco: Net Will Be a Great Equalizer," *PC World* 30 September, accessed June 2006 at http://www.pcworld.com/news/article/0,aid,13070,00.asp.

Steinberg, P.E. and McDowell, S.D. (2003), "Mutiny on the Bandwith: The Semiotics of Statehood in the Internet Domain Name Registries of Pitcarin and Niue," *New Media and Society* 5(1), 47–67.

Suellentrop, C. (2004), "Cheney Drops the Ball," *Slate* 6 October, accessed August 2006 at http://www.slate.com/id/2107809/.

Swartz, J. (2006), "Domain Names Become Premium Web Real Estate," *USA TODAY* 9 May, accessed August 2006 at http://www.usatoday.com/tech/news/2006–05–09–domainers_x.htm.

Tuan, Y. (1991), "Language and the Making of Place: A Narrative-Descriptive Approach," *Annals of the Association of American Geographers* 81(4), 684–96.

Valentine, G. and Holloway, S.L. (2002), "Cyberkids? Exploring Children Identities and Social Networks in On-line and Off-line Worlds," *Annals of the Association of American Geographers* 92(2), 302–19.

Walmsley, D.J. (2000), "Community, Place and Cyberspace," *Australian Geographer* 31(1), 5–19.

Wang, Y., Lai, P., and Sui, D. (2003), "Mapping the Internet Using GIS: The Death of Distance Hypothesis Revisited," *Journal of Geographical Systems* 5(4), 381–405.

Warf, B. (2001), "Segueways into Cyberspace: Multiple Geographies of the Digital Divide," *Environment and Planning B* 28(1), 3–19.

—— and Grimes, J. (1997), "Counterhegemonic Discourses and the Internet," *Geographical Review* 87, 259–74.

—— and Vincent, P. (2007), "Multiple Geographies of the Arab Internet," *Area* 39(1), 83–96.

Wass, E.S. (2003), *Addressing the World: National Identity and Internet Country Code Domains* (Lanham, MD: Rowman and Littlefield).

Weise, E. (2000), "Web Users to Get More Places to Visit: Domain-name Expansion Will Cast Wider Net," *USA TODAY* 5 December, 3D.

Williamson, L. and Pierson, E. (2003), "The Rhetoric of Hate on the Internet: Hateporn's Challenge to Modern Media Ethics," *Journal of Mass Media Ethics* 18(3/4), 250–67.

Wilson, M.I. (2001), "Location, Location, Location: The Geography of the Dot Com Problem," *Environment and Planning B* 28(1), 59–71.

Zook, M.A. (2000), "The Web of Production: The Economic Geography of Commercial Internet Production in the US," *Environment and Planning A* 32(3), 411–26.

Index